Plastics Fundamentals, Properties, and Testing

PLASTICS ENGINEERING

Founding Editor

Donald E. Hudgin

Professor
Clemson University
Clemson, South Carolina

1. Plastics Waste: Recovery of Economic Value, *Jacob Leidner*
2. Polyester Molding Compounds, *Robert Burns*
3. Carbon Black-Polymer Composites: The Physics of Electrically Conducting Composites, *edited by Enid Keil Sichel*
4. The Strength and Stiffness of Polymers, *edited by Anagnostis E. Zachariades and Roger S. Porter*
5. Selecting Thermoplastics for Engineering Applications, *Charles P. MacDermott*
6. Engineering with Rigid PVC: Processability and Applications, *edited by I. Luis Gomez*
7. Computer-Aided Design of Polymers and Composites, *D. H. Kaelble*
8. Engineering Thermoplastics: Properties and Applications, *edited by James M. Margolis*
9. Structural Foam: A Purchasing and Design Guide, *Bruce C. Wendle*
10. Plastics in Architecture: A Guide to Acrylic and Polycarbonate, *Ralph Montella*
11. Metal-Filled Polymers: Properties and Applications, *edited by Swapan K. Bhattacharya*
12. Plastics Technology Handbook, *Manas Chanda and Salil K. Roy*
13. Reaction Injection Molding Machinery and Processes, *F. Melvin Sweeney*
14. Practical Thermoforming: Principles and Applications, *John Florian*
15. Injection and Compression Molding Fundamentals, *edited by Avraam I. Isayev*
16. Polymer Mixing and Extrusion Technology, *Nicholas P. Cheremisinoff*
17. High Modulus Polymers: Approaches to Design and Development, *edited by Anagnostis E. Zachariades and Roger S. Porter*
18. Corrosion-Resistant Plastic Composites in Chemical Plant Design, *John H. Mallinson*
19. Handbook of Elastomers: New Developments and Technology, *edited by Anil K. Bhowmick and Howard L. Stephens*
20. Rubber Compounding: Principles, Materials, and Techniques, *Fred W. Barlow*

21. Thermoplastic Polymer Additives: Theory and Practice, *edited by John T. Lutz, Jr.*
22. Emulsion Polymer Technology, *Robert D. Athey, Jr.*
23. Mixing in Polymer Processing, *edited by Chris Rauwendaal*
24. Handbook of Polymer Synthesis, Parts A and B, *edited by Hans R. Kricheldorf*
25. Computational Modeling of Polymers, *edited by Jozef Bicerano*
26. Plastics Technology Handbook: Second Edition, Revised and Expanded, *Manas Chanda and Salil K. Roy*
27. Prediction of Polymer Properties, *Jozef Bicerano*
28. Ferroelectric Polymers: Chemistry, Physics, and Applications, *edited by Hari Singh Nalwa*
29. Degradable Polymers, Recycling, and Plastics Waste Management, *edited by Ann-Christine Albertsson and Samuel J. Huang*
30. Polymer Toughening, *edited by Charles B. Arends*
31. Handbook of Applied Polymer Processing Technology, *edited by Nicholas P. Cheremisinoff and Paul N. Cheremisinoff*
32. Diffusion in Polymers, *edited by P. Neogi*
33. Polymer Devolatilization, *edited by Ramon J. Albalak*
34. Anionic Polymerization: Principles and Practical Applications, *Henry L. Hsieh and Roderic P. Quirk*
35. Cationic Polymerizations: Mechanisms, Synthesis, and Applications, *edited by Krzysztof Matyjaszewski*
36. Polyimides: Fundamentals and Applications, *edited by Malay K. Ghosh and K. L. Mittal*
37. Thermoplastic Melt Rheology and Processing, *A. V. Shenoy and D. R. Saini*
38. Prediction of Polymer Properties: Second Edition, Revised and Expanded, *Jozef Bicerano*
39. Practical Thermoforming: Principles and Applications, Second Edition, Revised and Expanded, *John Florian*
40. Macromolecular Design of Polymeric Materials, *edited by Koichi Hatada, Tatsuki Kitayama, and Otto Vogl*
41. Handbook of Thermoplastics, *edited by Olagoke Olabisi*
42. Selecting Thermoplastics for Engineering Applications: Second Edition, Revised and Expanded, *Charles P. MacDermott and Aroon V. Shenoy*
43. Metallized Plastics, *edited by K. L. Mittal*
44. Oligomer Technology and Applications, *Constantin V. Uglea*
45. Electrical and Optical Polymer Systems: Fundamentals, Methods, and Applications, *edited by Donald L. Wise, Gary E. Wnek, Debra J. Trantolo, Thomas M. Cooper, and Joseph D. Gresser*
46. Structure and Properties of Multiphase Polymeric Materials, *edited by Takeo Araki, Qui Tran-Cong, and Mitsuhiro Shibayama*
47. Plastics Technology Handbook: Third Edition, Revised and Expanded, *Manas Chanda and Salil K. Roy*
48. Handbook of Radical Vinyl Polymerization, *edited by Munmaya K. Mishra and Yusef Yagci*

49. Photonic Polymer Systems: Fundamentals, Methods, and Applications, *edited by Donald L. Wise, Gary E. Wnek, Debra J. Trantolo, Thomas M. Cooper, and Joseph D. Gresser*
50. Handbook of Polymer Testing: Physical Methods, *edited by Roger Brown*
51. Handbook of Polypropylene and Polypropylene Composites, *edited by Harutun G. Karian*
52. Polymer Blends and Alloys, *edited by Gabriel O. Shonaike and George P. Simon*
53. Star and Hyperbranched Polymers, *edited by Munmaya K. Mishra and Shiro Kobayashi*
54. Practical Extrusion Blow Molding, *edited by Samuel L. Belcher*
55. Polymer Viscoelasticity: Stress and Strain in Practice, *Evaristo Riande, Ricardo Díaz-Calleja, Margarita G. Prolongo, Rosa M. Masegosa, and Catalina Salom*
56. Handbook of Polycarbonate Science and Technology, *edited by Donald G. LeGrand and John T. Bendler*
57. Handbook of Polyethylene: Structures, Properties, and Applications, *Andrew J. Peacock*
58. Polymer and Composite Rheology: Second Edition, Revised and Expanded, *Rakesh K. Gupta*
59. Handbook of Polyolefins: Second Edition, Revised and Expanded, *edited by Cornelia Vasile*
60. Polymer Modification: Principles, Techniques, and Applications, *edited by John J. Meister*
61. Handbook of Elastomers: Second Edition, Revised and Expanded, *edited by Anil K. Bhowmick and Howard L. Stephens*
62. Polymer Modifiers and Additives, *edited by John T. Lutz, Jr., and Richard F. Grossman*
63. Practical Injection Molding, *Bernie A. Olmsted and Martin E. Davis*
64. Thermosetting Polymers, *Jean-Pierre Pascault, Henry Sautereau, Jacques Verdu, and Roberto J. J. Williams*
65. Prediction of Polymer Properties: Third Edition, Revised and Expanded, *Jozef Bicerano*
66. Fundamentals of Polymer Engineering, *Anil Kumar and Rakesh K. Gupta*
67. Handbook of Polypropylene and Polymer, *Harutun Karian*
68. Handbook of Plastic Analysis, *Hubert Lobo and Jose Bonilla*
69. Computer-Aided Injection Mold Design and Manufacture, *J. Y. H. Fuh, Y. F. Zhang, A. Y. C. Nee, and M. W. Fu*
70. Handbook of Polymer Synthesis: Second Edition, *Hans R. Kricheldorf and Graham Swift*
71. Practical Guide to Injection Blow Molding, *Samuel L. Belcher*
72. Plastics Technology Handbook: Fourth Edition, *Manas Chanda and Salil K. Roy*
73. Industrial Polymers, Specialty Polymers, and Their Applications, *Manas Chanda and Salil K. Roy*
74. Plastics Fundamentals, Properties, and Testing, *Manas Chanda and Salil K. Roy*
75. Plastics Fabrication and Recycling, *Manas Chanda and Salil K. Roy*

Plastics Fundamentals, Properties, and Testing

Manas Chanda
Salil K. Roy

CRC Press
Taylor & Francis Group
Boca Raton London New York

CRC Press is an imprint of the
Taylor & Francis Group, an **informa** business

The material was previously published in *Plastics Technology Handbook, Fourth Edition* © Taylor & Francis, 2007.

CRC Press
Taylor & Francis Group
6000 Broken Sound Parkway NW, Suite 300
Boca Raton, FL 33487-2742

ISBN 13: 978-1-4200-8060-5 (hbk)

Library of Congress Cataloging-in-Publication Data

Chanda, Manas, 1940-
Plastics fundamentals, properties, and testing / Manas Chanda and Salil K. Roy.
p. cm. -- (Plastics engineering ; 74)
Includes bibliographical references and index.
ISBN 978-1-4200-8060-5 (alk. paper)
1. Plastics. 2. Plastics--Testing. I. Roy, Salil K., 1939- II. Title. III. Series.

TA455.P5C45 2008
620.1'923--dc22 2008011623

Visit the Taylor & Francis Web site at
http://www.taylorandfrancis.com

and the CRC Press Web site at
http://www.crcpress.com

Contents

Preface

Plastics today represent one of the fastest changing fields with new polymers being synthesized, new uses being found, and existing processes and products being modified and improved for ecological needs, better economics, and better values. Plastics touch every one of us in our daily lives as many of the things we use—our clothing, our food, and our bodies—are made of polymers. In one form or other, plastics also have some role to play in every realm of human activity, endeavor, and experience. Today the use of plastics ranges from being minor to major and from being an alternative to indispensable in such diverse areas of applications as packaging and transportation, building and construction, domestic appliances and business machines, agriculture and horticulture, electrical and electronic, medical and biomedical, automotive and aerospace, as well as sports, marine, and corrosion prevention. The remarkable versatility of the nonmetallic class of materials called *polymers* (often loosely referred to as *plastics*) poses many questions, even to a lay user — What makes plastics so distinctly different from metals? How and why different types of plastics differ from each other? What gives rise to the characteristic properties of plastics that we are familiar with? Which plastics melt on heating and which ones harden on heating? Which plastics should be used for any given application? What are the limitations of plastics? How stable and resistant are plastics to different use conditions, such as high or low temperature, sunlight, humidity, microorganisms, solvents, acid, alkali, radiation, high voltage, load, and impact? How can the properties of plastics be modified to meet specific needs? Answers to these questions can be found in this book, written in a lucid style so as to be comprehensible even to a non-technical layperson.

Chapter 1, which is the largest chapter in the book, looks at polymers at the molecular level to explain their inherent properties and how through changes in their molecular architecture or incorporation of various fillers and additives their end-use properties can be influenced. The chapter covers a wide spectrum of topics, coherently presented with focus on macromolecular basis of polymer behavior. Starting with basic definitions commonly used in polymer science, the reader is introduced, in a graded and sequential manner, to various aspects of polymers with the emphasis gradually shifting from structural aspects to properties and onto applications. Thus the main topics on polymers comprising this chapter are molecular weights, polymerization processes, structural shapes, configuration and conformation, amorphous and crystalline states, thermal transitions, cross-linking, thermal stability, deterioration and stabilization, diffusion and permeability, toxicity, and biodegradability. Also discussed are the functions of compounding agents, metal deactivators, light stabilizers, plasticizers, antistatic agents, flame retardants, smoke suppressants, colorants, and antimicrobials in the use of plastics.

While a qualitative concept regarding the genesis of polymer behavior and properties is gained from reading Chapter 1, the reader gets a more focused and quantitative picture of polymer behavior, evaluation, and characterization in Chapter 2. What gives polymers the dominant place among all materials is the wide range of properties inherent in them or imparted by various means. Chapter 2 discusses polymer properties under four main headings—mechanical, electrical, optical, and thermal—giving theoretical derivations where necessary and providing explanations on a molecular and structural basis. A major focus is given on mechanical properties of polymers as these often constitute the most important consideration and dictate the use of specific polymers in diverse industrial applications and myriad other uses. A wide range of topics have been covered under mechanical properties including stress–strain behavior, stress–strain–time behavior, mechanical models, effect of temperature, time–temperature superposition, dynamic mechanical properties, rheological behavior, impact behavior, fatigue, and hardness.

Reinforced plastics form an important area of structural application of plastics since the modulus and strength of plastics can be increased significantly through reinforcement. In reinforced plastics, the polymer (popularly called the *resin*) forms the matrix and a filler (mostly used in the form of fibers, but particles, for example glass spheres, are also used) provides the reinforcing effect. In view of their distinctive nature and extensive use as materials of construction in load-bearing applications, a special focus has been on analysis of properties of reinforced plastics, especially those reinforced by continuous or discontinuous fibers, as well as their deformation, fracture, fatigue, and impact behaviors.

Electrical and thermal insulation properties of plastics make these materials indispensable in many applications. These properties are discussed in detail and with a constant focus on the molecular origin of these properties. The discussion on electrical properties covers, specifically, dielectric strength, insulation resistance, arc resistance, dielectric constant in relation to use conditions, and dielectric losses. Optical properties of plastics, including optical clarity, index of refraction, piped lighting effects, and stress-optical characteristics, are discussed highlighting application aspects and relation of these properties to polymer composition and structure. The useful thermal properties of plastics, which include specific heat, thermal expansion, thermal conductivity, and thermal softening, are discussed in a lucid manner and their significance with regard to various applications, some of which are as commonplace as the teapot handle or the insulating connection to an electric iron, is explained.

Standard test methods have been developed for measuring the aforesaid properties. Since the property values are highly dependent on the specimen preparation, equipment, and testing techniques used, it is essential to refer to the appropriate official standard test methods when executing the work. Chapter 2 presents schematically (in simplified form) the bases of many of these standard test methods, according to ASTM and BS 2782 specifications, to indicate broadly the principles involved.

As plastics have become an inalienable part of our daily lives, it is important to be able to identify them and to know their characteristic properties so as to put them to better use. Various test methods are available to identify plastics some of which are very simple, such as observing behavior on heating and holding in a gas flame to take note of one or more of the following effects: (1) if the plastic burns and if so, how readily; (2) the nature and color of the flame as the material burns; (3) whether the material is self-extinguishing or continues to burn after removal of the flame; and (4) the nature of the residue. A section is therefore provided toward the end of the book dealing with identification of common plastics. For example, polyethylene burns with a luminous flame producing vapor–like paraffin wax (extinguished candles) and melts forming droplets that continue to burn, whereas PVC produces yellow–orange, green-bordered flame, producing strongly acidic fumes and black residue. Such observations on heating and ignition of common polymers are listed in a tabular form for easy reference. Tests for characteristic elements, such as nitrogen, sulfur, and halogens, may also serve to roughly indicate the nature of the base material and additives, if present. The different polymers have been classified into several groups according to the elements present, and the focus of identification has been further narrowed on the basis of other preliminary observations, e.g., fusibility or otherwise, melting point, heat distortion temperature, flame tests, thermal degradation, and solubility or extractability in water or different organic solvents. The solubility behaviors of common polymers have been compared in a tabular form for ready reference and use. When the observations and results of preliminary tests have been considered and most of the possible structures for the polymer base eliminated, an exact identification can then be made by carrying out specific tests. Some such tests for ready identification of specific polymers are described in Section 2.7. These will be useful to those who are interested in analysis and identification of unknown plastic samples.

The stimulus to bring out this book as a spin-off of our *Plastics Technology Handbook* has come from the constantly excellent response and acceptance shown by readers to the handbook ever since the publication of its first edition in 1982. Being much smaller in size and more convenient to use, the present book with its lucidity of approach and wealth of valuable information and specific data storage should find a wider reach and satisfy the knowledge thirst of a wide spectrum of people in diverse walks of life, ranging from a layman to a college student, an industry worker, and a scientist engaged in R&D activities. This book is an outcome of the initiative taken by Allison Shatkin, Materials Science and Chemical Engineering Editor at CRC Press/Taylor & Francis, who was the first to come up with the idea of producing this book for the convenience of readers. We thank her for the initiative and interest she has shown in the project.

Manas Chanda
Salil K. Roy

Authors

Manas Chanda has been a professor and is presently an emeritus professor in the Department of Chemical Engineering, Indian Institute of Science, Bangalore, India. He also worked as a summer-term visiting professor at the University of Waterloo, Ontario, Canada with regular summer visits from 1980 to 2000. A five-time recipient of the International Scientific Exchange Award from the Natural Sciences and Engineering Research Council, Canada, Dr. Chanda is the author or coauthor of nearly 100 scientific papers, articles, and books, including *Introduction to Polymer Science and Chemistry* (CRC Press/Taylor & Francis). A fellow of the Indian National Academy of Engineers and a member of the Indian Plastics Institute, he received a BS (1959) and MSc (1962) from Calcutta University, and a PhD (1966) from the Indian Institute of Science, Bangalore, India.

Salil K. Roy is a professor in the Postgraduate Program in Civil Engineering of the Petra Christian University, Surabaya, Indonesia. Earlier he worked as lecturer, senior lecturer, and associate professor at the National University of Singapore. Prior to that he was a research scientist at American Standard, Piscataway, New Jersey.

Dr. Roy is a fellow of the Institution of Diagnostic Engineers, U.K., and has published over 250 technical papers in professional journals and conference proceedings; he also holds several U.S. Patents. He received a BSc (1958) and MSc (Tech.) (1961) from the University of Calcutta, India, and a ScD (1966) from the Massachusetts Institute of Technology, Cambridge, Massachusetts. Dr. Roy is a subject of biographical record in the prestigious *Great Minds of the 21st Century* published by the American Biographical Institute, *Who's Who in the World* published by the Marquis Who's Who in the World, and *2000 Outstanding Intellectuals of the 21st Century* published by the International Biographical Centre, Cambridge, England.

Authors

1

Characteristics of Polymers

1.1 What Is a Polymer?

A molecule has a group of atoms which have strong bonds among themselves but relatively weak bonds to adjacent molecules. Examples of small molecules are water (H_2O), methanol (CH_3OH), carbon dioxide, and so on. Polymers contain thousands to millions of atoms in a molecule which is large; they are also called macromolecules. Polymers are prepared by joining a large number of small molecules called monomers. Polymers can be thought of as big buildings, and monomers as the bricks that go into them.

Monomers are generally simple organic molecules containing a double bond or a minimum of two active functional groups. The presence of the double bond or active functional groups act as the driving force to add one monomer molecule upon the other repeatedly to make a polymer molecule. This process of transformation of monomer molecules to a polymer molecule is known as polymerization. For example, ethylene, the prototype monomer molecule, is very reactive because it has a double bond. Under the influence of heat, light, or chemical agents this bond becomes so activated that a chain reaction of self-addition of ethylene molecules is generated, resulting in the production of a high-molecular-weight material, almost identical in chemical composition to ethylene, known as polyethylene, the polymer of ethylene (Figure 1.1).

The difference in behavior between ordinary organic compounds and polymeric materials is due mainly to the large size and shape of polymer molecules. Common organic materials such as alcohol, ether, chloroform, sugar, and so on, consist of small molecules having molecular weights usually less than 1,000. The molecular weights of polymers, on the other hand, vary from 20,000 to hundreds of thousands.

The name polymer is derived from the Greek poly for many and *meros* for parts. A polymer molecule consists of a repetition of the unit called a *mer*. Mers are derived from monomers, which, as we have seen for ethylene, can link up or polymerize under certain conditions to form the polymer molecule. The number of mers, or more precisely the number of repetitions of the mer, in a polymer chain is called the degree of polymerization (DP). Since the minimum length or size of the molecule is not specified, a relatively small molecule composed of only, say, 3 mers might also be called a polymer. However, the term polymer is generally accepted to imply a molecule of large size (macromolecule). Accordingly, the lower-molecular-weight products with low DP should preferably be called oligomers (*oligo*=few) to distinguish them from polymers. Often the term high polymer is also used to emphasize that the polymer under consideration is of very high molecular weight.

Because of their large molecular size, polymers possess unique chemical and physical properties. These properties begin to appear when the polymer chain is of sufficient length—i.e., when the molecular weight exceeds a threshold value—and becomes more prominent as the size of the molecule increases. The dependence of the softening temperature of polyethylene on the degree of polymerization is shown

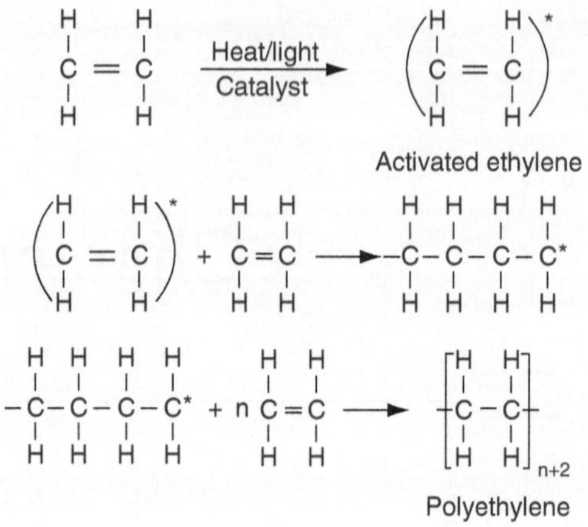

FIGURE 1.1 Intermediate steps during formation of polyethylene.

in Figure 1.2a. The dimer of ethylene is a gas, but oligomers with a DP of 3 or more (that is, C_6— or higher paraffins) are liquids, with the liquid viscosity increasing with the chain length. Polyethylenes with DPs of about 30 are greaselike, and those with DPs around 50 are waxes. As the DP value exceeds 400 or the molecular weight exceeds about 10,000, polyethylenes become hard resins with softening points about 100°C. The increase in softening point with chain length in the higher-molecular-weight range is small. The relationship of such polymer properties as tensile strength, impact strength, and melt viscosity with molecular weight is indicated in Figure 1.2b. Note that the strength properties increase rapidly first as the chain length increases and then level off, but the melt viscosity continues to increase rapidly, Polymers with very high molecular weights have superior mechanical properties but are difficult to process and fabricate due to their high melt viscosities. The range of molecular weights chosen for commercial polymers represents a compromise between maximum properties and processability.

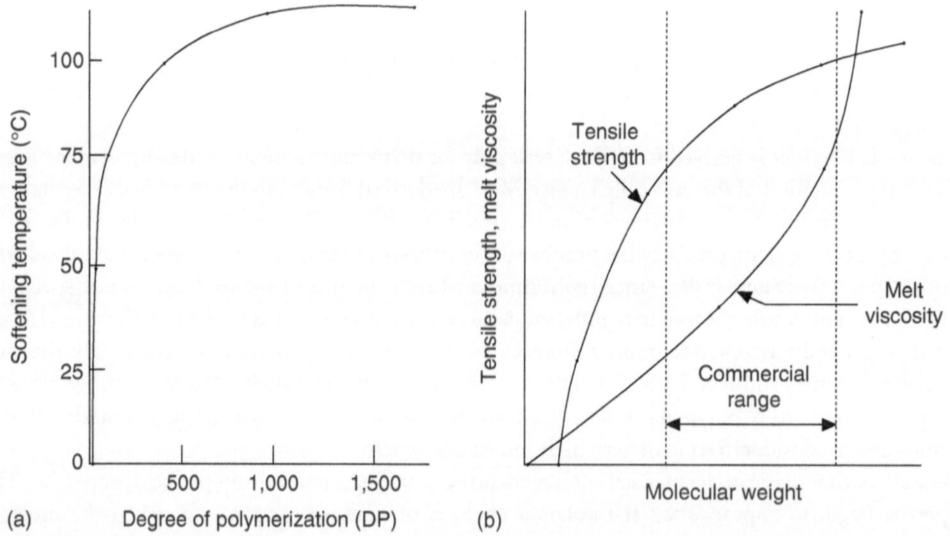

FIGURE 1.2 Polymer properties versus polymer size. (a) Softening temperature of polyethylene. (b) Tensile strength, and melt viscosity. (Adapted from Seymour, R. B. and Carraher, C. E. Jr., 1992. *Polymer Chemistry. An Introduction.* Marcel Dekker, New York.)

1.2 Molecular Weight of Polymers

In ordinary chemical compounds such as sucrose, all molecules are of the same size and therefore have identical molecular weights (M). Such compounds are said to be monodisperse. In contrast, most polymers are polydisperse. Thus a polymer does not contain molecules of the same size and, therefore, does not have a single molecular weight. In fact, a polymer contains a large number of molecules—some big, some small. Thus there exists a variation in molecular size and weight, known as *molecular-weight* distribution (MWD), in every polymeric system, and this MWD determines to a certain extent the general behavior of polymers. Since a polymer consists of molecules of different sizes and weights, it is necessary to calculate an average molecular weight ($\bar{M}$) or an *average degree of polymerization* ($\overline{\mathrm{DP}}$).

The molecular weights commonly used in the characterization of a polydisperse polymer are the number average, weight average, and viscosity average.

Consider a sample of a polydisperse polymer of total weight W in which N = total number of moles; N_i = number of moles of species i (comprising molecules of the same size); n_i = mole fraction of species i; W_i = weight of species i; w_i = weight fraction of species i; M_i = molecular weight of species i; x_i = degree of polymerzation of species i.

1.2.1 Number-Average Molecular Weight ($\bar{M}_{\mathrm{n}}$)

From the definition of molecular weight as the weight of sample per mole, we obtain

$$\bar{M}_{\mathrm{n}} = \frac{W}{N} = \frac{\sum N_i M_i}{N} = \sum n_i M_i \tag{1.1}$$

$$= \frac{\sum W_i}{\sum W_i/M_i} = \frac{\sum w_i}{\sum w_i/M_i} = \frac{1}{\sum w_i/M_i} \tag{1.2}$$

Dividing $\bar{M}_{\mathrm{n}}$ by the mer weight M_0, we obtain a number-average degree of polymerization, $\overline{\mathrm{DP}}_{\mathrm{n}}$, where

$$\overline{\mathrm{DP}}_{\mathrm{n}} = \frac{\bar{M}_{\mathrm{n}}}{M_0} = \frac{\sum N_i x_i}{\sum N_i} \tag{1.3}$$

The quantity $\bar{M}_{\mathrm{n}}$ is obtained by end-group analysis or by measuring a colligative property such as elevation of boiling point, depression of freezing point, or osmotic pressure [1,2].

1.2.2 Weight-Average Molecular Weight ($\bar{M}_{\mathrm{w}}$)

Equation 1.1 indicates that in the computation of $\bar{M}_{\mathrm{n}}$, the molecular weight of each species is weighted by the mole fraction of that species. Similarly, in the computation of weight-average molecular weight the molecular weight of each species is weighted by the weight fraction of that species:

$$\bar{M}_{\mathrm{w}} = \sum w_i M_i = \frac{\sum W_i M_i}{\sum W_i} \tag{1.4}$$

$$= \frac{\sum N_i M_i^2}{\sum N_i M_i} \tag{1.5}$$

The weight-average degree of polymerization, $\overline{\text{DP}}_w$, is obtained by dividing $\bar{M}_w$, by the mer weight:

$$\overline{\text{DP}}_w = \frac{\bar{M}_w}{M_0} = \frac{\sum W_i x_i}{\sum W_i} \tag{1.6}$$

$\bar{M}_w$ can be determined by measuring light scattering of dilute polymer solution [3,4]. $\bar{M}_w$ is always higher than $\bar{M}_n$. Thus for a polymer sample containing 50 mol% of a species of molecular weight 10,000 and 50 mol% of species of molecular weight 20,000, Equation 1.1 and Equation 1.5 give $\bar{M}_n = 0.5(10{,}000 + 20{,}000) = 15{,}000$ and $\bar{M}_w = [(10{,}000)^2 + (20{,}000)^2]/[10{,}000 + 20{,}000] = 17{,}000$.

1.2.3 Viscosity-Average Molecular Weight ($\bar{M}_v$)

The viscosity-average molecular weight is defined by the equation

$$\bar{M}_v = \left[\sum w_i M_i^a\right]^{1/a} = \left[\sum N_i M_i^{1+a} / \sum N_i M_i\right]^{1/a} \tag{1.7}$$

For $a = 1$, $\bar{M}_v = \bar{M}_w$ and for $a = -1$, $\bar{M}_v = \bar{M}_n$. Thus, $\bar{M}_v$ falls between $\bar{M}_w$ and $\bar{M}_n$, and for many polymers it is 10%–20% below $\bar{M}_w$. $\bar{M}_v$ is calculated from the intrinsic viscosity $[\eta]$ by the empirical relation

$$[\eta] = K\bar{M}_v^{\alpha} \tag{1.8}$$

where K and α are constants. $[\eta]$ is derived from viscosity measurements by extrapolation to "zero" concentration [5,6].

In correlating polymer properties (such as reactivity) which depend more on the number of molecules in the sample than on the sizes of the molecules, $\bar{M}_n$ is a more useful parameter than $\bar{M}_w$ or $\bar{M}_v$. Conversely, for correlating polymer properties (such as viscosity) which are more sensitive to the size of the polymer molecules, $\bar{M}_w$ or $\bar{M}_v$ is more useful.

Because it is easy to determine, *melt index* often is used instead of molecular weight in routing characterization of polymers. It is defined as the mass rate of polymer flow through a specified capillary under controlled conditions of temperature and pressure. The index can often be related empirically to some average molecular weight, depending on the specific polymer. A lower melt index indicates a higher molecular weight, and vice versa.

1.2.4 Polydispersity Index

The ratio of weight-average molecular weight to number-average molecular weight is called the dispersion or polydispersity index (I). It is a measure of the width of the molecular-weight distribution curve (Figure 1.3) and is used as such for characterization purposes. Normally I is between 1.5 and 2.5, but it may range to 15 or greater. The higher the value of I is, the Greater is the spread of the molecular-weight distribution of the polymer. For a monodisperse system (e.g., pure chemicals), $I = 1$.

There is usually a molecular size for which a given polymer property will be optimum for a particular application. So a polymer sample containing the greatest number of molecules of that size will have the optimum property. Since samples with the same average molecular weight may possess different molecular-weight distributions, information regarding molecular-weight distribution is necessary for a proper choice of polymer for optimum performance. A variety of fractionation techniques, such as fractional precipitation, precipitation chromatography, and gel permeation chromatography (GPC), based on properties such as solubility and permeability, which vary with molecular weight, may be used for separating polymers of narrow size ranges.

Example 1

A sample of poly(vinyl chloride) is composed according to the following fractional distribution (Figure 1.3).

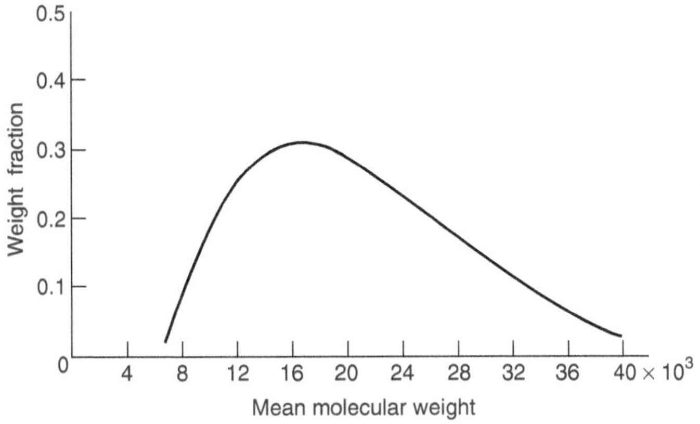

FIGURE 1.3 Molecular-weight distribution of a polymer.

Wt fraction	0.04	0.23	0.31	0.25	0.13	0.04
Mean mol. wt $\times 10^{-3}$	7	11	16	23	31	39

(a) Compute $\bar{M}_n$, $\bar{M}_w$, $\overline{DP}_n$, and $\overline{DP}_w$.

(b) How many molecules per gram are there in the polymer?

Answer. (a)

Wt fraction (w_i)	Mean mol. wt (M_i)	$w_i \times M_i$	w_i/M_i
0.04	7,000	280	0.57×10^{-5}
0.23	11,000	2,530	2.09×10^{-5}
0.31	16,000	4,960	1.94×10^{-5}
0.25	23,000	5,750	1.90×10^{-5}
0.13	31,000	4,030	0.42×10^{-5}
0.04	39,000	1,560	0.10×10^{-5}
Σ		19,110	6.21×10^{-5}

From Equation 1.2

$$\bar{M}_n = \frac{1}{6.21 \times 10^{-5}} = 16{,}100 \text{ g/mole}$$

From Equation 1.4

$$\bar{M}_w = 19{,}110 \text{ g/mole}$$

1 mer weight of vinyl chloride (C_2H_3Cl) = (2)(12) + (3)(1) + 35.5 = 62.5 g/mer

$$\overline{DP}_n = \frac{16{,}000 \text{ g/mole}}{62.5 \text{ g/mer}} = 258 \text{ mers/mole}$$

$$\overline{DP}_w = \frac{19{,}110\ \text{g/mole}}{62.5\ \text{g/mer}} = 306\ \text{mers/mole}$$

(b) Number of molecules per gram

$$= \sum \frac{w_i}{M_i}(\text{Avogadro number})$$

$$= (6.21 \times 10^{-5})(6.02 \times 10^{23})$$

$$= 3.74 \times 10^{19}\ \text{molecules/g}$$

1.3 Polymerization Processes

There are two fundamental polymerization mechanisms. Classically, they have been differentiated as addition polymerization and condensation polymerization. In the addition process, no by-product is evolved, as in the polymerization of vinyl chloride (see below); whereas in the condensation process, just as in various condensation reactions (e.g., esterification, etherification, amidation, etc.) of organic chemistry, a low-molecular-weight by-product (e.g., H_2O, HCl, etc.) is evolved. Polymers formed by addition polymerization do so by the successive addition of unsaturated monomer units in a chain reaction promoted by the active center. Therefore, addition polymerization is called chain polymerization. Similarly, condensation polymerization is referred to as step polymerization since the polymers in this case are formed by stepwise, intermolecular condensation of reactive groups. Another polymerization process that has now appeared as a new research area of considerable interest is supramolecular polymerization (see later).

1.3.1 Addition or Chain Polymerization

In chain polymerization, a simple, low-molecular-weight molecule possessing a double bond, referred to in this context as a monomer, is treated so that the double bond opens up and the resulting free valences join with those of other molecules to form a polymer chain. For example, vinyl chloride polymerizes to poly(vinyl chloride):

$$\underset{\text{Vinyl chloride}}{H_2C{=}\underset{\displaystyle Cl}{\underset{|}{CH}}} \xrightarrow{\text{Polymerization}} \underset{\text{Poly(vinyl chloride)}}{\left(CH_2{-}\underset{\displaystyle Cl}{\underset{|}{CH}} \right)_n} \qquad (1.9)$$

It is evident that no side products are formed; consequently the composition of the mer or repeating unit of the polymer ($-CH_2-CHCl-$) is identical to that of the monomer ($CH_2{=}CHCl$). The identical composition of the repeating unit of a polymer and its monomer(s) is, in most cases, an indication that the polymer is an addition polymer formed by chain polymerization process. The common addition polymers and the monomers from which they are produced are shown in Table 1.1.

Chain polymerization involves three processes: chain initiation, chain propagation, and chain termination. (A fourth process, chain transfer, may also be involved, but it may be regarded as a combination of chain termination and chain initiation.) Chain initiation occurs by an attack on the monomer molecule by a free radical, a cation, or an anion; accordingly, the chain polymerization processes are called free-radical polymerization, cationic polymerization, or anionic polymerization. (In coordination addition or chain polymerization, described below separately, the chain initiation step is,

TABLE 1.1 Typical Addition Polymers (Homopolymers)

	Monomer	Polymer	Comments
1.	Ethylene $CH_2{=}CH_2$	Polyethylene (PE) $-\!\left(CH_2-CH_2\right)_n\!-$	High density polyethylene (HDPE) and low density polyethylene (LDPE); molded objects, tubing, film, electrical insulation, used for household products, insulators, pipes, toys, bottles, e.g., Alkathene, Lupolan, Hostalen, Marlex.
2.	Propylene $CH_2{=}CH(CH_3)$	Polypropylene (PP) $-\!\left(CH_2-CH(CH_3)\right)_n\!-$	Lower density, stiffer, and higher temperature resistance than PE; used for water pipes, integral hinges, sterilizable hospital equipment, e.g., Propathene, Novolen, Moplen, Hostalen, Marlex.
3.	Styrene 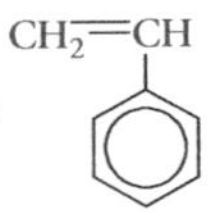	Polystyrene (PS) 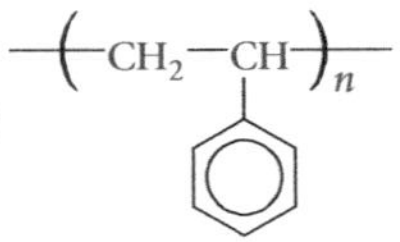	Transparent and brittle; used for cheap molded objects, e.g., Styron, Carinex, Hostyren, Lustrex. Modified with rubber to improve toughness, e.g., High impact Polystyrene (HIPS) and acrylonitrile–butadiene–styrene copolymer (ABS). Expanded by volatilization of a blended blowing agent (e.g., pentane) to make polystyrene foam, e.g., Styrocell, Styrofoam
4.	Acrylonitrile $H_2C{=}CH-CN$	Polyacrylonitrile $-\!\left(CH_2-CH(CN)\right)_n\!-$	Widely used as fibers; best alternative to wool for sweaters, e.g., Orlon, Acrilan.
5.	Vinylacetate $CH_2{=}CH-O-C(=O)-CH_3$	Poly(vinyl acetate) $-\!\left(CH_2-CH(O-C(=O)-CH_3)\right)_n\!-$	Emulsion paints, adhesives, sizing, chewing gum, e.g., Flovic, Mowilith, Mowicoll.
6.	Vinyl chloride $CH_2{=}CH-Cl$	Poly(vinyl chloride) (PVC) $-\!\left(CH_2-CH(Cl)\right)_n\!-$	Water pipes, bottles, gramophone records, plasticized to make PVC film, leather cloth, raincoats, flexible pipe, tube, hose, toys, electrical cable sheathing, e.g., Benvic, Darvic, Geon, Hostalit, Solvic, Vinoflex, Welvic.

(continued)

TABLE 1.1 (Continued)

	Monomer	Polymer	Comments
7.	Tetrafluoroethylene $CF_2{=}CF_2$	Polytetrafluoroethylene $-\!\!(CF_2{=}CF_2)\!\!-$	High Temperature resistance, chemically inert, excellent electrical insulator, very low coefficient of friction, expensive; moldings, films, coatings, used for non-stick surfaces, insulation, gaskets; e.g. Teflon, Fluon.
8.	Methyl methacrylate $CH_2{=}C(CH_3)-C(=O)-OCH_3$	Poly(methyl methacrylate) (PMMA) $-\!\!(CH_2-C(CH_3)(C{=}O\,OCH_3))_n-$	Transparent sheets and moldings; more expensive than PS; known as *organic* glass, used for aeroplane windows; e.g., Perspex, Plexiglass, Lucite, Diakon, Vedril.
9.	Isobutylene $CH_2{=}C(CH_3)_2$	Polyisobutylene (PIB) $-\!\!(CH_2-C(CH_3)_2)_n-$	Lubricating oils, sealants, copolymerized with 0.5–2.5 mol% isoprene to produce Butyl rubber for tire inner tubes and inner liners of tubeless tires.
10.	Isoprene $CH_2{=}C(CH_3)-CH{=}CH_2$	*cis*-1,4-Polyisoprene $-\!\!(CH_2-C(CH_3){=}CH-CH_2)_n-$	Tires, mechanical goods, footwear, sealants, caulking compounds, e.g., Coral, Natsyn, Clariflex I.
11.	Butadiene $CH_2{=}CH-CH{=}CH_2$	*cis*-1,4-Polybutadiene $-\!\!(CH_2-CH{=}CH-CH_2)_n-$	Tires and tire products, e.g., *Cis*-4, Ameripol-CB, Diene.

however, assumed to be the insertion of the first monomer molecule into a transition metal–carbon bond.) A free radical is a reactive substance having an unpaired electron and is usually formed by the decomposition of a relatively unstable material called an *initiator.* Benzoyl peroxide is a common free-radical initiator and can produce free radicals by thermal decomposition as

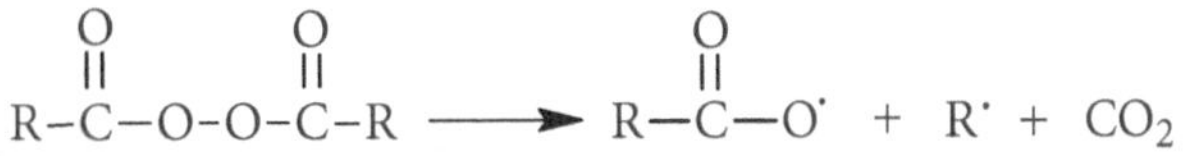

(1.10)

(R = Phenyl group for benzoyl peroxide initiator)

Free radicals are, in general, very active because of the presence of unpaired electrons (denoted by dot). A free-radical species can thus react to open the double bond of a vinyl monomer and add to one side of the broken bond, with the reactive center (unpaired electron) being transferred to the other side of the broken bond:

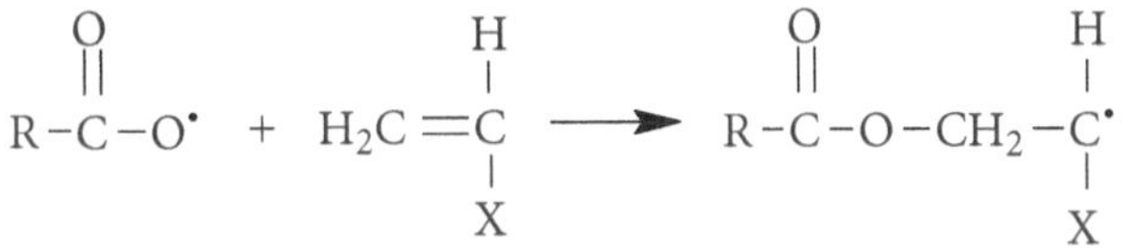

(1.11)

(X = CH_3, C_6H_5, Cl, etc.)

The new species, which is also a free radical, is able to attack a second monomer molecule in a similar way, transferring its reactive center to the attacked molecule. The process is repeated, and the chain continues to grow as a large number of monomer molecules are successively added to propagate the reactive center:

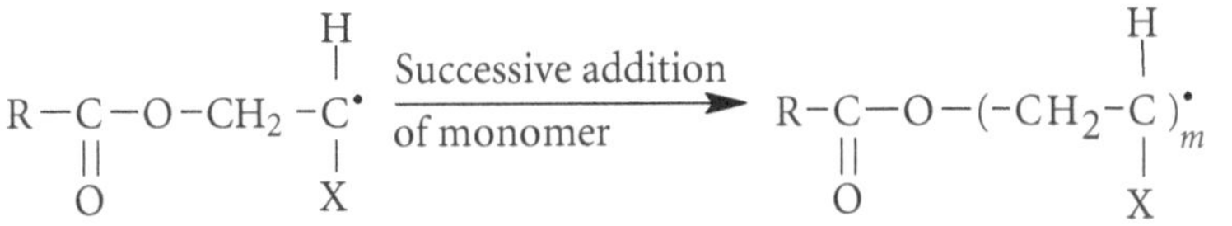

(1.12)

This process of propagation continues until another process intervenes and destroys the reactive center, resulting in the termination of the polymer growth. There may be several termination reactions depending on the type of the reactive center and the reaction conditions. For example, two growing radicals may combine to annihilate each other's growth activity and form an inactive polymer molecule; this is called termination by combination or coupling:

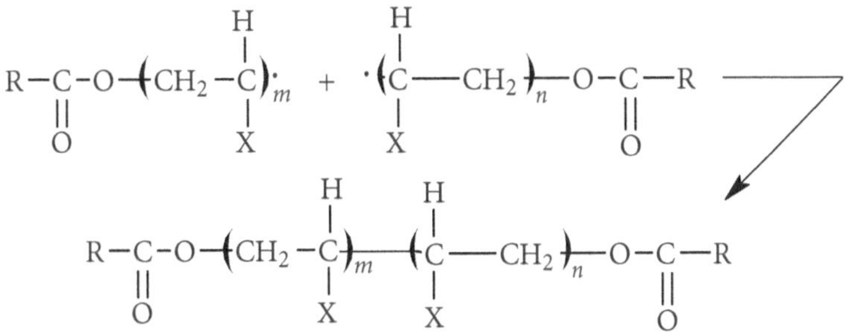

(1.13)

A second termination mechanism is disproportionation, shown by the following equation:

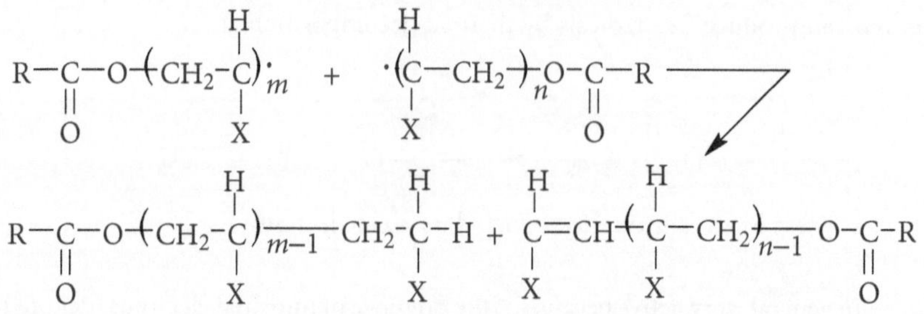

(1.14)

In chain polymerization initiated by free radicals, as in the previous example, the reactive center, located at the growing end of the molecule, is a free radical. Similarly, in chain polymerizations initiated by ionic systems, the reactive center is ionic, i.e., a carbonium ion (in cationic initiation) or a carbanion (in anionic initiation). Regardless of the chain initiation mechanism—free radical, cationic, or anionic—once a reactive center is produced it adds many more molecules in a chain reaction and grows quite large extremely rapidly, usually within a few seconds or less. (However, the relative slowness of the initiation stage causes the overall rate of reaction to be slow and the conversion of all monomers to polymers in most polymerizations requires at least 30 min, sometimes hours.) Evidently, at any time during a chain polymerization process the reaction mixture will consist only of unreacted monomers, high polymers and unreacted initiator species, but no intermediate sized molecules. The chain polymerization will thus show the presence of high-molecular-weight polymer molecules at all extents of conversion (see Figure 1.4). In certain ionic chain polymerizations, which feature a fast initiation process coupled with the absence of reactions that terminate the propagating reactive centers, molecular weight increases linearly with conversion. This is known as "living" ionic chain polymerization.

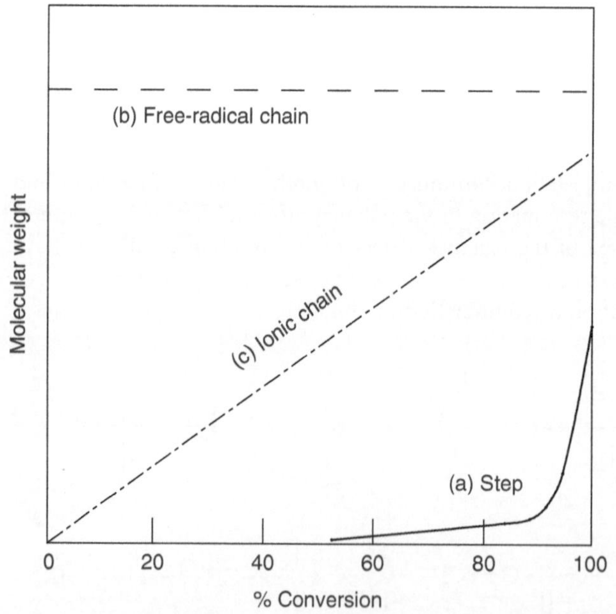

FIGURE 1.4 Variation of molecular weight with conversion in (a) step polymerization, (b) free-radical polymerization, and (c) ionic chain polymerization.

1.3.2 Coordination Addition Polymerization

Many polymers are now manufactured on a commercial scale using Ziegler–Natta catalysts, an outstanding example being polypropylene of high molecular weight which cannot be made by commercial processes of free-radical or ionic chain polymerization. Perhaps the best known Ziegler–Natta systems are those derived from $TiCl_4$ or $TiCl_3$ and an aluminum alkyl. The catalyst systems appear to function by formation of a coordination complex between the catalyst, growing chain, and incoming monomer. Hence the process is referred to as coordination addition polymerization and the catalysts as coordination catalysts. Polymers with stereoregular structures (see later) can be produced with these catalysts.

The efficiency or activity of the early Ziegler–Natta catalyst systems was low. The term activity usually refers to the rate of polymerization, expressed in terms of kilograms of polymer formed per gram of catalyst. Thus a low activity meant that large amounts of catalyst were needed to obtain reasonably high yields of polymer, and the spent catalyst had then to be removed from the product to avoid contamination. This problem effectively disappeared with the advent of subsequent generations of catalysts leading to large increases in activity without loss of stereospecificity. This was achieved by increasing the effective surface area of the active component by more than two orders of magnitude through impregnation of the catalyst on a solid support such as $MgCl_2$ or MgO. For example, in contrast to a typical $TiCl_3$–AlR_3 catalyst which yields about 50–200 g of polyethylene per gram of catalyst per hour per atmosphere of ethylene, as much as 200,000 g of polyethylene and over 40,000 g of polypropylene per gram titanium per hour may be produced using a $MgCl_2$-supported catalyst, thus obviating the need to remove the spent catalyst (a costly step) from the product. Such catalyst systems are often referred to as high-mileage catalysts. Stereospecificity of the catalyst is kept high (>90%–98% isotactic dyads) by adding electron-donor additives such as ethyl benzoate.

The catalyst complex of the $TiCl_3/AlR_3$ system essentially acts as a template for the successive orientation and isotactic placement of the incoming monomer units. Though a number of structures have been proposed for the active species, they fall into either of two general categories: monometallic and bimetallic, depending on the number of metal centers. The two types can be illustrated by the structures (I) and (II) for the active species from titanium chloride ($TiCl_4$ or $TiCl_3$) and alkylaluminum (AlR_3 or AlR_2Cl).

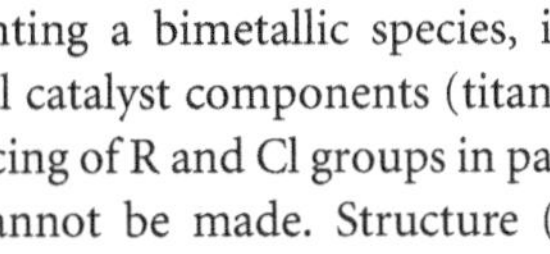

(I)

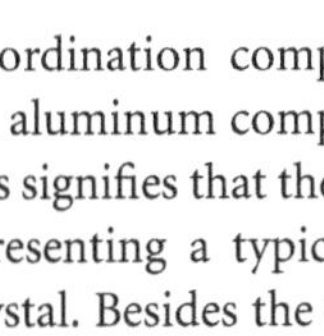

(II)

Structure (**I**), representing a bimetallic species, is the coordination complex that arises from the interaction of the original catalyst components (titanium and aluminum compounds) with exchange of R and Cl groups. The placing of R and Cl groups in parentheses signifies that the exact specification of the ligands on Ti and Al cannot be made. Structure (**II**), representing a typical monometallic species, constitutes an active titanium site at the surface of a $TiCl_3$ crystal. Besides the four chloride ligands that the central Ti atom shares with its neighboring Ti atoms, it has an alkyl ligand (received through exchange reactions with alkyl aluminum) and a vacant orbital (□).

The truly active bimetallic catalysts are complexes that have an electron-deficient bond, e.g., Ti···C···Al in (**I**). Chain propagation by the bimetallic mechanism [7] occurs at two metal centers of the bridge complex as shown in Figure 1.5. With the chain growth taking place always from the metal end, the incoming monomer is oriented, for steric reasons, with the $=CH_2$ group pointing into the lattice and the CH_3 group to one side, with the result that the process leads to the formation of an isotactic polymer.

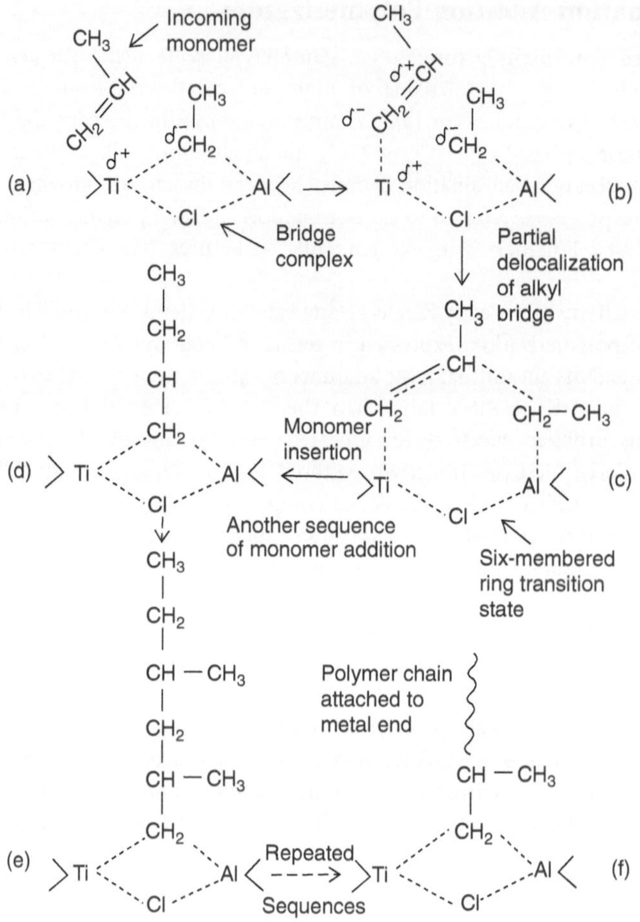

FIGURE 1.5 Bimetallic mechanism for stereospecific polymerization. (Adapted from Patat, F. and Sinn, H. 1958. *Angew. Chem.*, 70, 496.)

While a limited amount of experimental evidence does lend support to the bimetallic concept, majority opinion, however, favors the second and simpler alternative, the monometallic mechanism (described next).

It is generally accepted that the d-orbital in the transition element are the main source of catalytic activity and that it is the Ti-alkyl bond that acts as the polymerization center where chain growth occurs. For α-$TiCl_3$ catalyst the active center [8] is formed by the interaction of aluminum alkyl with an octahedral vacancy around Ti, as shown in Figure 1.6. To elaborate, the five-coordinated Ti^{3+} on the surface has a vacant d-orbital, represented by -□, which facilitates chemisorption of the aluminum alkyl and this is followed by alkylation of the Ti^{3+} ion by an exchange mechanism to form the active center $TiRCl_4$-□. The vacant site at the active center can accommodate the incoming monomer unit, which forms a π-complex with the titanium at the vacant d-orbital and is then inserted into the Ti-alkyl bond. The sequence of steps is shown in Figure 1.7 using propylene as the monomer.

After the monomer is inserted into the Ti-alkyl bond, the polymer chain migrates back to its initial position, while the vacant site migrates to its original position to accept another monomer molecule. This migration is necessary, as otherwise an alternating position would be offered to the monomer leading to the formation of a syndiotactic polymer instead of an isotactic polymer.

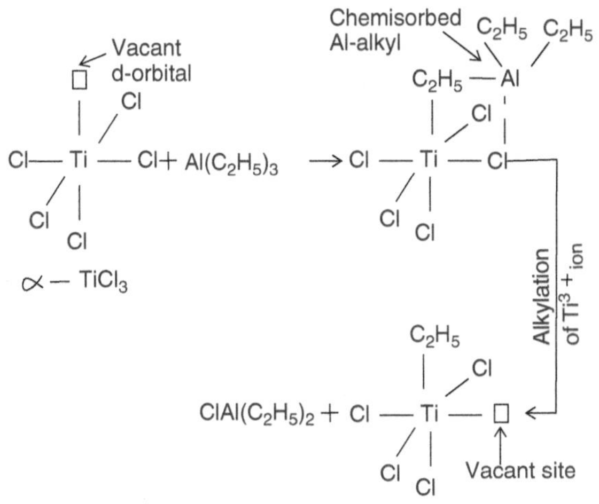

FIGURE 1.6 Interaction of aluminum alkyl with an octahedral vacancy around Ti in the first stage of monometallic mechanism. (After Cossee, P. 1967. *The Stereochemistry of Macromolecules*, A. D. Ketley, ed., *Vol. 6*. Marcel Dekker, New York.)

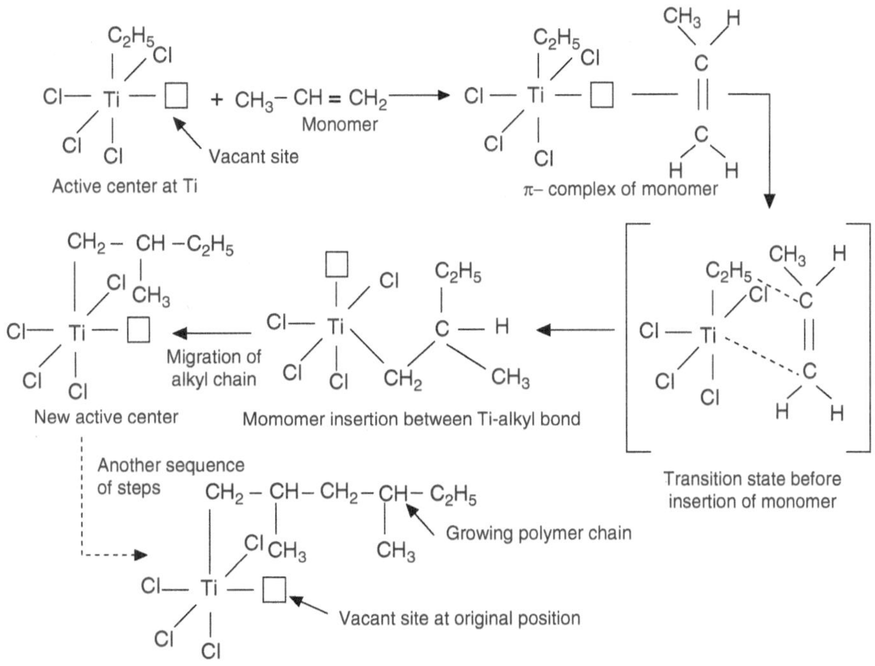

FIGURE 1.7 Monometallic mechanism for stereospecific polymerization. (After Cossee, P. 1967. *The Stereochemistry of Macromolecules*, A. D. Ketley, ed., *Vol. 6*. Marcel Dekker, New York.)

The termination of a polymer chain growing at an active center may occur by various reactions, as shown below with propylene as the example.

1. *Chain Transfer to Monomer:*

$$\bullet\!-\!\mathrm{Ti{-}CH_2{-}CH(CH_3)}\!\sim\!\sim\!\sim + \mathrm{CH_3CH{=}CH_2} \xrightarrow{k_{tr,M}} \bullet\!-\!\mathrm{Ti{-}CH_2CH_2CH_3} + \mathrm{CH_2{=}C(CH_3)}\!\sim\!\sim\!\sim \quad (1.15)$$

$$\bullet\!-\!\mathrm{Ti{-}CH_2{-}CH(CH_3)}\!\sim\!\sim\!\sim + \mathrm{CH_3CH{=}CH_2} \xrightarrow{k_{tr,M}} \bullet\!-\!\mathrm{Ti{-}CH{=}CH{-}CH_3} + \mathrm{CH_3{-}CH(CH_3)}\!\sim\!\sim\!\sim \quad (1.16)$$

   where ●-Ti represents the transition metal active center on the catalyst site at which chain propagation takes place. Note that it is the methylene carbon atom from the monomer that is bonded to the transition metal atom (cf. Figure 1.5).

2. *Chain transfer to the Group I–III metal alkyl:*

$$\bullet\!-\!\mathrm{Ti{-}CH_2{-}CH(CH_3)}\!\sim\!\sim\!\sim + \mathrm{Al(C_2H_5)_3} \longrightarrow \bullet\!-\!\mathrm{Ti{-}CH_2CH_3} + \mathrm{(C_2H_5)_2Al{-}CH_2{-}CH(CH_3)}\!\sim\!\sim\!\sim \quad (1.17)$$

3. *Spontaneous intramolecular hydride transfer:*

$$\bullet\!-\!\mathrm{Ti{-}CH_2{-}CH(CH_3)}\!\sim\!\sim\!\sim \xrightarrow{k_s} \bullet\!-\!\mathrm{Ti{-}H} + \mathrm{CH_2{=}CH(CH_3)}\!\sim\!\sim\!\sim \quad (1.18)$$

4. *Chain transfer to an active hydrogen compound such as molecular hydrogen (external agent):*

$$\bullet\!-\!\mathrm{Ti{-}CH_2{-}CH(CH_3)}\!\sim\!\sim\!\sim + \mathrm{H_2} \xrightarrow{k_{tr,H_2}} \bullet\!-\!\mathrm{Ti{-}H} + \mathrm{CH_3{-}CH(CH_3)}\!\sim\!\sim\!\sim \quad (1.19)$$

The above reactions terminating the growth of polymer chains are indeed chain transfer reactions since in each case a new propagating chain is initiated. The relative extents of these reactions depend on various factors such as the monomer, the initiator components, temperature, concentrations, and other reaction conditions. Under normal conditions of polymerization, intramolecular hydride transfer is negligible and termination of propagating chains occurs mostly by chain transfer processes. Being a highly effective chain transfer agent, molecular hydrogen is often used for polymer molecular weight control.

1.3.3 Step Polymerization

Step polymerization occurs by stepwise reaction between functional groups of reactants. The reaction leads successively from monomer to dimer, trimer, tetramer, pentamer, and so on, until finally a polymer molecule with large DP is formed. Note, however, that reactions occur at random between the intermediates (e.g., dimers, trimers, etc.) and the monomer as well as among the intermediates themselves. In other words, reactions of both types, namely,

$$n - \text{mer} + \text{monomer} \rightarrow (n+1) - \text{mer}$$

and

$$n - \text{mer} + m - \text{mer} \rightarrow (n+m) - \text{mer}$$

occur equally. Thus, at any stage the product consists of molecules of varying sizes, giving a range of molecular weights. The average molecular weight builds up slowly in the step polymerization process, and a high-molecular-weight product is formed only after a sufficiently long reaction time when the conversion is more than 98% (see Figure 1.4a).

Since most (though not all) of the step polymerization processes involve poly-condensation (repeated condensation) reactions, the terms step polymerization and condensation polymerization are often used synonymously. Consider, for example, the synthesis of a polyamide, i.e., a polymer with amide (–CONH–) as the characteristic linkage. If we start with, say, hexamethylenediamine and adipic acid as reactants, the first step in the formation of the polymer (nylon) is the following reaction producing a monoamide:

$$\begin{array}{l} H_2N{-}(CH_2)_6{-}NH_2 + HO{-}\underset{\underset{O}{\|}}{C}{-}(CH_2)_4{-}\underset{\underset{O}{\|}}{C}{-}OH \longrightarrow \\ H_2N{-}(CH_2)_6{-}NH{-}\underset{\underset{O}{\|}}{C}{-}(CH_2)_4{-}\underset{\underset{O}{\|}}{C}{-}OH + H_2O \end{array} \tag{1.20}$$

The reaction continues step-by-step to give the polyamide nylon-6,6. The overall reaction may thus be represented as

$$\begin{array}{l} n\,H_2N{+}CH_2{+}_6NH_2 + n\,HO{-}\underset{\underset{O}{\|}}{C}{+}CH_2{+}_4\underset{\underset{O}{\|}}{C}{-}OH \longrightarrow \\ H{-}\!\left[NH{+}CH_2{+}_6NH{-}\underset{\underset{O}{\|}}{C}{+}CH_2{+}_4\underset{\underset{O}{\|}}{C}\right]_n\!{-}OH + H_2O \end{array} \tag{1.21}$$

Poly(hexamethylene adipamide)

We see that the composition of the repeating unit (enclosed in square brackets) equals that of two monomer molecules minus two molecules of water. Thus a condensation polymer may be defined as one whose synthesis involves elimination of small molecules or whose repeating unit lacks certain atoms present in the monomer(s).

With the development of polymer science and the synthesis of new polymers, the previous definition of condensation polymer is inadequate. For example, in polyurethanes (Table 1.2), which are classified as condensation polymers, the repeating unit has the same net composition as the two monomers (i.e., a diol and a diisocyanate), which react without eliminating any small molecule. To overcome such problems, chemists have introduced a definition which describes condensation polymers as consisting of structural units joined by internal functional groups such as

ester $(-\underset{\underset{O}{\|}}{C}{-}O{-})$, amide $(-\underset{\underset{O}{\|}}{C}{-}NH{-})$, imide $\left(-N{<}{}^{CO-}_{CO-}\right)$, urethane $(-O{-}\underset{\underset{O}{\|}}{C}{-}NH{-})$, sulfide $(-S-)$, ether $(-O-)$, carbonate $(-O{-}\underset{\underset{O}{\|}}{C}{-}O{-})$, and sulfone $(-\underset{\underset{O}{\|}}{\overset{\overset{O}{\|}}{S}}-)$ linkages.

TABLE 1.2 Typical Condensation Polymers

	Polymerization Reaction[a]	Comments
Polyamide (PA)	$n\,H_2N{-}R{-}NH_2 + n\,HOC(=O){-}R'{-}C(=O)OH \longrightarrow H{-}\!\left(NH{-}R{-}NHC(=O){-}R'{-}C(=O)\right)_n\!{-}OH + (2n-1)\,H_2O$ $n\,H_2N{-}R{-}NH_2 + n\,Cl{-}C(=O){-}R'{-}C(=O){-}Cl \longrightarrow H{-}\!\left(NH{-}R{-}NHC(=O){-}R'{-}C(=O)\right)_n\!{-}Cl + (2n-1)\,HCl$ $n\,H_2N{-}R{-}C(=O)OH \longrightarrow H{-}\!\left(NH{-}R{-}C(=O)\right)_n\!{-}OH + (n-1)\,H_2O$	Moldings, fibers, tirecord; poly(hexamethylene adipamide) (Nylon 6,6)e.g., Ultramid A; polycaprolactam (nylon-6), e.g., Ultramid B, Akulon, Perlenka, poly(hexamethylene sebacamide) (Nylon-6,10), e.g., Ultramid S, Zytel.
Polyester	$n\,HO{-}R{-}OH + n\,HOC(=O){-}R'{-}C(=O)OH \longrightarrow HO{-}\!\left(R{-}OC(=O){-}R'{-}C(=O)O\right)_n\!{-}H + (2n-1)\,H_2O$ $n\,HO{-}R{-}OH + n\,R''OC(=O){-}R'{-}C(=O)OR'' \longrightarrow HO{-}\!\left(R{-}OC(=O){-}R'{-}C(=O)O\right)_n\!{-}R'' + (2n-1)\,R''OH$ $n\,HO{-}R{-}C(=O)OH \longrightarrow HO{-}\!\left(R{-}C(=O)O\right)_n\!{-}H + (n-1)\,H_2O$	Textile fibers, film, bottles; poly(ethylene terephthalate) (PET) e.g., Terylene, Dacron, Melinex, Mylar.
Polyurethane (PU)	$n\,HO{-}R{-}OH + n\,OC{:}N{-}R'{-}NCO \longrightarrow H{-}\!\left(O{-}R{-}OC(=O){-}NH{-}R'{-}NHC(=O)\right)_{(n-1)}\!{-}O{-}R{-}OC(=O){-}NH{-}R'{-}NCO$	Rubbers, foams, coatings; e.g., Vulkollan, Adiprene C, Chemigum SL, Desmophen A, Moltopren.
Polysulphide	$n\,Cl{-}R{-}Cl + n\,Na_2S_x \longrightarrow {-}\!\left(R{-}S_x\right)_n\!{-} + 2n\,NaCl$	Adhesives, sealants, binders, hose, e.g., Thiokol.

Polysiloxane	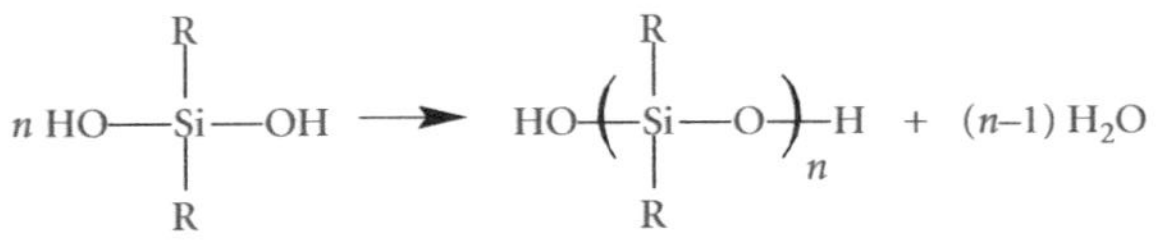	Elastomers, sealants, fluids, e.g., Silastic, Silastomer, Silopren.
Phenol–formaldehyde (PF)	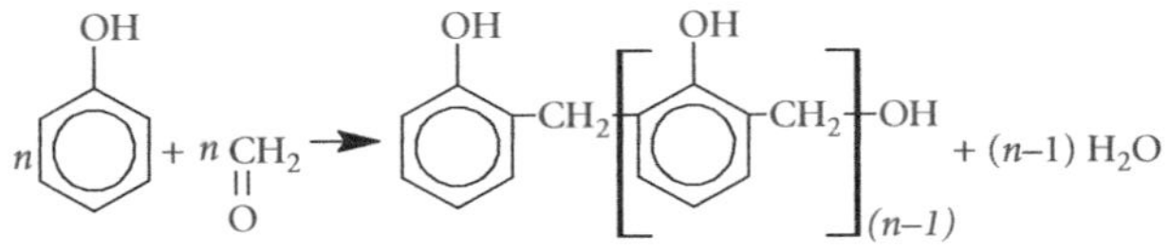	Plywood adhesives, glass-fiber insulation, molding compound, e.g., Hitanol, Sirfen, Trolitan.
Urea–formaldehyde (UF)		Particle-board binder resin, paper and textile treatment, molding compounds, coatings, e.g., Beetle, Resolite, Cibanoid.
Melamine–formaldehyde (MF)	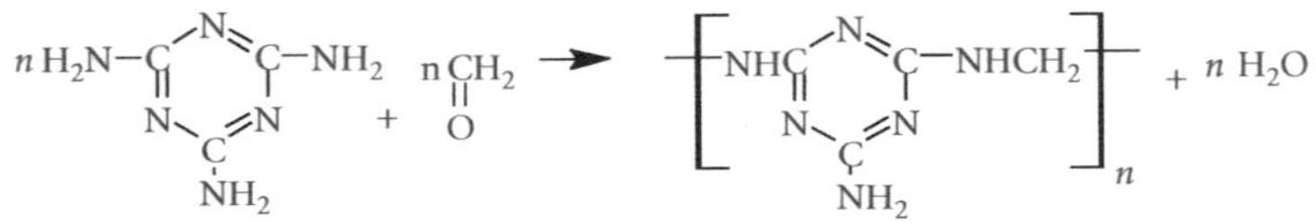	Dinnerware, table tops, coatings, e.g., Formica, Melalam, Cymel.

[a] R, R′, R″ represent aliphatic or aromatic ring. The repeating unit of the polymer chain is enclosed in parentheses.

A polymer satisfying either or both of the above definitions is classified as a condensation polymer. Phenol–formaldehyde, for example, satisfies the first definition but not the second. Some condensation polymers along with their repeating units and condensation reactions by which they can be synthesized are shown in Table 1.2. Some high-performance polymers prepared by polycondensation are listed in Table 1.3.

The ring-opening polymerizations of cyclic monomers, such as propylene oxide,

$$H_3C-\overbrace{CH-CH_2}^{O} \longrightarrow \left[CH_2-\underset{\substack{|\\CH_3}}{CH}-O\right]_n \qquad (1.22)$$

or –caprolactam

$$\left(CH_2\right)_5\!\!\!\overbrace{}^{\overset{O}{\overset{\|}{C}}}\!\!\!NH \longrightarrow \left[NH-\left(CH_2\right)_5-CO\right]_n \qquad (1.23)$$

proceed either by chain or step mechanisms, depending on the particular monomer, reaction conditions, and initiator employed. However, the polymers produced in Equation 1.22 and Equation 1.23 will be structurally classified as condensation polymers, since they contain functional groups (e.g., ether, amide) in the polymer chain. Such polymerizations thus point out very clearly that one must distinguish between the classification based on polymerization mechanism and that based on polymer structure. The two classifications cannot always be used interchangeably. Both structure and mechanism are usually needed in order to clearly classify a polymer.

1.3.4 Supramolecular Polymerization

Supramolecular polymers are a relatively new class of polymers in which monomeric repeating units are held together with directional and reversible (noncovalent) secondary interactions, unlike conventional macromolecular species in which repetition of monomeric units is mainly governed by covalent bonding. A schematic comparison of a covalent polymer and a supramolecular polymer is shown in Figure 1.8.

The directionality and strength of the supramolecular bonding, such as hydrogen bonding, metal coordination, and π–π interactions, are important features resulting in polymer properties in dilute and concentrated solutions, as well as in the bulk. It should be noted that supramolecular interactions are not new to polymer science, where hydrogen bonding and other weak reversible interactions are important in determining polymer properties and architectures. However, for linear supramolecular polymers to form, it is a prerequisite to have strong and highly directional interactions as a reversible alternative for the covalent bond. Hydrogen bonds between neutral organic molecules, though they hold a prominent place in supramolecular chemistry because of their directionality and versatility, are not among the strongest noncovalent interactions. Hence, either multiple hydrogen bonds with cooperativity must be used or hydrogen bonds should be supported by additional forces like excluded volume interactions [9]. Though the concept has been known for years, it was not known how to incorporate such sufficiently strong but still reversible interactions. However, in the past decade following the development of strong hydrogen-bonding dimers, several research groups have applied these dimers for the formation of hydrogen-bonded supramolecular polymers. Thus the finding by Sijbesma et al. [10] that derivatives of 2-ureido-4[1H]-pyrimidinone (UPy, **1** in Figure 1.9) are easy to synthesize and they dimerize strongly (dimerization constant $>10^6$ M^{-1} in $CHCl_3$) by self-complimentary quadrupole (array of four) hydrogen bonding (**2** in Figure 1.9) prompted them to use this functionality as the associating end group in reversible self-assembling polymer systems.

TABLE 1.3 Some High-Performance Condensation Polymers

Polymer Type and Polycondensation Reaction		Comments
Polycarbonate (PC)	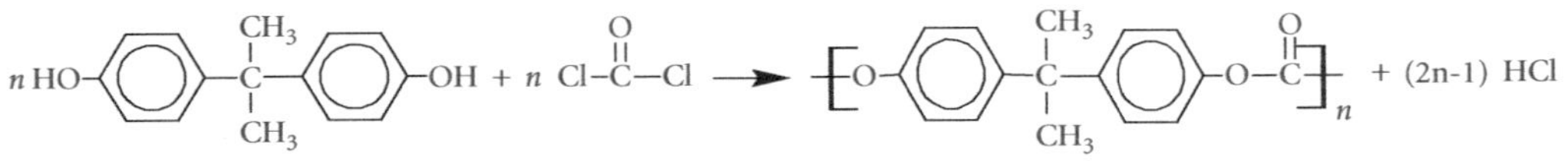	Moldings and sheets; transparent, tough and physiologically inert: used for safety glasses, lenses, screens and glazings, electrical and electronics, appliances, compact discs, e.g. Merlon, Baylon, Jupilon.
Polyether sulfone (PES)	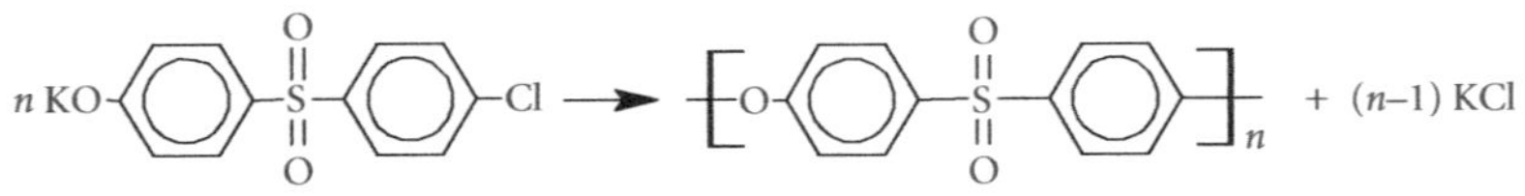	Moldings, coatings, membranes; rigid, transparent, self-extinguishing, resistant to heat deformation: used for electrical components, molded circuit boards, appliances operating at high temperatures, e.g., Victrex PES.
Polyetherether ketone (PEEK)		Moldings, composites, bearings, coatings; very high continuous use temperature (260°C): used in coatings and insulation for high performance wiring, composite prepregs with carbon fibers, e.g., Victrex PEEK.
Poly(phenylene sulphide) (PPS)	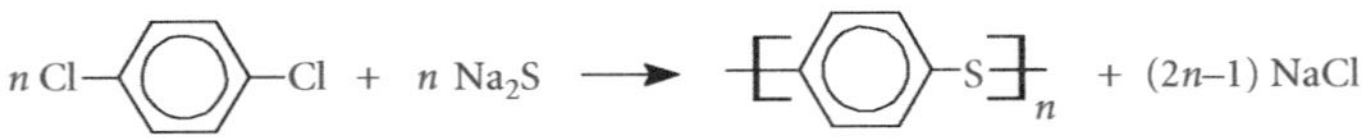	Moldings, composites, coatings; outstanding in heat resistance, flame resistance, chemical resistance and electrical insulation resistance: used for electrical components, mechanical parts, e.g., Ryton, Tedur, Fortron.

(continued)

TABLE 1.3 (Continued)

Polymer Type and Polycondensation Reaction		Comments
Poly(*p*-phenylene terephthalamide)		High modulus fibers; as strong as steel but have one-fifth of weight, ideally suited as tire cord materials and for ballistic vests, e.g., Kevlar, Twaron.
Polyimide	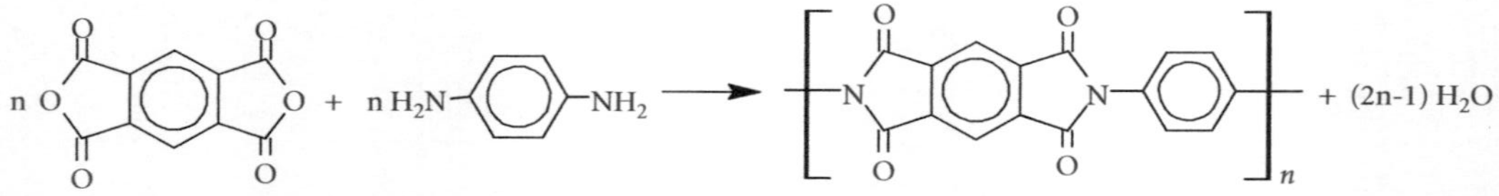	Films, coatings, adhesives, laminates; outstanding in heat resistance, flame resistance, abrasion resistance, electrical insulation resistance, resistance to oxidative degradation, high energy radiation and most chemicals (except strong bases): used in specialist applications, e.g., Kapton, Vespel.

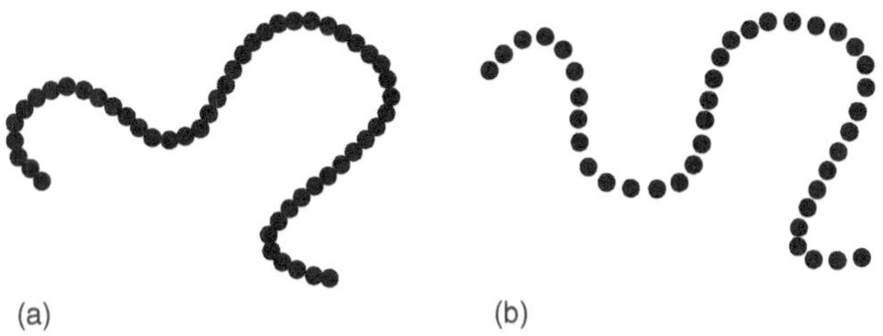

FIGURE 1.8 Schematic representation of (a) a covalent polymer and (b) a supramolecular polymer. (After Brunsveld, L., Folmer, B. J. B., Meijer, E. W., and Sijbesma, R. P. 2001. *Chem. Rev.*, 101, 4071. With permission.)

A difunctional UPy compound, **4** in Figure 1.10, possessing two UPy units can be easily made in a one step procedure, from commercially available compounds, methylisocytosine (R=CH_3) and hexyldiisocyanate (R=C_6H_{12}). The compound forms very stable and long polymer chains (5 in Figure 1.10) in solution as well as in the bulk [9,11]. Dissolving a small amount of the compound in chloroform gives solutions with high viscosities, while calculations show that polymers with molecular weights of the order of 10^6 can be formed. Deliberate addition of small amounts of monofunctional compounds (**1** in Figure 1.9) results in a sharp drop in viscosity, proving that linkages between the building blocks are reversible and unidirectional and that the monofunctional compounds act as chain stoppers. For the same reason, the supramolecular polymers show polymer-like viscoelastic behavior in bulk and solution, whereas at elevated temperatures they exhibit liquid-like properties [9].

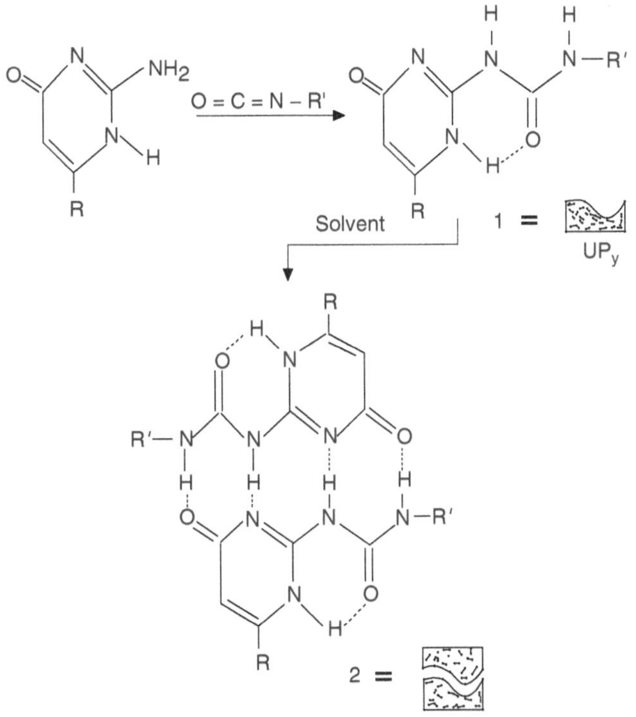

FIGURE 1.9 Synthesis of a monofunctional 2-ureido-4[1H]–pyrimidinone (UPy) (**1**) and dimerization of **1** in solution forming a quadrupole hydrogen-bonded unit. (After Sijbesma, R. P., Beijer, F. H., Brunsveld, L., Folmer, B. J. B., Hirschberg, J. H. K., Lange, R. F. M., Lowe, J. K. L., and Meijer, E. W. 1997. *Science*, 278, 1601. With permission.)

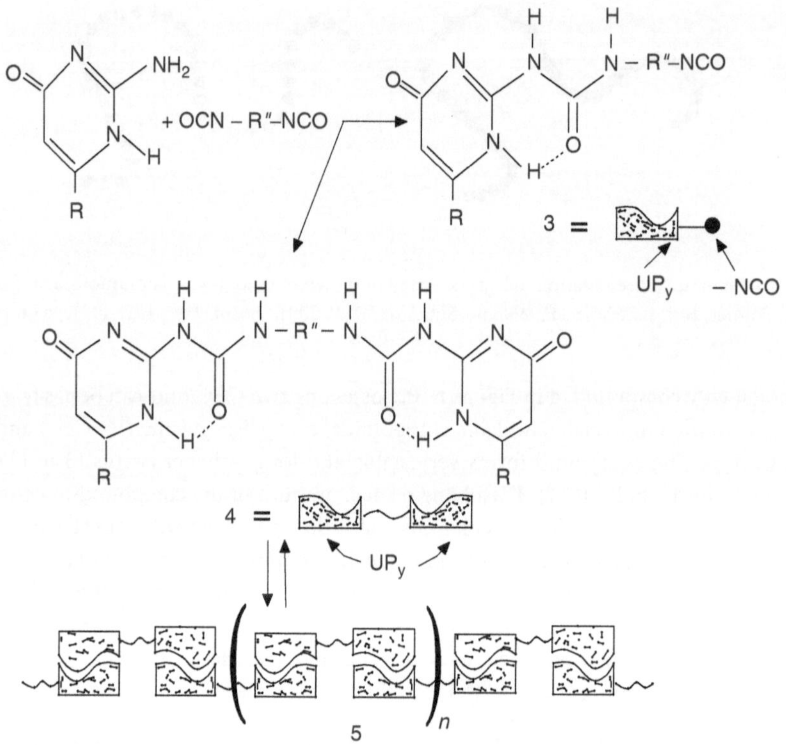

FIGURE 1.10 Preparation of (a) UPy possessing an isocyanate functional group (**3**) and (b) a difunctional UPy compound (**4**) which forms a supramolecular polymer (**5**) by hydrogen bonding (cf. Figure 1.9). (After Brunsveld, L., Folmer, B. J. B., Meijer, E. W., and Sijbesma, R. P. 2001. *Chem. Rev.*, 101, 4071 and Folmer, B. J. B., Sijbesma, R. P., Versteegen, R. M., van der Rijt, J. A. J., and Meijer, E. W. 2000. *Adv. Mater.*, 12, 12, 874. With permission.)

The quadrupole hydrogen-bonded unit can be employed in the chain extension of telechelic oligomers such as polysiloxanes, polyethers, polyesters, and polycarbonates [11]. Thus the electrophilic isocyanate group (–NCO) of "synthon" (**3** in Figure 1.10) can be reacted with common nucleophilic end groups (–OH or $-NH_2$) of telechelic oligomers, resulting in supramolecular polymers by chain extension (Figure 1.11). Thus the material properties of telechelic polymers have been shown to improve dramatically upon functionalization with synthon, and materials have been obtained that combine many of the mechanical properties of conventional macromolecules with the low melt viscosity of oligomers [9]. In contrast to conventional high-molecular-weight polymers, supramolecular (reversible) polymers with a high "virtual" molecular weight show excellent processability due to the strong temperature dependency of the melt viscosity [11]. Moreover, hybrids between blocks of covalent macromolecules and supramolecular polymers can be easily made.

1.3.5 Copolymerization

All the addition polymers we have considered so far (Table 1.1) contain only one type of repeating unit or mer in the chain. Polymers can also be synthesized by the aforesaid processes with more than one type of mer in the chain. Such polymers are called *copolymers.* They are produced by polymerizing a mixture of monomers (copolymerization) [12] or by special methods. Copolymers can be of different types, depending on the monomers used and the specific method of synthesis. The copolymer with a relatively random distribution of the different mers in its structure is referred to as a random copolymer. Representing, say, two different mers by A and B, a random copolymer can be depicted as

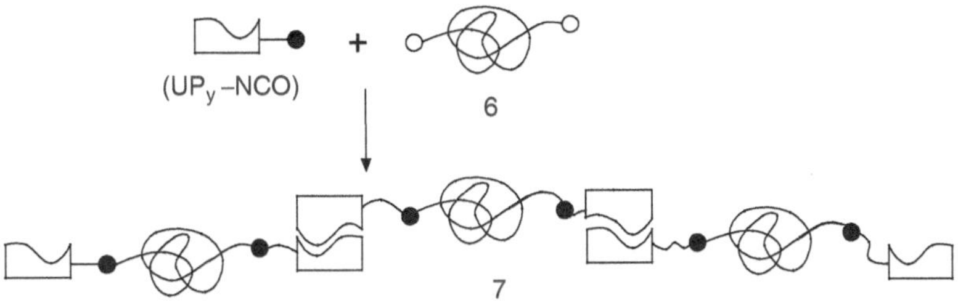

FIGURE 1.11 Schematic representation of the formation of supramolecular polymer (**7**) by chain extension of reactive telechelic oligomer with UPy. (After Folmer, B. J. B., Sijbesma, R. P., Versteegen, R. M., van der Rijt, J. A. J., and Meijer, E. W. 2000. *Adv. Mater.*, 12, 12, 874. With permission.)

ABBABBBAABBAABAAABBA

There other copolymer structures [13] are known: alternating, block, and graft copolymer structures (Figure 1.12). In the alternating copolymer the two mers alternate in a regular fashion along the polymer chain:

ABABABABABABABABABAB

A block copolymer is a linear copolymer with one or more long uninterrupted sequences of each mer in the chain:

AAAAAAAAAABBBBBBBBBB

A graft copolymer, on the other hand, is a branched copolymer with a backbone of one type of mer to which are attached one or more side chains of another mer.

AAAAAAAAAAAAAAAA
B
B
B
B
B

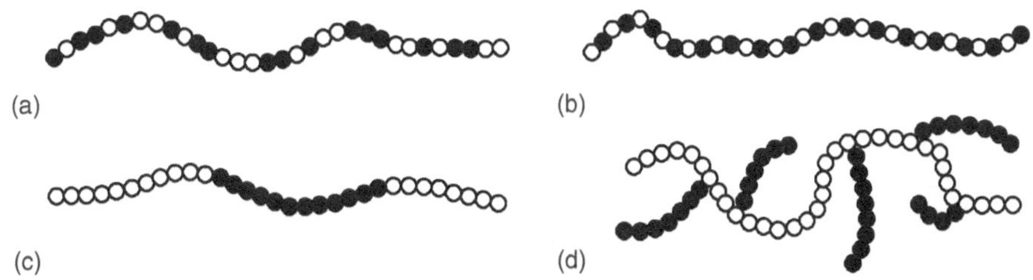

FIGURE 1.12 Copolymer arrangements. (a) Two different types of mers (denoted by open and filled circles) are randomly placed. (b) The mers are alternately arranged. (c) A block copolymer. (d) A graft copolymer.

Copolymerization, which may be compared to alloying in metallurgy, is very useful for synthesizing polymers with the required combination of properties. For example, polystyrene is brittle and polybutadiene is flexible; therefore copolymers of styrene and butadiene should be more flexible than polystyrene but tougher than polybutadiene. The general-purpose rubber GRS (or SBR), the first practical synthetic rubber, is a copolymer of styrene and butadiene.

1.4 Configurations of Polymer Molecules

The long threadlike shape of polymer molecules induces generally a random arrangement leading to inter- and intramolecular entanglement, somewhat like a bowl of cooked spaghetti. A typical molecule of polyethylene, for example, might be represented by a cylindrical chain with a length of 50,000 Å and a diameter of less than 5 Å. This is similar to a rope that is 45 m long and 4.5 mm in diameter. A molecule such as this can easily get knotted and entangled with surrounding molecules. In the structure some parts of the molecular chains can be more ordered than others. The ordered regions are termed micelles or *crystallites.* These regions are embedded in the unordered or amorphous matrix.

It is quite logical that if we want a molecule to go into some kind of ordered, repetitive pattern, then its structure must also have a regularly repeating pattern. The degree of crystallinity of the polymer will thus increase with the linearity and steric regularity of the molecules and also with interchain attractive forces.

Linear polymers are found in nature, or they may be formed by polymerization of simple monomers. When monosubstituted ethylene monomers ($CH_2{=}CHR$) polymerize, the addition reaction may be head-to tail, head-to-head/tail-to-tail, or a random mixture of the two:

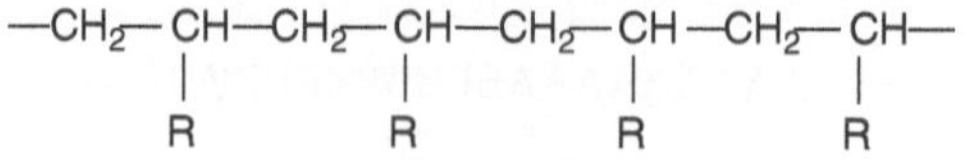

Head-to-tail

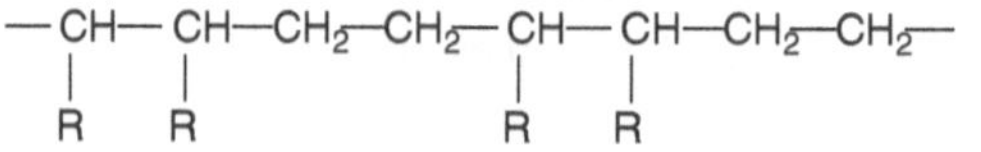

Head-to-head/tail-to-tail

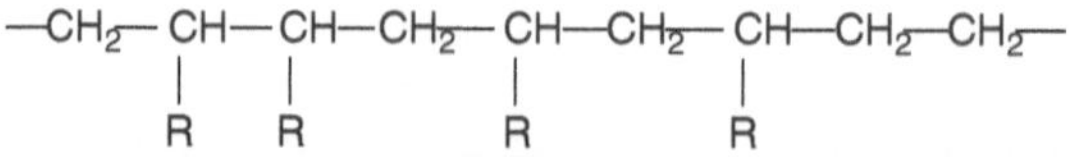

Random

The head-to-tail configuration is preferred almost to the exclusion of the other two. An important reason for this is steric hindrance, which favors head-to-tail reaction, especially if R is bulky.

Another aspect of stereoregularity is tacticity. Figure 1.13 show a polymer chain in which all of the chain carbons are in the same plane. Three configurations can be obtained: A polymer molecule is isotactic if all the substituted groups lie on the same side of the main chain. In a syndiotactic polymer molecule the substituted groups regularly alternate from one side to the other. The molecule is *atactic* if the positioning of substituted groups is random.

The relative arrangement of groups and atoms in successive monomer units in a polymer chain not only affects the crystallinity but also induces completely different properties in polymers. One example of this effect is found in polypropylene steroisomers. (The three stereoisomers of polypropylene can be obtained by replacing R by CH_3 in Figure 1.13.) The structural difference results in profound variations in the properties of polypropylene isomers. As is evident from Table 1.4, the three polypropylene isomers appear to be three altogether different materials.

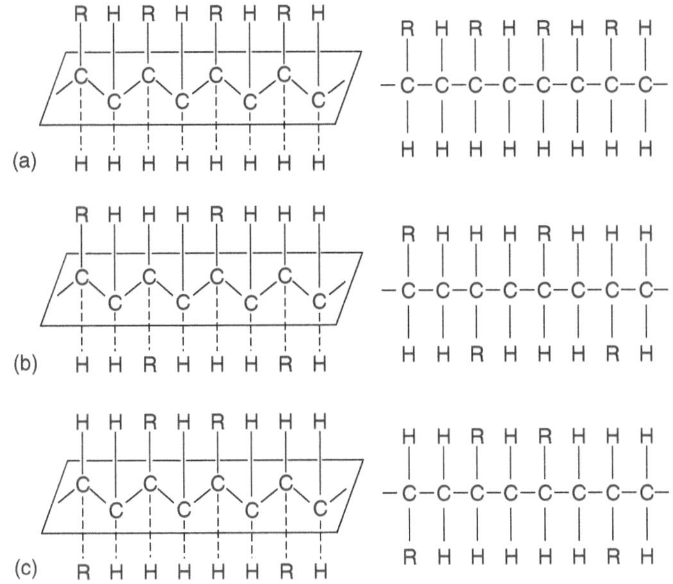

FIGURE 1.13 Diagrams of (a) isotactic, (b) syndiotactic, and (c) atactic configuration in a vinyl polymer. The corresponding Fischer projections are shown on the right.

1.5 Conformations of a Polymer Molecule

Just as a rope can be stretched, folded back on itself, curled up into a ball, entangled, knotted, and so forth, a polymer molecule can take on many conformations. Consider a molecule as a chain of N links, each link of length l_0 and attached to the preceding link by a rotating joint.

For a free rotating polymer chain the average conformation is characterized by the mean square distance $\langle r^2 \rangle$ between the ends of the chain and is given by

$$\langle r^2 \rangle = N l_0^2 \tag{1.24}$$

If we assume that adjacent links in the chain form fixed angles θ with each other but rotate freely about that angle, then

$$\langle r^2 \rangle = \frac{1 - \cos\theta}{1 + \cos\theta} N l_0^2 = 2N l_0^2 \tag{1.25}$$

since bonds in a tetrahedral carbon unit are at 109.5° to each other.

Thus, for a polyethylene molecule comprising, say, 40,000 freely rotating $-CH_2-$ units ($l_0 = 1.54$ Å), the end-to-end distance would only be $\sqrt{2N}\, l_0$ or approximately 435 Å. In contrast, if fully extended, the molecule will be in all-*trans* conformation with a linear zigzag structure (like corrugated sheets) shown in (III). (The dotted lines denote bonds and the wedges signify bonds above the plane of the page.)

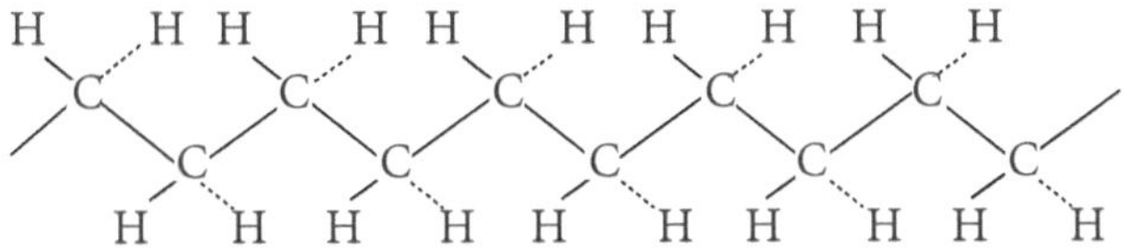

(III)

TABLE 1.4 Properties of Polypropylene Stereoisomers

Property	Stereoisomers		
	Isotactic	Syndiotactic	Atactic
Appearance	Hard solid	Hard solid	Soft rubbery
Melting temperature (°C)	175	131	<100
Density (g/cc)	0.90–0.92	0.89–0.91	0.86–0.89
Tensile strength [psi (N/m^2)]	5,000 (3.4×10^7)	—	—
Solubility	Insoluble in most organic solvents	Soluble in ether and aliphatic hydrocarbon	Soluble in common organic solvents
Crystallinity (%)	<70	—	—
Glass transition temperature (°C)	0 to −35	—	−11 to −35

The end-to-end distance of this fully extended polyethylene molecule will be $Nl_0 \sin(109.5°/2)$, or approximately 50,000 Å, for a molecule comprising 40,000 $-CH_2-$ units, and the contour length is Nl_0 or 60,000 Å.

1.6 Polymer Crystallinity

X-ray scattering and electron microscopy have shown that the crystallites are made up of lamellae which, in turn, are built-up of folded polymer chains as explained below.

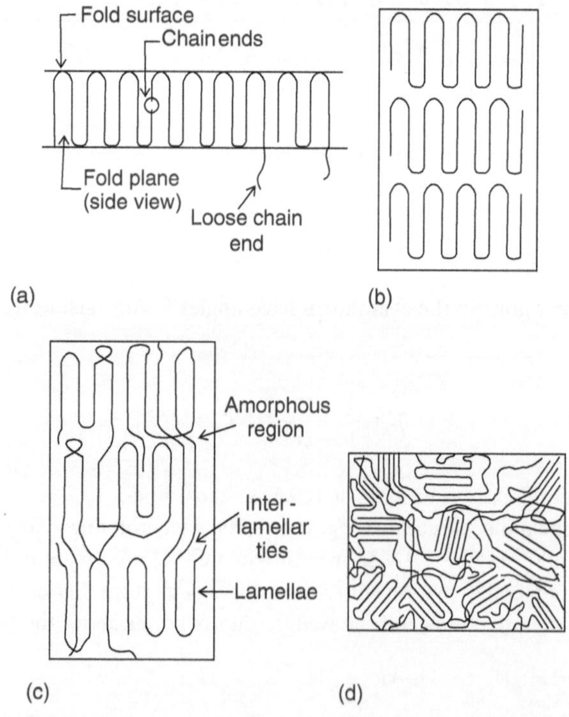

FIGURE 1.14 Schematic representation of (a) fold plane showing regular chain folding, (b) ideal stacking of lamellar crystals, (c) interlamellar amorphous model, and (d) fringed micelle model of randomly distributed crystallites.

Lamellae are thin, flat platelets on the order of 100–200 Å (0.01–0.02 μm) thick and several microns in lateral dimensions, while polymer molecules are generally on the order of 1,000–10,000 Å long. Since the polymer chain axis is perpendicular to the plane of the lamellae, as revealed by electron diffraction, the polymer molecules must therefore be folded back and forth within the crystal. This arrangement has been shown to be sterically possible. In polyethylene, for example, the molecules can fold in such a way that only about five chain carbon atoms are required for the fold, that is, for the chain to reverse its direction. Each molecule folds up and down in a regular fashion to establish a fold plane. As illustrated in Figure 1.14a, a single fold plane may contain many polymer chains. The height of the fold plane is known as the fold period. It corresponds to the thickness of the lamellae.

Figure 1.14b shows an idealized model of lamellae structure with ideal stacking of lamellar crystals. A more useful model, however, is that of stacks of lamellae interspersed with and connected by amorphous regions that consist of disordered chain segments of polymer molecules. Such a model, referred to as interlamellar amorphous model (Figure 1.14c), helps explain the ductility and strength of polymers as a direct consequence of the molecular links between the lamellae forming interlamellar ties. For semicrystalline polymers with amorphous regions to the tune of 20%–50%, it is often more advantageous to adopt a fringed micelle or fringed crystalline model (Figure 1.14d). It pictures polymers as two-phase systems in which the amorphous regions are interspersed between the randomly distributed crystallites.

1.6.1 Determinants of Polymer Crystallinity

The extent to which polymer molecules will crystallize depends on their structures and on the magnitudes of the secondary bonds forces among the polymer chains: the greater the structural regularity and symmetry of the polymer molecule and the stronger the secondary forces, the greater the tendency toward crystallization. We give a few examples:

1. Linear polyethylene has essentially the best structure for chain packing. Its molecular structure is very simple and perfectly regular, and the small methylene groups fit easily into a crystal lattice. Linear polyethylene (high density) therefore crystallizes easily and to a high degree (over 90%) even though its secondary forces are small. Branching impairs the regularity of the structure and makes chain packing difficult. Branched polyethylene (low density) is thus only partially (50%–60%) crystalline. Most of the differences in properties between low-density and high-density polyethylenes can be attributed to the higher crystallinity of the latter. Thus, linear polyethylenes have higher density than the branched material (density range of 0.95–0.97 versus 0.91–0.94 g/cm^3), higher melting point (typically 135 versus 115°C), greater stiffness (modulus of 100,000 versus 20,000 psi), greater tensile strength, greater hardness, and less permeability to gases and vapors.
2. Substituents hanging off polymer chains lead to difficulties in packing and generally decrease the tendency toward crystallization. Moreover, crystallization does not take place easily when polymer molecules have a low degree of symmetry. Thus, polymers such as polystyrene, poly(methyl methacrylate), poly(vinyl acetate), etc., all of which have bulky side groups oriented at random with respect to the main carbon chain (in atactic polymers), show very poor crystallization tendencies and tend to have amorphous structures. However, crystallinity would result if the side groups could be arranged in a regular orientation. Indeed, this can be done by controlled polymerization with properly chosen catalysts.
3. Copolymerization reduces the structural symmetry of a polymer. Thus it is a very effective method of decreasing the crystallization tendency of a polymer.
4. Chain flexibility also affects the crystallizability of a polymer. Excessive flexibility in a polymer chain, as in natural rubber and polysiloxanes, gives rise to difficulty in chain packing, with the result that such polymers remain almost completely in the amorphous state. In the other extreme, excessive rigidity in polymers due to extensive cross-linking, as in thermosetting resins like phenol–formaldehyde and urea–formaldehyde, also results in an inability to crystallize.

5. The presence of polar groups—such as amide, carboxyl, hydroxyl, chlorine, fluorine, and nitrile—along the polymer chains greatly increases the intermolecular or secondary attraction forces, which is favorable for crystallization. However, high secondary forces alone may not give rise to high crystallinity unless the chain segments are aligned. Mechanical stretching of the polymer makes this alignment easier. For example, nylon-6,6 (a polyamide) has less than the expected degree of crystallinity in the unstretched condition and is used as a plastic. Highly crystalline strong fibers are produced by stretching (cold-drawing) the polyamide polymer 400%–500%. Mechanical stretching also makes it possible to develop a degree of order and crystallinity in several other thermoplastic resins that do not ordinarily crystallize. An unusual example of alignment and crystallization on stretching is rubber.
6. The degree of crystallinity of polymeric materials is reduced by adding plasticizers. Crystallization in many synthetic resins is not always desirable because it makes shaping more difficult and reduces transparency by closely packing neutralization of the intermolecular forces of attraction by coming between polymer molecules, thus enhancing flexibility and plasticity, are often added to the polymeric mass before shaping. The oldest example is celluloid, made by plasticizing nitrocellulose (ordinarily a crystalline material) with camphor. Cellophane (regenerated cellulose film produced by a *viscose* process) is plasticized with glycerine to prevent crystallization and loss of transparency. Polyvinyl chloride (PVC) is made flexible by adding plasticizers, such as dioctyl phthalate, for use as wire coating, upholstery, film and tubing. The unplasticized rigid PVC is used for the production of pipe, sheet, and molded parts. The disadvantage of plasticizers is that they reduce the tensile strength and chemical resistance of the material.

1.7 The Amorphous State

The amorphous state is the characteristic of all polymers at temperatures above their melting points (except under special circumstances where liquid crystals may form). If a molten polymer retains its amorphous nature on cooling to the solid state, the process is called vitrification. In the vitrified amorphous state, the polymer resembles a glass. It is characteristic of those polymers in the solid state that, for reasons of structure, exhibit no tendency toward crystallization. The amorphous solid state is characterized by glass transition (T_g), which is described in a later section. We consider below only the behavior of polymer melt.

When an amorphous polymer achieves a certain degree of rotational freedom, it can be deformed. If there is sufficient freedom, the molecules begin to move past one another and the polymer flows. The science of deformation and flow is called rheology. It is of fundamental importance in industrial applications since it is usually in the molten state that polymers are molded into useful objects.

To cause a polymer to deform or flow requires the application of a force. If a force is applied and then withdrawn quickly, the polymer molecules tend to revert to their previous undeformed configuration, a process called relaxation. In other words, the polymer melt exhibits a certain elastic quality. This elasticity comes about because the molecules were disturbed from what was a thermodynamically favorable arrangement. If, however, the force is applied gradually and consistently, the molecules begin to flow irreversibly. (Silly putty, a siloxane polymer, is ideal for demonstrating this effect. If dropped, it bounces; but it can be shaped by the slow application of pressure.) Because of entanglement of the polymer chains and frictional effects, the flowing liquid will be very viscous. This combination of properties, namely elasticity and viscous flow, is why polymers are referred to as viscoelastic materials.

A consequence of viscoelasticity of polymer melt is die swell, which refers to the fact that the thickness of the melt emerging through a narrow orifice or die is greater than the width of the die opening. This is explained as follows: as the molecules flow rapidly through the die opening, they are compressed, and when they emerge, the resultant reduction in pressure causes the molecules to rebound to a degree. This dimensional increase of the extrudate must be taken into account by engineers who design polymer processing machinery (Chapter 1 of *Plastics Fabrication and Recycling*).

Chain entanglement that contributes to the high viscosity of polymer melt is clearly going to change as the molecular weight increases. Molecular weight is thus a critical variable in polymer rheology. Studies have shown that with flexible chain polymers, there exists a critical molecular weight ($\bar{M}_c$) for entanglement to begin. For most common polymer, $\bar{M}_c$ falls in the range 4,000–15,000. Although $\bar{M}_c$ varies from one polymer to another, it has been shown that elimination of mass effects arising from substituents on the chains and calculation of chain length leads to a value that is remarkably constant from one polymer to another and corresponds to a $\overline{\text{DP}}$ of about 600. In other words, a critical chain length, rather than a critical molecular weight per se, is necessary for entanglement.

1.8 Structural Shape of Polymer Molecules

Polymers can be classified, based on the structural shape of polymer molecules, as linear, branched, or cross-linked. Schematic representations are given in Figure 1.15. Linear polymers have repeating units linked together in a continuous length (Figure 1.15a). When branches protrude from the main polymer chain at irregular intervals, the polymer is termed as branched polymer. Branches may be long or short, forming a comblike structure (Figure 1.15b), or divergent (Figure 1.15c), forming a dendritelike structure. [Regularly repeating side groups which are a part of the monomer structure are not considered as branches. Thus polypropylene is a linear polymer, as are polystyrene and poly(methyl methacrylate).]

Both linear and branched polymers are thermoplastic; that is, they can be softened and hardened reversibly by changing the temperature. Fabricating processes like injection molding, extrusion molding, casting, and blowing take advantage of this feature to shape thermoplastic resins. The rigidity of thermoplastic resins at low temperatures is attributed to the existence of secondary bond forces between the polymer chains. These bonds are destroyed at higher temperatures, thereby causing fluidity of the resin.

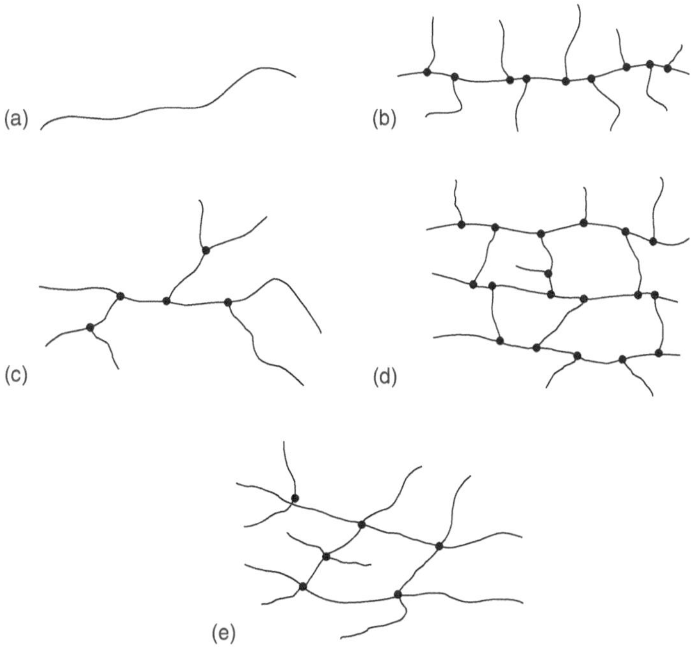

FIGURE 1.15 Schematic representation of (a) linear, (b and c) branched, and (d and e) cross-linked polymers. The branch points and junction points are indicated by heavy dots.

Polymers used as textile fibers are linear. However, they must satisfy two additional requirements: (1) high molecular weight and (2) a permanent orientation of the molecules parallel to the fiber axis. The molecules must have a high degree of order and/or strong secondary forces to permit orientation and crystallization. The chain orientation necessary to develop sufficient strength by crystallization is achieved by a process known as cold drawing, in which the initially formed filaments (unoriented or only slightly oriented) are drawn at a temperature above the glass transition temperature (discussed later), which is the temperature at which sufficient energy is available to the molecular segments to cause them to begin to rotate.

Elastomeric materials, like thermoplastic resins and fibers, are essentially linear polymers. But certain distinctive features in their molecular structure give rise to rubberlike elasticity. Elastomeric polymers have very long chain molecules occurring in randomly coiled arrangements in the unstressed condition. A large deformation is thus possible merely by reorienting the coiled molecules. When elongated, the molecular coils partially open up and become aligned more or less parallel to the direction of elongation. The aligned configuration represents a less probable state or a state of lower entropy than a random arrangement. The aligned polymer chains therefore have a tendency to return to their original randomly coiled state. The large deformability of elastomeric materials is due to the presence of a certain internal mobility that allows rearranging the chain orientation, the absence of which in linear chain plastic materials (at normal temperatures) constitutes the essential difference between the two groups.

Although the aforesaid requirements are necessary conditions for ensuring a large extent of deformability, the remarkable characteristic of the rubbery state—namely, nearly complete recovery—cannot be obtained without a permanent network structure, since permanent deformation rather than elastic recovery will occur. A small amount of cross-linkage is necessary to provide this essential network structure [14,15]. Natural rubber (polyisoprene), for example, simply flows like an extremely viscous liquid at room temperature if it is not cross-linked. Cross-links are introduced into a rubber by heating raw rubber with sulfur (1%–2% by weight) and accelerating agents. Sulfur reacts with the double-bonded carbon atoms to produce a network structure, as shown schematically in Figure 1.16. The amount of cross-linkage must be as small as possible to retain the structure; excessive cross-linkages will make the internal structure too stiff to permit even the required rearrangement of chain orientation during both deformation and recovery—in other words, it will destroy the rubbery state. An example of this is best furnished by ebonite, which is a rigid plastic made by vulcanizing natural rubber with large quantities of sulfur.

Example 2

(a) How much sulfur is required to fully cross-link natural rubber? (b) What is the sulfur content of vulcanized natural rubber that is 50% cross-linked? (Assume that each cross-link contains one sulfur atom).

Answer. Mer weight of isoprene (Figure 1.16a):

$$C_5H_8 = (5)(12) + (8)(1) = 68 \text{ g/mer}$$

Assuming $x=1$ in Figure 1.16b, one sulfur atom, on the average, is required for cross-linking per mer of isoprene. Therefore,

(a) Amount of sulfur = (32/68)100 = 47 g/100 g of raw rubber

(b) Sulfur content = 100(0.5)(32)/[0.5(32) + 68] = 19%

Example 3

A rubber contains 60% butadiene, 30% isoprene, 5% sulfur, and 5% carbon black. What fraction of possible cross-links are joined by vulcanization? (Assume that all the sulfur is used in cross-linking.)

Answer. 1 mer weight of butadiene (C_4H_6) = (4)(12) + (6)(1) = 54 g/mer

1 mer weight of isoprene (C_5H_8) = (5)(12) + (8)(1) = 68 g/mer

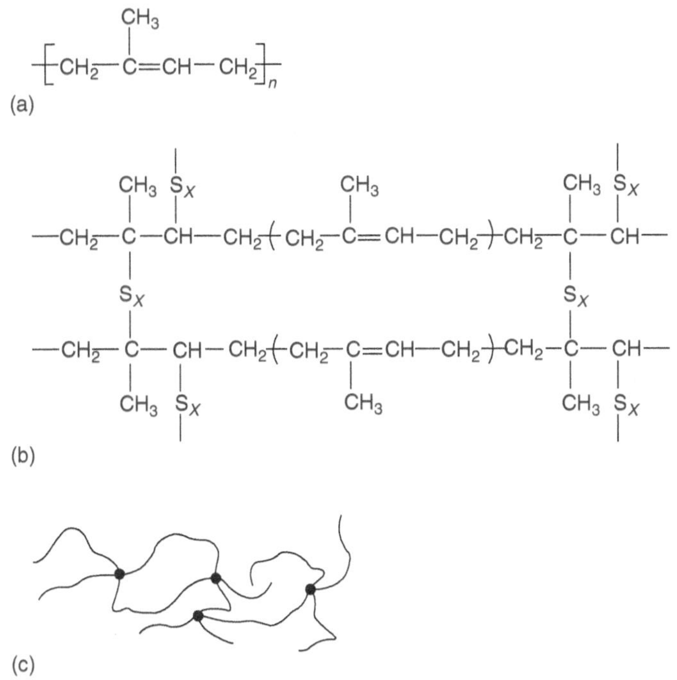

FIGURE 1.16 Vulcanization of natural rubber with sulfur. (a) Linear polyisoprene (natural rubber). (b) Vulcanized structure (idealized) showing cross-links of sulfide groups. The number, x, of sulfur atoms in a cross-link is 1 or 2 in efficient vulcanization systems but may be as high as 8 under conditions where cyclic and other structures are also formed in the reaction. (c) The effect of cross-linking is to introduce points of linkage or anchor points between chain molecules, restricting their slippage. The cross-links in elastomers are typically a few hundred carbon atoms apart.

1 atomic weight of sulfur = 32. Assuming that, on the average, one sulfur atom per mer is required for cross-linking, we get

$$\text{Fraction of cross-links} = \frac{5/32}{60/54 + 30/68} = 0.101 \text{ or } 10.1\%$$

1.9 Thermal Transitions in Polymers

The term "transition" refers to a change of state induced by changing the temperatures or pressure. Two major thermal transitions are the glass transition and the melting, the respective temperatures being called T_g and T_m.

1.9.1 T_g and T_m

All polymers are hard rigid solids at sufficiently low temperatures, but as the temperature rises a thermoplastic polymer eventually acquires sufficient thermal energy to enable its chains to move freely enough for it to behave like a viscous liquid (assuming there is no degradation). There are several ways in which a polymer can pass from the solid to the liquid phase, depending on the structure of the polymer chains and their arrangement in the sample. The different types of thermal response in this transition from a rigid solid to an eventually liquid state can be illustrated in several ways. One of the simplest and most satisfactory is to trace the change in specific volume, as shown schematically in Figure 1.17.

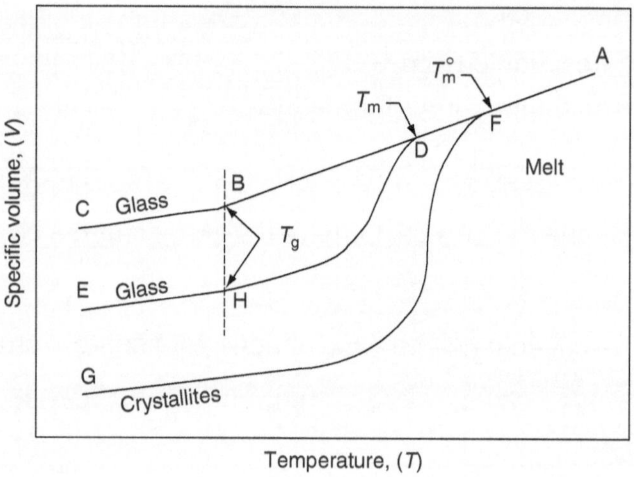

FIGURE 1.17 Schematic representation of the change of specific volume of a polymer with temperature for a completely amorphous sample (A–B–C), a semicrystalline sample (A–D–E), and a perfectly crystalline material (A–F–G).

In first-order transitions, such as melting, there is a discontinuity in the volume–temperature plot (Figure 1.17) or enthalpy–temperature plot at the transition temperature. In second-order transitions, only a change in slope occurs and thus there is a marked change in the first derivative or temperature coefficients, as illustrated in Figure 1.18. The glass transition is not a first-order transition, as no discontinuities are observed at T_g when the specific volume or entropy of the polymer is measured as a

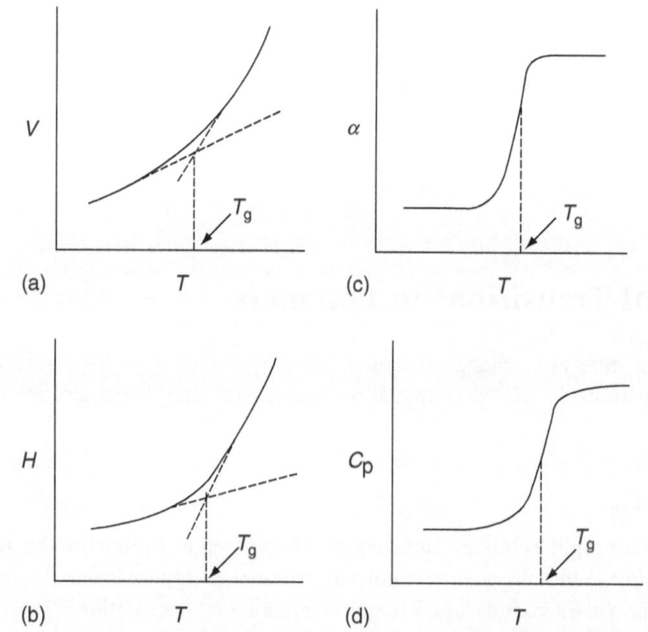

FIGURE 1.18 Idealized variations (a,b) in volume (*V*) and enthalpy (*H*). Also shown (c,d) are α, the volume coefficient of expansion, and C_p, the heat capacity, which are, respectively, the first derivatives of *V* and *H* with respect to temperature (*T*).

function of temperature. However, the first derivative of the property–temperature curve, i.e., the temperature coefficient of the property, exhibits a marked change in the vicinity of T_g; for this reason it is sometimes called a second-order transition.

$$\alpha = \frac{1}{V}\left(\frac{\partial V}{\partial T}\right)_P \tag{1.26}$$

where V is the volume of the material, and α has the units K^{-1}. While this quantity increases rather sharply at T_g, the increase actually occurs over a range (10°C–30°C). Similar changes occur in the heat capacity at constant pressure (C_p), which is the first derivative of enthalpy H with respect to temperature.

A thermoplastic polymer may be completely amorphous in the solid state, which means that the polymer molecular chains in the specimen are arranged in a totally random fashion. Referring to Figure 1.17, the volume change in amorphous polymers follows the curve ABC. In the region C–B, the polymer is a glassy solid and has the characteristics of glasses, including hardness, stiffness and brittleness. In the glassy region, the available thermal energy (RT energy units/mol) is insufficient to allow rotation about single bonds in the polymer backbone overcoming intramolecular energy barriers, and movements of large (some 10–50 consecutive chain atoms) segments of the polymer chain cannot take place. But as the sample is heated, it passes through a temperature T_g, called the glass transition temperature, above which it softens and becomes rubberlike. This is an important temperature and marks the onset of extensive molecular motion which is reflected in marked changes in properties, such as specific volume, refractive index, stiffness, and hardness. Above T_g, the material may be more easily deformed or become ductile. A continuing increase in temperature along B–A leads to a change of the rubbery polymers to a viscous liquid without any sharp transition.

In a perfectly crystalline polymer, all the chains would be contained in regions of three-dimensional order, called crystallites, and no glass transition would be observed, because of the absence of disordered chains in the sample. A perfectly crystalline polymer, on heating, would follow curve G–F–A, melting at T_m° to become a viscous liquid.

Perfectly crystalline polymers are, however, not encountered in practice and real polymers may instead contain varying proportions of ordered and disordered regions in the sample. These semicrystalline polymers usually exhibit both T_g and T_m (not T_m°) corresponding to the disordered and ordered regions, respectively, and follow curves similar to E–H–D–A. As T_m° is the melting temperature of a perfectly crystalline polymer of high molecular weight T_m is lower than T_m° and more often represents a melting range, because the semicrystalline polymer contains crystallites of various sizes with many defects which act to depress the melting temperature.

Both T_g and T_m are important parameters that serve to characterize a given polymer. While T_g sets an upper temperature limit for the use of amorphous thermoplastics like poly(methyl methacrylate) or polystyrene and a lower temperature limit for rubbery behavior of an elastomer like SBR rubber or 1,4-*cis*-polybutadiene, T_m or the onset of the melting range determines the upper service temperature. Between T_m and T_g, semicrystalline polymers tend to behave as a tough and leathery material. As a general rule, however, semicrystalline polymers are used at temperatures between T_g and a practical softening temperature which lies above T_g and below T_m. (The onset of softening is usually measured as the temperature required for a particular polymer to deform a given amount under a specified load. These values are known as heat deflection temperatures. Such data do not have any direct relations with T_m, but they are widely used in designing with plastics.)

The T_g and T_m values for some polymers are shown in Table 1.5. In general, both T_g and T_m are affected in the same manner by considerations of polymer structure [16]. Thus, both T_g and T_m increase with higher molecular symmetry, structural rigidity, and secondary forces of polymer chains.

The T_g and T_m values of a polymer determine the temperature range in which it can employed. Amorphous elastomeric polymers, for example, must be used at temperatures (region B–D in Figure 1.17) well above T_g to permit the high, local segmental mobility required in such materials. Thus styrene–butadiene (25/75) copolymer ($T_g = -57$°C), polyisoprene ($T_g = -73$°C), and

TABLE 1.5 Glass Transition Temperatures (T_g) and Crystalline Melting Temperatures (T_m) of Polymers

Polymer	T_g (°C)	T_m (°C)
Polyethylene (high density)	−115	137
Polyoxymethylene	−85	181
Polyisoprene (natural rubber)	−73	28
Polyisobutylene	−73	44
Polypropylene	−20	176
Poly(vinylidene chloride)	−19	190
Poly(chlorotrifluoroethylene) (kel-F)	45	220
Poly(hexamethylene adipamide) (nylon-6,6)	53	265
Poly(ethylene terephthalate) (Terylene, Dacron)	69	265
Poly(vinyl chloride)	81	212
Polystyrene	100	240
Poly(methyl methacrylate) (Perspex, Lucite)	105	200
Cellulose triacetate	105	306
Polytetrafluoroethylene (Teflon)	127	327

polyisobutylene ($T_g = -73$°C) can be used as rubbers at ambient temperatures. Amorphous structural polymers, such as polystyrene and poly(methyl methacrylate), depend on their glasslike rigidity below T_g for their utility; they should therefore have high T_g values so that under ambient conditions they are well below T_g.

Tough, leatherlike polymers are limited for use in the immediate vicinity of their T_g. Such behavior is observed in vinyl chloride-based plastics, which are used as substitutes for leather in automobile seat covers, travel luggage, and ladies' handbags. Highly crystalline fiber-forming polymers must be used at temperatures substantially below T_m (about 100°C), since changes in crystal structure can occur as T_m is approached. The T_m of a fiber must therefore be above 200°C to remain unaffected at use temperatures encountered in cleaning and ironing. (T_m should not, however, be excessively high—not more than 300°C; otherwise spinning of the fiber by melt-spinning processes may not be possible.) The T_g of a fiber, on the other hand, should have an intermediate value, because too high a value of T_g would interfere with the stretching operation as well as with ironing, and too low a value of T_g would not permit crease retention in fabrics. Nylon and terylene, as may be seen from Table 1.5, therefore have optimal values of T_m and T_g. Semicrystalline polymers with about 50% crystallinity are used at temperatures between T_g and T_m, since in this range the material exhibits moderate rigidity and a high degree of toughness, somewhat analogous to reinforced rubber. Branched polyethylene (low density), with $T_g = -120$°C and $T_m = 115$°C, used at ambient temperatures is a typical example.

1.9.2 Regions of Viscoelastic Behavior

As we have seen above, the transition that separates the glassy state from the viscous state is known as the glass–rubber transition. This transition attains the properties of a second-order transition at very slow rates of heating or cooling. In order to clearly locate the region of this transition and to provide a broader picture of the temperature dependence of polymer properties, the principal regions of viscoelastic behavior of polymers will be briefly discussed.

Broadly, there are five regions of viscoelastic behavior for linear amorphous polymers as shown in Figure 1.19. In region I (a–b), the polymer is glassy and frequently brittle. Typical examples are polystyrene and poly(methyl methacrylate) at room temperature. Young's modulus in this region just below the glass transition temperature is approximately 3×10^{10} dyne/cm^2 (3×10^9 Pa) and it is nearly the same for a wide range of polymers.

Range II (b–c) is the glass transition region. Typically, the modulus drops a factor of about a thousand in a 20°C–30°C range. The behavior of polymers in this region is best described as leathery. For static or quasistatic measurements, such as illustrated in Figure 1.19, the glass transition temperature, T_g, is often

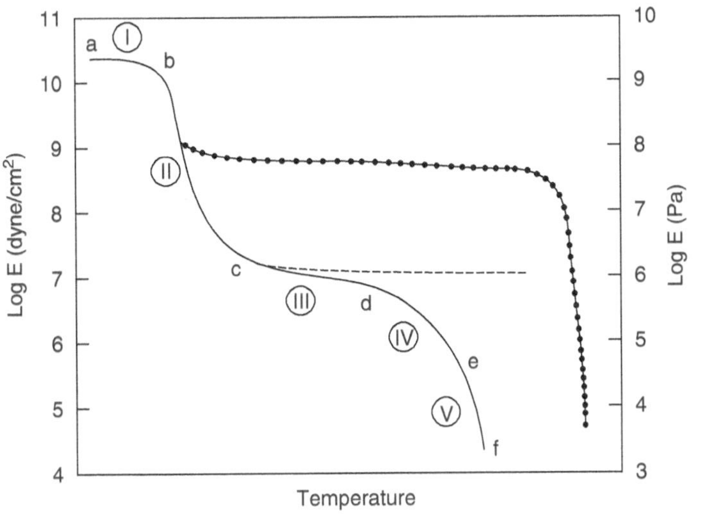

FIGURE 1.19 Five regions of viscoelastic behavior for a linear, amorphous polymer: I (a–b), II (b–c), III (c–d), IV (d–e), and V (e–f). Also illustrated are effects of crystallinity (dotted line) and cross-linking (dashed line).

taken at the maximum rate of turndown of the modulus at the elbow, i.e., where d^2E/dT^2 is at a maximum. A few T_g values are shown in Table 1.5.

Region III (c–d) in Figure 1.19 is the rubbery plateau region. After a sharp drop that the modulus exhibits in the glass transition region, it again becomes almost constant in the rubbery plateau region, with typical values of 2×10^7 dyne/cm^2 (2×10^6 Pa). In this region, polymers exhibit high rubber elasticity, so much so that an elastomer in this region can be stretched perhaps several hundred percent and substantially snap back to its original length on being released.

In region III (c–d) of Figure 1.19, two cases need be distinguished. First, if the polymer is linear, the solid line is followed on which the modulus drops off slowly with increasing temperature. The width of the plateau is governed primarily by the molecular weight of the polymer; the higher the molecular weight, the longer the plateau. Second, if the polymer is cross-linked, the dashed line in Figure 1.19 is followed, and improved rubber elasticity is observed, with the creep portion suppressed. A common example of a cross-linked polymer above its glass transition temperature is the ordinary rubber band. For cross-linked polymers, region III remains in effect at higher temperatures up to the decomposition temperature, and regions IV and V do not occur.

[The discussion above has been limited to amorphous polymers. However, if the polymer is semicrystalline, the dotted line in Figure 1.19 is followed. Since the crystalline regions in the polymer matrix tend to behave as a filler phase and also as a type of physical cross-link between the chains, the height of the plateau (i.e., the modulus) will be governed by the degree of crystallinity.]

In the case of linear amorphous polymers, raising the temperature past the rubbery plateau region brings them to the rubbery flow region—region IV. In this region, the polymer exhibits both rubber elasticity and flow properties depending on the time scale of the experiment. For experiments performed in a short time, the physical entanglements of polymer chains are not able to relax and the material still behaves rubbery. For longer-duration experiments, the increased molecular motion imparted by the increased temperature cause chains to move, resulting in visible flow. An example of a material in the rubbery flow region at ambient temperature is Silly Putty which bounces like a ball when thrown (quick experiment) but deforms like a taffy when pulled (a much slower experiment).

Region V (e–f) in Figure 1.19 is the liquid flow region which is reached at still higher temperatures where the increased kinetic energy of the chains permits them wriggle out through entanglements rapidly and move as individual molecules, often producing highly viscous flow. This is the melting temperature and it is always above the glass transition temperature.

Example 4

A new polymer was reported to soften at 60°C but it is not known for sure whether the softening was a glass transition or a melting point. Describe a simple experiment to distinguish between the two possibilities. It is known however, that the new polymer is essentially a linear polymer.

Answer. If 60°C is a glass transition, then heating the polymer slowly past 60°C would take it to the rubbery plateau region (region III in Figure 1.19), where the modulus *E*, and hence hardness would remain fairly constant with increase of temperature. For a melting transition, however, the modulus would drop rapidly and the polymer would become increasingly softer in a similar experiment.

Pressing one's thumb in an object is a simple way to gauze the object's hardness. Its scientific analog is the measurement of hardness by indentation. In practice, the point of a weighted needle is allowed to rest on the polymer surface as the temperature is raised. The movement of the needle as it penetrates the surface can be monitored by means of an amplification gauge. Though less accurate than other more sophisticated methods, it is useful for the preliminary engineering-oriented examination of systems.

1.9.3 Factors Affecting T_g

As T_g marks the onset of molecular motion, a number of factors which affect rotation about links (necessary for movement of polymer chains) will also influence the T_g of a polymer. These include (a) chain flexibility, (b) molecular structure (steric effects), (c) molecular weight, and (d) branching and cross-linking.

1.9.3.1 Chain Flexibility

The flexibility of the chain is undoubtedly the most important factor influencing the T_g of a polymer. The chain flexibility depends more on the rotation or torsion of skeletal bonds than on changes in bond angles. When a randomly coiled chain is pulled out into an elongated conformation, the skeletal bonds "unwind" rather than undergo angular distortion (see Figure 1.20). Thus, flexibility on a macroscopic scale depends on torsional mobility at the molecular level. If a highly flexible chain is present, T_g will generally be low and if the chain is rigid, the T_g value will be high.

For symmetrical polymers, the chemical nature of the backbone chain is the important factor determining the chain flexibility and hence T_g. Chains made up of bond sequences that are able to rotate easily are flexible, and hence polymers containing –(–CH_2–CH_2–)–,–(–CH_2–O–CH_2–)–, or –(–Si–O–Si–)– links will have correspondingly low values of T_g. For example, poly(dimethyl siloxane) has one of the lowest T_g values known (−123°C) presumably because the Si–O bonds have considerable torsional mobility.

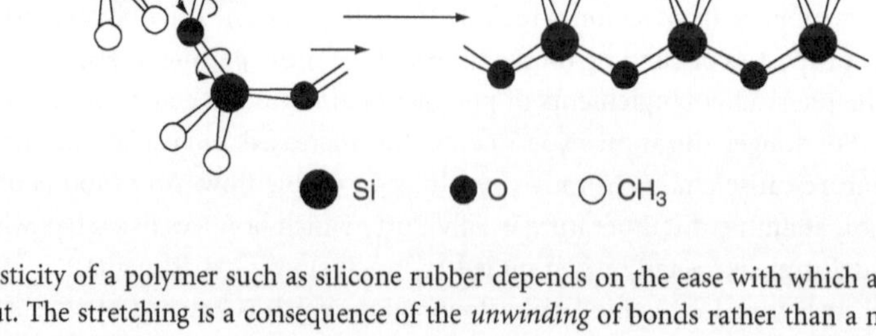

FIGURE 1.20 Elasticity of a polymer such as silicone rubber depends on the ease with which a random coil chain can be stretched out. The stretching is a consequence of the *unwinding* of bonds rather than a marked widening of bond angles.

The value of T_g is raised markedly by the insertion of groups that stiffen the chain by impeding rotation, so that more thermal energy is required to set the chain in motion. Particularly effective in this respect is the *p*-phenylene ring. Thus, a chain consisting entirely of *p*-phenylene rings, namely, poly(*p*-phenylene) (IV)

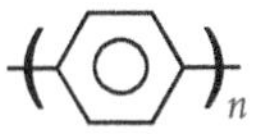

(IV)

has a highly intractable, rigid structure with no softening point. This structure can be modified by introducing flexible groups in the chain to produce tractable polymers with high values of T_g. Some examples are poly(phenylene oxide) (V), $T_g=83°C$ and poly(*p*-xylylene) (VI), $T_g=280°C$ as compared to polyethylene, $T_g=-93°C$, and poly(ethylene oxide), $T_g=-67°C$.

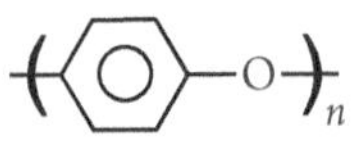

(V)

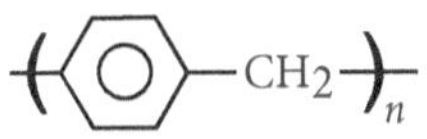

(VI)

1.9.3.2 Steric Effects

When the polymer chains are unsymmetrical, with repeat units of the type $-(-CH_2-CHX-)-$, an additional restriction is imposed by steric effects depending on the size of the pendant group X. Bulky pendant groups hinder the rotation about the backbone and cause T_g to increase. There is some evidence of a correlation between T_g and the size of the pendant group measured by its molar volume V_X. For example, T_i (°C) increases with increasing V_X (cm^3/mol) in the progressive series, polyethylene ($T_g=-93$, $V_X=3.7$), polypropylene ($T_g=-20$, $V_X=25.9$) polypropylene ($T_g=100$, $V_X=92.3$), and poly(vinylnaphthalene) ($T_g=135$, $V_X=143.9$).

Superimposed on the pendant group size factor are the effects of polarity and the intrinsic flexibility of the pendant group itself. Greater intermolecular interactions due to polar groups hinder molecular motion and increase T_g. Thus polar groups tend to encourage a higher T_g than nonpolar groups of similar size, as seen when comparing T_g (°C) and V_X (cm^3/mol) of polypropylene ($T_g=-20$, $V_X=25.9$), poly(vinyl chloride) ($T_g=81$, $V_X=22.1$), and polyacrylonitirile ($T_g=105$, $V_X=30.0$). On the other hand, greater flexibility of side chain leads to lower T_g as is evident on examination of the polyacrylate series from methyl through butyl: poly(ethylacrylate) ($T_g=-24$, $V_X=82.3$), poly(propyl acrylate) ($T_g=-48$, $V_X=104.5$), and poly(butyl acrylate) ($T_g=-55$, $V_X=126.7$), where T_g is in °C and V_X is in cm^3/mol.

A further increase in steric hindrance is caused by the presence of an α-methyl group, which restricts rotation even further, thus causing T_g to increase. Typical examples are the pair polystyrene ($T_g=100°C$)–poly(α-methylstyrene) ($T_g=172°C$) and the pair poly(methyl acrylate) ($T_g=6°C$)–poly(methyl methacrylate) ($T_g=105°C$).

1.9.3.3 Configurational Effects

It should be noted that the steric effects of the pendant groups considered above are simply additional contributions to the main chain effects. Similarly cis–trans isomerism in polydienes and tacticity variations in certain α-methyl substituted polymers alter chain flexibility and hence affect T_g. Well-known examples of cis–trans variations are polybutadiene *cis*($T_g=-108°C$) and *trans*($T_g=-18°C$) or polyisoprene *cis*($T_g=-73°C$) and *trans*($T_g=-53°C$). An example of tacticity variation is poly(methyl methacrylate) for which the isotactic, atactic, and syndiotactic stereostructures are associated with T_g values of 45, 105, and 115°C, respectively.

1.9.3.4 Effect of Cross-Linking

When cross-links are introduced into a polymer, the molecular motion in the sample is restricted and T_g rises. This transition is ill-defined for a high cross-link density, but at lower values T_g is found to increase linearly with the number of cross-links.

1.9.4 Factors Affecting T_m

The application of macroscopic thermodynamics leads to some useful generalizations of factors affecting T_m. At T_m, the free energy change is zero, i.e.,

$$\Delta G_m = \Delta H_m - T_m \Delta S_m = 0$$

where $T_m = \Delta H_m / \Delta S_m$. This expression for T_m predicts that a high melting point can be the result of a high value of the enthalpy change ΔH_m and/or a small value of the entropy change ΔS_m in melting. The former corresponds to stronger binding of adjacent but unbonded units in the polymer lattice and thus to higher degree of crystallinity. The factors that affect crystallinity and T_m can be classified as symmetry, intermolecular bonding, tacticity, branching, and molecular weight. These are discussed below. If ΔS_m is small, melting does not result in a large gain of conformational entropy and, to some degree, the structure of the solid must persist in the melt. For example, molecules of isotactic polypropylene crystallize in the form of helices and these are thought to occur in the melt as well, thus making ΔS_m small and T_m high.

1.9.4.1 Symmetry

The formation of stable crystalline regions (crystallites) in a polymer requires (a) that an efficient close-packed arrangement of the polymer chains can be achieved in three dimensions and (b) that a favorable change in internal energy is obtained during this process. This imposes restrictions on the type of chain that can form crystallites easily and one would expect linear chains with high degree of symmetry such as polyethylene, polytetrafluoroethylene, various polyesters, and polyamides to crystallize most readily.

Linear symmetrical molecules such as polyethylene, polytetrafluoroethylene, and other linear molecules with more complex backbones containing –(–O–)–, –(–COO–)–, and –(–CONH–)– groups, such as polyethers, polyesters, and polyamides, all posses a suitable symmetry for crystallite formation and usually assume extended zigzag conformations when aligned in the lattice. On the other hand, chains containing irregular units, which impair the symmetry, reduce the ability of the polymer to crystallize. Thus cis double bonds (VII), *o*-phenylene groups (VIII), and *m*-phenylene groups (IX) all encourage bending and twisting in the chains and make close packing very difficult.

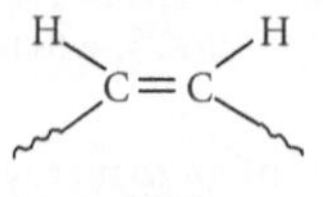

(VII)

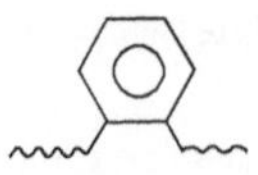

(VIII)

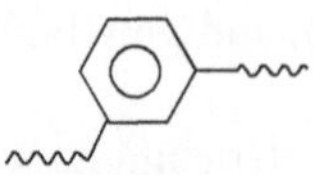

(IX)

If, however, the phenylene rings are para-oriented, the chains retain their axial symmetry and can crystallize more readily. Similarly, double bonds in trans configuration maintain the chain symmetry thus allowing for crystallite formation. This is highlighted by a comparison of the amorphous elastomeric *cis*-polyisoprene ($T_m = 28°C$) with highly crystalline *trans*-polyisoprene ($T_m = 74°C$), which is a nonelastomeric rigid polymer, or *cis*-1,4-polybutadiene ($T_m = -11°C$) with *trans*-1,4-polybutadiene ($T_m = 148°C$).

Another aspect of high chain symmetry is the possibility of molecular motion within the crystal lattice contributing to higher T_m. For example, polyethylene and polytetrafluoroethylene are both sufficiently symmetrical to be considered as smooth, stiff cylindrical rods. In the crystal, these rods tend to roll over each other and change position when thermally agitated. This motion within the crystal lattice, called premelting, effectively stabilizes the lattice. Consequently, more thermal energy is required to break down the crystallite into a disordered melt, ant T_m is raised. Irregularly shaped polymers with bends and bumps in the chain cannot move in this way without disrupting the crystal lattice, and so have lower T_m values.

1.9.4.2 Intermolecular Bonding

Any interaction between polymer chains in the crystal lattice serves to hold the structure more firmly and raise the melting temperature. In polyethylene, the close packing in crystallites achieved due to high chain symmetry and lack of substituents on the chains allows the van der Walls forces to act cooperatively and provide additional stability to the lattice. In polymers containing polar substituents, for example, Cl, CN, or OH, the chains can be aligned and held rigidly by the strong dipole–dipole interactions between the substituents. This effect is more obvious in the symmetrical polyamides. These polymers can form intermolecular hydrogen bonds which greatly enhance crystallite stability and raise the T_m significantly.

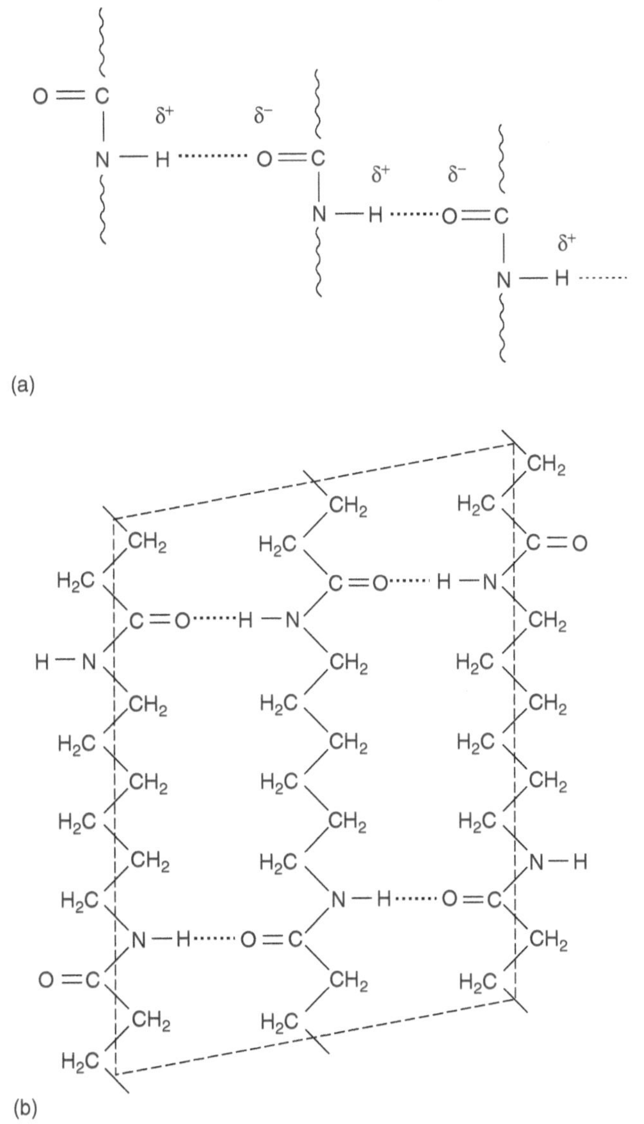

FIGURE 1.21 (a) Hydrogen bonds between neighboring chains of polyamide. (b) Arrangement of chains in hydrogen-bonded sheets in the crystal structure of nylon-6,6.

This is illustrated in Figure 1.21 for nylon-6,6, where the extended zigzag structure greatly facilitates the formation of regular intermolecular hydrogen bonding. The increased stability is reflected in the value of T_m, which for nylon-6,6 is 267°C, compared with 137°C for high-density polyethylene.

1.9.4.3 Tacticity

Stiffening of the polymer chain due to the presence of large pendant groups tends to raise T_m, but this will also increase the difficulty of close packing to form a crystalline array and so tend to lower T_m. However, this latter problem can be overcome if the pendant groups are arranged in a regular fashion along the chain. Thus isotactic polymers tend to form helices to accommodate the substituents in the most stable steric positions, and being of regular forms these helices are capable of regular alignment promoting crystallite formations. Thus isotactic polystyrene is semicrystalline ($T_m = 240°C$), though atactic polystyrene is amorphous. Syndiotactic polymers are also sufficiently regular to crystallize, not necessarily as a helix, but rather in glide planes.

1.9.4.4 Branching, Chain Flexibility, and Molecular Weight

Chain flexibility has a direct bearing on the melting point. Insertion of groups that stiffen the chain increases T_m, while introducing flexible groups into the chain lowers the value of T_m (see "Factors Affecting T_m"). Branching in the side group tends to stiffen the chain and raise T_m, as shown in the series poly(but-1-ene) ($T_m = 126°C$), poly(3-methyl but-1-ene) ($T_m = 145°C$), poly(3,3′-dimethyl but-1-ene) ($T_m > 320°C$). However, if the side groups is flexible and nonpolar, T_m is lowered. Also, if the chain is substantially branched to reduce the packing efficiency, the crystalline content is lowered, as is the melting point. A good example is low-density polyethylene where extensive branching lowers the density and T_m of the polymer.

Molecular weight in relatively lower order can have significant effect on T_m. Since chain ends are relatively free to move and the number of chain ends increases with the decrease in molecular weight T_m is lowered because less energy is then required to stimulate chain motion and melting. For example, polypropylene with molecular weight 2,000 has $T_m = 114°C$, whereas a sample with molecular weight 30,000 has $T_m = 170°C$.

1.9.5 Relation between T_m and T_g

While T_m is a first-order transition, T_g is a second-order transition and this precludes the possibility of a simple relation between them. There is, however, a crude relation between T_m and T_g. Thus the ratio T_g/T_m range from 0.5 to 0.75 when the temperatures are expressed in Kelvin. The relation is represented in Figure 1.22 where a broad band covers most of the results for linear hompolymers and the ratio (T_g/T_m) lies between 0.5 and 0.75 for about 80% of these [17,18]. The ratio is closer to 0.5 for symmetrical polymers such as polyethylene and polybutadiene, but closer to 0.75 for unsymmetrical polymers, such as polystyrene and polyisoprene. The difference in these values may be related to the fact that in unsymmetrical chains with repeat units of the type –(CH_2–CHX–)– an additional restriction to rotation is imposed by steric effects causing T_g to increase, and conversely, an increase in symmetry lowers T_g.

1.10 Designing a Polymer Structure for Improved Properties

The three principles [19] applied to give strength and resistance to polymers are (1) crystallization, (2) cross-linking, and (3) increasing inherent stiffness of polymer molecules. Combinations of any two or all of the three strengthening principles have proved effective in achieving various properties with polymers (Figure 1.23). For polymers composed of inherently flexible chains, crystallization and cross-linking are the only available means to enhance polymer properties. The factors affecting the crystallinity of polymers have been discussed previously, and the methods of introducing cross-links between polymer molecules are discussed later.

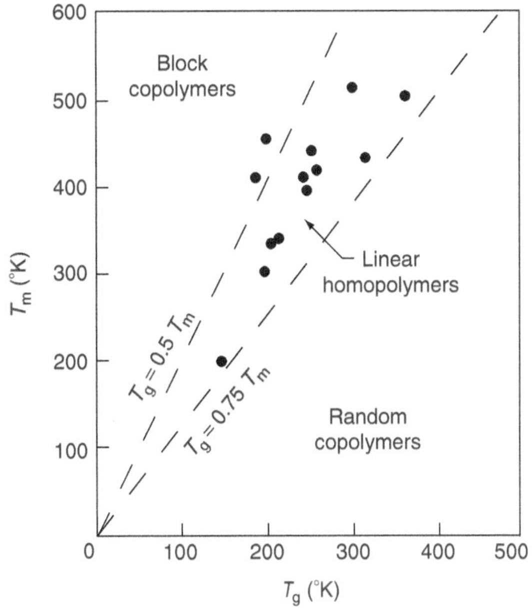

FIGURE 1.22 Plot of T_m against T_g for linear homopolymers with T_g/T_m in the range 0.5 and 0.75 broadly demarcating $T_m - T_g$ domains covered by homopolymers and copolymers.

The third, and relatively new, strengthening principle is to increase chain stiffness. One possible way of stiffening a polymeric chain is to hang bulky side groups on the chain to restrict chain bending. For example, in polystyrene, benzene rings are attached to the carbon backbone of the chain; this causes stiffening of the molecule sufficient to make polystyrene a hard and rigid plastic with a softening temperature higher than that of polyethylene, even though the polymer is neither cross-linked nor crystalline. The method is advantageous because the absence of crystallinity makes the material completely transparent, and the absence of cross-linking makes it readily moldable. A similar example is poly(methyl methacrylate).

However, the disadvantage of attaching bulky side groups is that the material dissolves in solvents fairly easily and undergoes swelling, since the bulky side groups allow ready penetration by solvents and swelling agents. This problem can be eliminated by stiffening the backbone of the chain itself. One way to do this is to introduce rigid ring structures in the polymer chain. A classic example of such a polymer is cellulose, which is the structural framework of wood and the most abundant organic material in the world. Its chain molecule consisting of a string of ring-shaped condensed glucose molecules has an intrinsically stiff backbone. Cellulose therefore has a high tensile strength. In poly/ethylene terephthalate fiber the chains are only moderately stiff, but the combination of chain stiffness and crystallization suffices to give the fiber high strength and a high melting point (265°C). The new plastic polycarbonate containing benzene rings in the backbone of the polymer chain (Table 1.3) is so tough that it can withstand the blows of a hammer. Ladder polymers based on aromatic chains consist of double-stranded chains made up of benzene-type rings [20]. These hard polymers are completely unmeltable and insoluble.

The combination of all three principles has led to the development of new and interesting products with enhanced properties. Composites based on expoxy and urethane-type polymers may be cited as an example. Thus, the stiff polymeric chains of epoxy and urethane types are cross-linked by curing reactions, and fillers are added to produce the equivalent of crystallization.

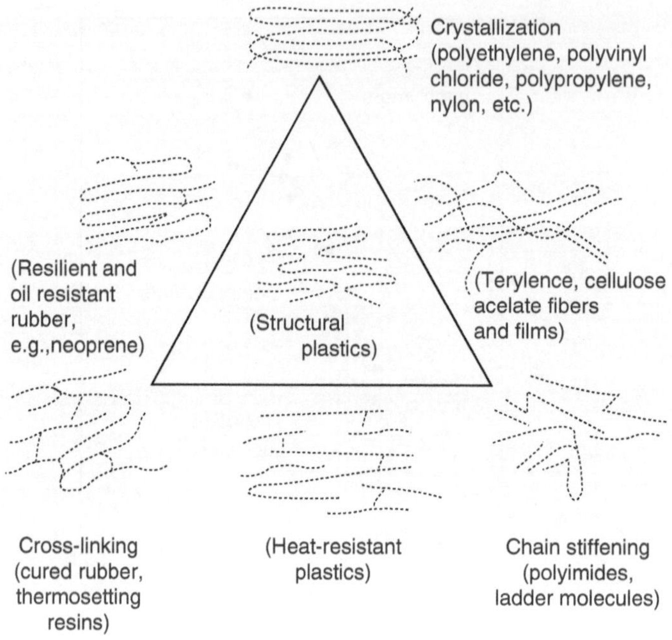

FIGURE 1.23 Three basic principles—crystallization, cross-linking, and chain stiffening—for making polymers strong and temperature resistant are represented at the three corners of the triangle. The sides and the center of the triangle indicate various combinations of the principles. (After Mark, H. F. 1967. *Sci. Am.*, 217, 3, Sept. 19, 148, see also p. 156.)

1.11 Cross-Linking of Polymer Chains

Cold flow can be prevented by cross-links between the individual polymer chains. The structure of polymer chains present in the cross-linked polymers is similar to the wire structure in a bedspring, and chain mobility, which permits one chain to slip by another (which is responsible for cold flow) is prevented. Natural rubber, for example, is a sticky product with no cross-linking, and its polymer chains undergo unrestricted slippage; the product has limited use. However, as we have seen, when natural rubber is heated with sulfur, cross-linking takes place (Figure 1.16). Cross-linking by sulfur at about 5% of the possible sites gives rubber enough mechanical stability to be used in automobile tires but still enables it to retain flexibility. Introducing more sulfur introduces more cross-links and makes rubber inflexible and hard.

Cross-linking of polymer chains can be brought about by (1) reactions of functional groups; (2) vulcanization, using peroxides, sulfur, or sulfur-containing compounds; (3) free-radical reactions caused by ionizing radiation; (4) photolysis involving photosensitive functional groups; or (5) coulombic interactions of ionic species.

1.11.1 Reactions of Functional Groups

Cross-linking by chemical reactions is an important process in polymer technology. A few common examples of cross-linking involving prepolymers are illustrated in Figure 1.24 through Figure 1.30. The cross-links are usually formed between prepolymer molecules by foreign molecules [e.g., styrene in unsaturated polyesters (Figure 1.28)] or by small chain segments, as in phenolic resin (Figure 1.24) and glyptal resin (Figure 1.30).

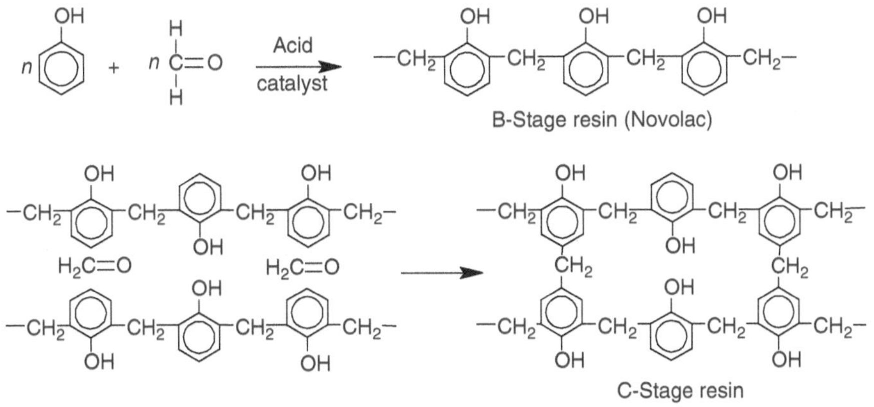

FIGURE 1.24 Equations (idealized) for the production of phenol-formaldehyde resins.

A high degree of cross-linking gives rise to three-dimensional or space network polymers in which all polymer chains are linked to form on giant molecule. Thus, instead of being composed of discrete molecules, a piece of highly cross-linked polymer constitutes, essentially, just one molecule. At high degree of cross-linking, polymers acquire rigidity, dimensional stability, and resistance to heat and chemicals [15]. Because of their network structure such polymers cannot be dissolved in solvents and cannot be softened by heat; strong heating only cause decomposition. Polymers or resins which are transformed into a cross-linked product, and thus take on a set on heating, are said to be of thermosetting type. Quite commonly, these materials are prepared, by intent, in only partially polymerized states (prepolymers), so that they may be deformed in the heated mold and then hardened by curing (cross-linking).

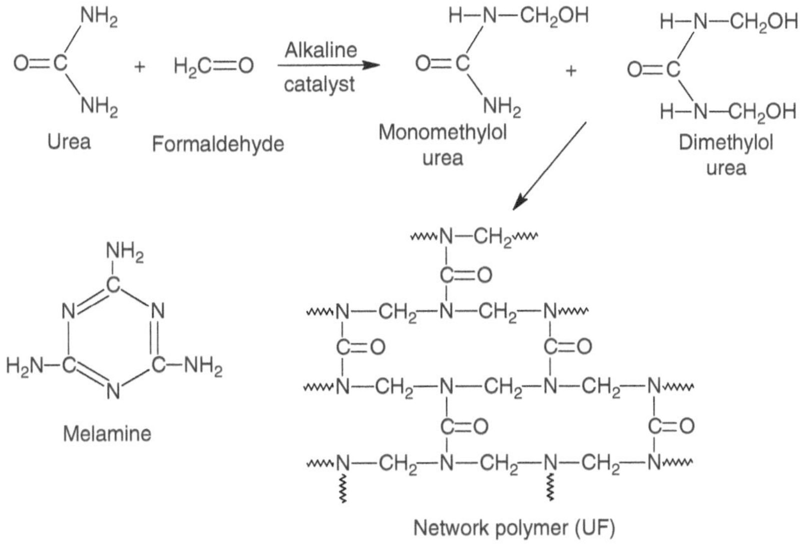

FIGURE 1.25 The two important classes of amino resins are the products of condensation reactions of urea and melamine with formaldehyde. Reactions for the formation of urea-formaldehyde amino resins (UF) are shown. Preparation of melamine-formaldehyde resins is similar.

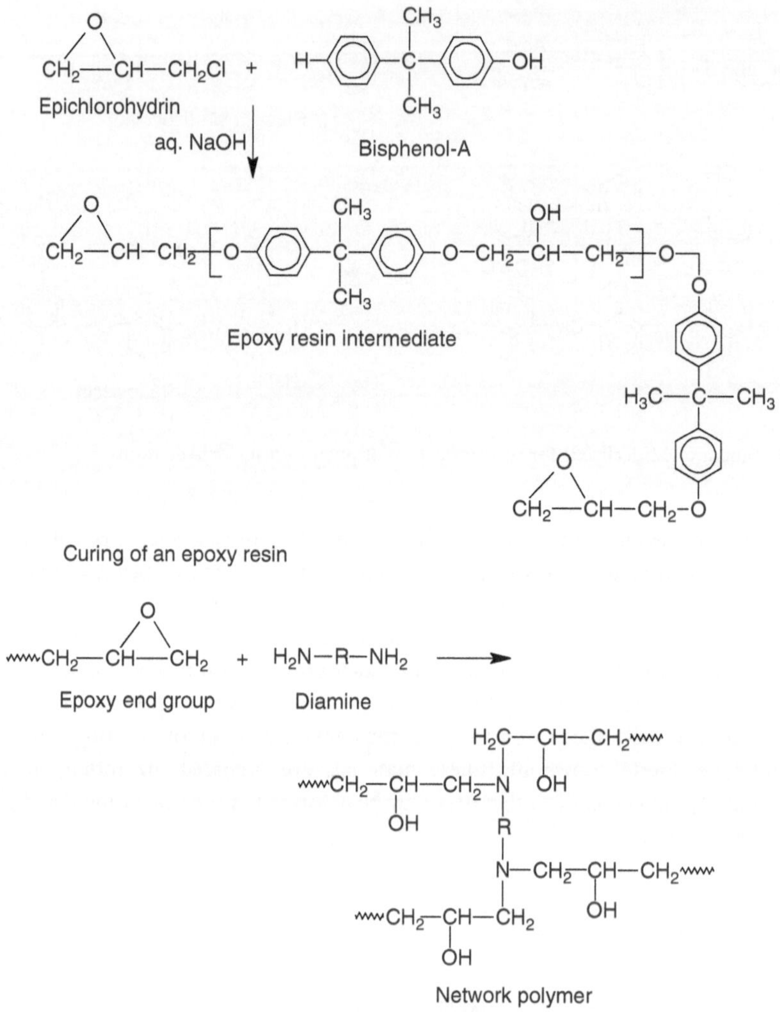

FIGURE 1.26 Epoxy monomers and polymer and curing of epoxy resins. Polyamines such as diethylenetriamine ($H_2NC_2H_4NHC_2H_4NH_2$) are widely used for the production of network polymers by room temperature curing.

The most important resins in commercial applications are phenolic resins (Figure 1.24), amino-resins (Figure 1.25), epoxy resins (Figure 1.26), vinylesters (Figure 1.27) unsaturated polyester resins (Figure 1.28), urethane foams (Figure 1.29), and the alkyds (Figure 1.30). The conversion of an uncross-linked thermosetting resin into a cross-linked network is called curing. For curing, the resin is mixed with an appropriate hardener and heated. However, with some thermosetting systems (e.g., epoxies and unsaturated polyesters), the cross-linked or network structure is formed even with little or no application of heat.

Linear polymers having functional groups can be cross-linked using suitable polyfunctional agents. Thus cellulosic fibers (cotton, rayon) are cross-linked by reaction of the hydroxyl groups of cellulose with formaldehyde, diepoxides, diisocyanates, and various methylol compounds such as urea-formaldehyde prepolymers, *N,N′*-dimethylol-*N,N′*-dimethylene urea, and trimethylomelamine. Cross-linking imparts crease and wrinkle resistance and results in crease-resistant fabrics.

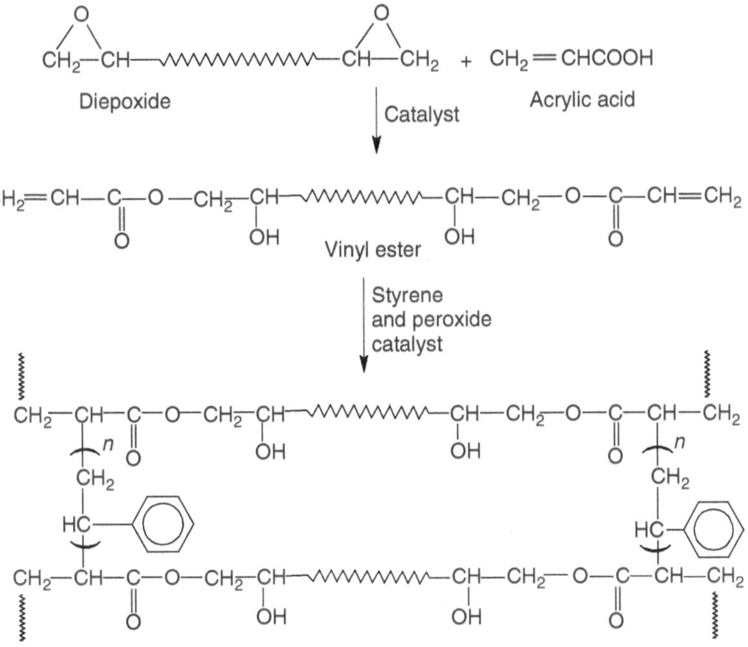

FIGURE 1.27 Equations for the preparation of a vinyl ester.

1.11.2 Vulcanization

Vulcanization is a general term applied to cross-linking of polymers, particularly elastomers. Many polymers are cross-linked by compounding with a peroxide such as dicumyl peroxide or di-*t*-butyl peroxide and then heating the mixture. Peroxide-initiated cross-linking of saturated polymers such as polyethylene proceeds by hydrogen abstraction (Equation 1.28) by radicals resulting from hemolytic cleavage of peroxide (Equation 1.27), followed by radical combination (Equation 1.29):

$$\mathrm{ROOR} \longrightarrow 2\mathrm{RO}\cdot \tag{1.27}$$

$$\mathrm{RO}\cdot + \sim\!\mathrm{CH_2CH_2}\!\sim \longrightarrow \sim\!\dot{\mathrm{C}}\mathrm{HCH_2}\!\sim + \mathrm{ROH} \tag{1.28}$$

$$\begin{matrix} \sim\!\dot{\mathrm{C}}\mathrm{H\,CH_2}\!\sim \\ + \\ \sim\!\dot{\mathrm{C}}\mathrm{H\,CH_2}\!\sim \end{matrix} \longrightarrow \begin{matrix} \sim\!\mathrm{CHCH_2}\!\sim \\ | \\ \sim\!\mathrm{CHCH_2}\!\sim \end{matrix} \tag{1.29}$$

With unsaturated polymers, hydrogen abstraction probably occurs at the allylic position (Equation 1.30) with subsequent cross-linking again resulting from radical combination (Equation 1.31).

$$\sim\!\mathrm{CH_2CH{=}CHCH_2}\!\sim + \mathrm{RO}\cdot \longrightarrow \sim\!\dot{\mathrm{C}}\mathrm{HCH{=}CHCH_2}\!\sim + \mathrm{ROH} \tag{1.30}$$

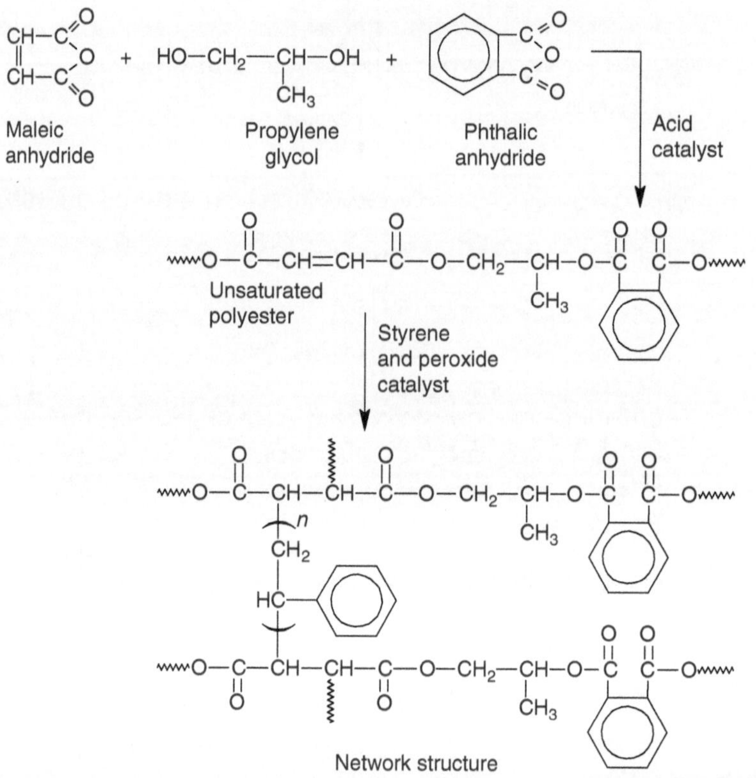

FIGURE 1.28 Equations for preparation and curing of an unsaturated polyester resin. The presence of ethylenic unsaturation provides sites for cross-linking by a chain reaction mechanism in the presence of styrene. Phthalic anhydride increases flexibility by increasing spacing of cross-links.

$$\begin{array}{ccc} \sim\dot{C}HCH{=}CHCH_2\sim & & \sim CHCH{=}CHCH_2\sim \\ + & \longrightarrow & | \\ \sim\dot{C}HCH{=}CHCH_2\sim & & \sim CHCH{=}CHCH_2\sim \end{array} \qquad (1.31)$$

Moreover, addition-transfer processes (Equation 1.32 and Equation 1.33) can also cause cross-linking, because in many instances considerably more cross-links are formed than would be expected on the basis of only abstraction–combination reactions described above.

$$\begin{array}{ccc} \sim\dot{C}HCH{=}CHCH_2\sim & & \sim CHCH{=}CHCH_2\sim \\ + & \longrightarrow & | \\ \sim CH_2CH{=}CHCH_2\sim & & \sim CH_2CH-\dot{C}HCH_2\sim \\ & & \downarrow \sim CH_2CH{=}CHCH_2\sim \end{array} \qquad (1.32)$$

$$\begin{array}{cc} & \sim CHCH{=}CHCH_2\sim \\ \sim\dot{C}HCH{=}CHCH_2\sim + & | \\ & \sim CH_2CH-CH_2CH_2\sim \end{array} \qquad (1.33)$$

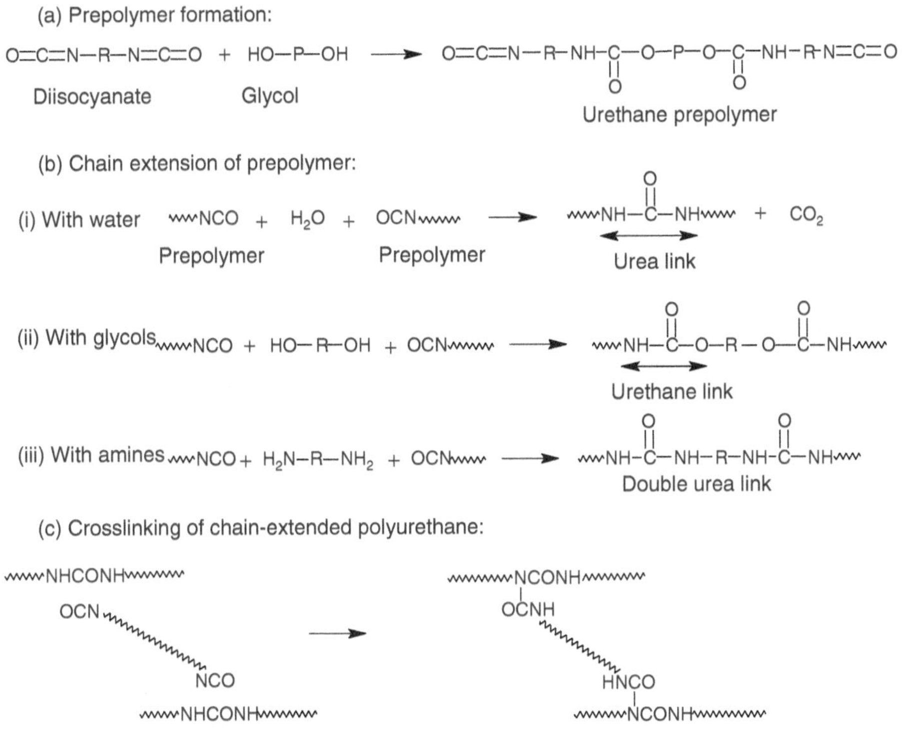

FIGURE 1.29 Equations for preparation, chain extension, and curing of polyurethane.

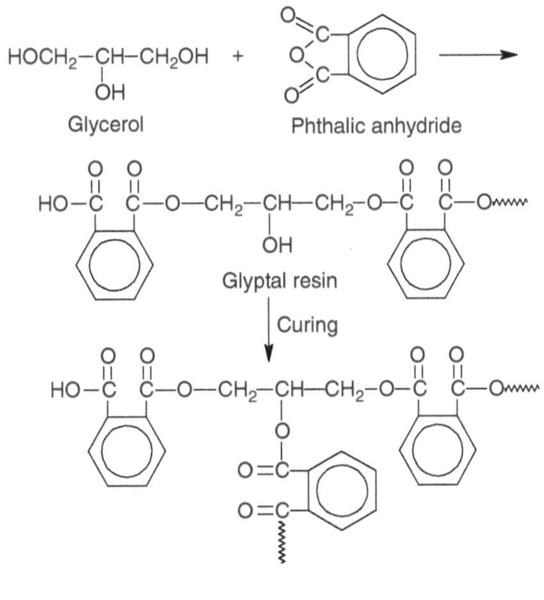

FIGURE 1.30 Equations for preparation of network glyptal resin.

Peroxide cross-linking of diene polymers works with all except butyl rubber, which undergoes chain scission. The cross-links formed via peroxides are more thermally stable than those formed via sulfur vulcanization. However, peroxide cross-linking is not economically competitive with sulfur cross-linking because of the high cost of peroxides. Peroxide cross-linking is primarily used for those polymers that cannot be easily cross-linked by sulfur, such as polyethylene and other polyolefins, ethylenepropylene (no diene) rubbers (EPM), and polysiloxanes. Cross-linking of polyethylene increases its strength properties and the upper limit of use temperature. Uses include electrical wire and cable insulation, pipe, and hose. For EPM and polysiloxanes, cross-linking is essential to their use as elastomers.

Cross-linking with elemental sulfur is the oldest method of vulcanization, discovered independently in 1839 by Goodyear in the United States and MacIntosh and Hancock in Great Britain. This results in many reactions, producing a variety of cross-links, as represented schematically in Figure 1.31. The rate of vulcanization with sulfur can be, and normally is, increased by addition of accelerators such as zinc salts of dithiocarbamic acids or organosulfur compounds such as tetramethylthiuram disulfide (see Chapter 1 of *Plastics Fabrication and Recycling*, Table 1.4). Other compounds, usually zinc oxide and stearic acid, are also added as activators. Although the mechanism of acceleration is not well understood, acceleration is known to increase the number of monosulfide and disulfide cross-links, and decrease the number of cyclic monosulfide groups (Figure 1.31).

Other cross-linking reactions are also used for elastomers. A variety of halogen-containing elastomers are cross-linked by heating with a basic oxide (e.g., MgO or ZnO) and a primary diamine. Such elastomers include poly(epichlorohydrin); various co-ad terpolymers of fluorinated monomers such as vinylidene fluoride, hexafluoropropylene, perfluoro(methyl vinyl ether), and tetrafluoroethylene (see "Fluoroelastomers"); and terpolymers of alkyl acrylate, acrylonitrile, and 2-chloroethyl vinyl ether (see "Polyacrylates"). Cross-linking involves dehydrohalogenation followed by addition of the diamine to the double bond (Equation 1.34) with the metal oxide acting as an acid acceptor. Some vulcanizations employa dithiol instead of a diamine. Elastomeric terpolymers of alkyl acrylate, ethylene, and a small amount of alkenoic acid (see "Acrylate Polymers") are vulcanized by addition of diamine. The alkenoic acid is added to introduce sites (C=C) for subsequent cross-linking via reaction with primary diamines.

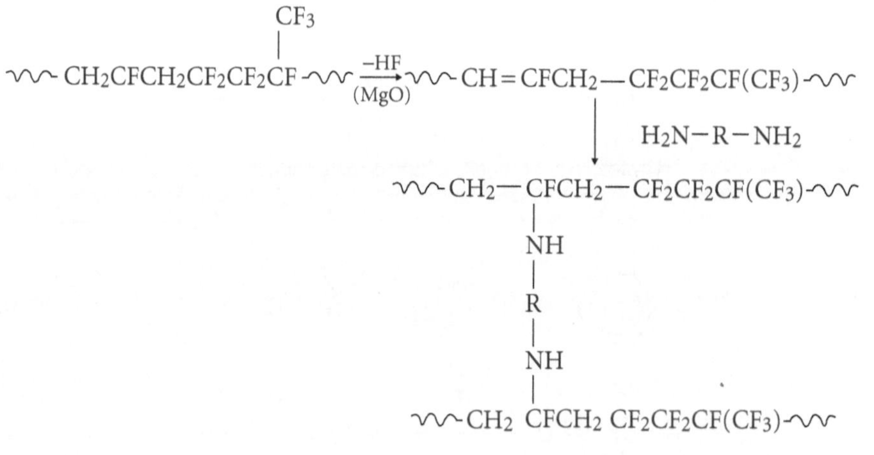

(1.34)

The vulcanization of polychloroprene (Neoprene) is carried out in different ways. Conventional sulfur vulcanization is not practiced to a large extent, since best physical properties of neoprene rubber are achieved by vulcanization with metal oxides (without diamine), either alone or in combination with sulfur (sometimes together with an accelerator). Halogenated butyl rubber is cross-linked in a similar manner. The mechanism for cross-linking with metal oxide alone is not established.

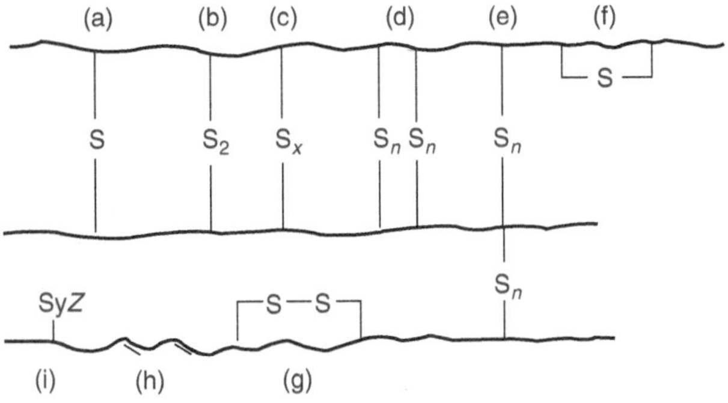

FIGURE 1.31 Typical chemical groupings in a sulfur-vulcanized natural rubber network. (a) Monosulfide cross-link; (b) disulfide cross-link; (c) polysulfide cross-link ($x=3$–6); (d) parallel vicinal cross-links ($n=1$–6) attached to adjacent main-chain atoms; (e) cross-links attached to common or adjacent carbon atoms; (f) intrachain cyclic monosulfide; (g) intrachain cyclic disulfide; (h) conjugated diene; (i) pendant sulfide group terminated by moiety Z derived from accelerator.

Aging of polymers is often accompanied by cross-linking due to the effect of the surroundings. Such cross-linking is undesirable because it greatly reduces the elasticity of the polymer, making it more brittle and hard. The well-known phenomenon of aging of polyethylene with loss of flexibility is due to cross-linking by oxygen under the catalytic action of sunlight (Figure 1.32a). Cheap rubber undergoes a similar loss of flexibility with time due to oxidative cross-linking (Figure 1.32b). This action may be discouraged by adding to the polymer an antioxidant, such as a phenolic compound and an opaque filler, such as carbon black, to prevent entry of light.

1.11.3 Radiation Cross-Linking

When vinyl polymers are subjected to ionizing radiation (whether it be photons, electrons, neutrons, or protons), two main types of reaction occur: cross-linking and degradation. Generally, both occur simultaneously, though degradation predominates with high doses of radiation. With low radiation doses, the polymer structure determines which will be the major reaction. Thus, geminally disubstituted polymers, such as poly(α-methylstyrene), poly(methylmethacrylate), and polyisobutylene tend to undergo chain scission, with monomer being formed as the major degradation product. Such polymers

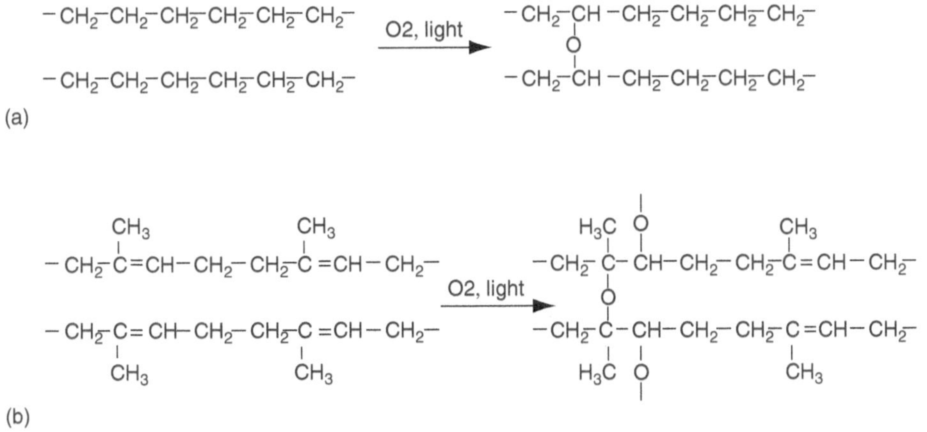

FIGURE 1.32 Aging of (a) polyethylene and (b) natural rubber by oxidative cross-linking.

TABLE 1.6 Behavior of Polymers Subjected to High Energy Radiation

Polymers that Cross-Link	Polymers that Degrade
Polyethylene	Polyisobutylene
Poly(acrylic acid)	Poly(α-methyl styrene)
Poly(methyl acrylate)	Poly(methyl methacrylate)
Polyacrylamide	Poly(methacrylic acid)
Polychloroprene	Poly(vinylidene chloride)
Polydimethylsiloxanes	Polypropylene
Styrene–acrylonitrile copolymers	Polytetrafluoroethylene

will decrease in molecular weight on exposure to radiation. Halogen-substituted polymers, such as poly (vinyl chloride), degrade with loss of halogen. With most other vinyl monomers, cross-linking predominates. Table 1.6 lists some polymers that cross-link and some that degrade (i.e., liable to chain scission).

The mechanism of cross-linking is free-radial in nature. The reaction is essentially the same as in peroxide cross-linking except that polymer radicals are formed by the interaction of ionizing radiation with polymer. The reaction probably involves initial rejection of a proton (Equation 1.35), which in turn removes another proton from an adjacent site on a neighboring chain Equation 1.36. Cross-linking then occurs by radical combination as by Equation 1.37. This assumption is reasonable because hydrogen is a major side product, and random formation of radicals would not give the efficiency of cross-linking that is generally observed.

$$\sim\!CH_2CH_2\!\sim \xrightarrow{h\nu} \sim\!\dot{C}HCH_2\!\sim + H\cdot \tag{1.35}$$

$$\underset{\text{(neighboring chain)}}{\sim\!CH_2CH_2\!\sim} + H\cdot \longrightarrow \sim\!\dot{C}HCH_2\!\sim + H_2 \tag{1.36}$$

$$\begin{matrix} \sim\!\dot{C}H\,CH_2\!\sim \\ + \\ \sim\!\dot{C}H\,CH_2\!\sim \end{matrix} \longrightarrow \begin{matrix} \sim\!CHCH_2\!\sim \\ | \\ \sim\!CHCH_2\!\sim \end{matrix} \tag{1.37}$$

Low-density polyethylene gives rise to significant amounts of gaseous hydrocarbons on irradiation. This probably involves the ejection of a chain branch following the ejection of a proton (Equation 1.38).

$$\sim\!CH_2\!-\!\underset{R}{\underset{|}{CH}}\!\sim \longrightarrow \sim\!\dot{C}H\!-\!\underset{R}{\underset{|}{CH}}\!\sim + H\cdot \longrightarrow \sim\!CH\!=\!CH\!\sim + RH \tag{1.38}$$

Fragmentation reactions of this type, as well as ejection of hydrogen, lead to double bonds in the polymer chains (Equation 1.38) in addition to cross-linking. Radiolysis effects on numerous vinyl polymers have been studied, but polyethylene is most important. Irradiated polyethylene film is used commercially because of its improved mechanical and thermal properties. Uses of radiation cross-linking include electrical wire and cable insulation and heat-shrinkable products (tubing, packaging film, and bags). Curing of coating and adhesives are other applications of radiation cross-linking. Polystyrene is quite resistant to radiation (a characteristic of aromatic polymers, in general), but it can be cross-linked with higher doses. A limitation of radiation cross-linking is that radiation does not penetrate deep into the polymer matrix. The method is therefore used primarily with films.

1.11.4 Photochemical Cross-Linking

Ultraviolet or visible light induced cross-linking (photocross-linking) has assumed increasing importance in recent years. Among the numerous applications are printed circuits for electronic equipment; printing inks; coatings for optical fibers; varnishes for paper arid carton board; and finishes for vinyl flooring, wood, paper, and metal. Photocross-linking applied to photoresist technology is described in Chapter 2 of *Industrial Polymers, Specialty Polymers, and Their Applications.*

There are two basic methods for bringing about photocross-linking of polymers; (1) incorporating photosensitizers which absorb photons and thereby induce formation of free radicals, and (2) incorporating groups into the polymer that undergo either light-initiated cross-linking polymerization or *photocycloaddition* reactions.

When photosensitizers such as benzophenone are added to polymer, absorption of ultraviolet radiation results in excitation of the sensitizer followed by hydrogen abstraction from the polymer to yield radical sites (Equation 1.39) available for cross-linking by combination reactions. Copolymers of vinyl esters and fluorinated monomers that can be cross-linked by ultraviolet radiation have been developed for use as weather-resistant wood coatings. In this application, the vinyl ester constitutes about 10% of the copolymer and benzophenone is added as a photosensitizer. The vinyl ester polymer also undergoes α-cleavage reaction (Equation 1.40) with subsequent cross-linking.

$$\mathrm{Ph{-}\overset{\overset{\displaystyle O}{\|}}{C}{-}Ph} \xrightarrow{h\nu} \mathrm{(Ph{-}\overset{\overset{\displaystyle O}{\|}}{C}{-}Ph)^*} \xrightarrow{RH} \mathrm{Ph{-}\overset{\overset{\displaystyle OH}{|}}{\underset{\bullet}{C}}{-}Ph} + \mathrm{R\cdot} \qquad (1.39)$$

$$[\mathrm{Ph} = \mathrm{phenyl}(C_6H_5)\mathrm{group}]$$

$$\sim\!\mathrm{CH_2{-}\underset{\underset{\displaystyle \underset{\displaystyle |}{C{=}O}}{|}}{\underset{}{CH}}}\!\!\!\!\!\!\sim \xrightarrow{h\nu} \sim\!\mathrm{CH_2\dot{C}H}\!\sim + \mathrm{O{=}\dot{C}{-}OR} \qquad (1.40)$$

$$\text{(side group on CH: } \mathrm{C{=}O} \text{ bearing } \mathrm{OR}\text{)}$$

A wide variety of functional groups has been used to effect photocycloaddition or light-induced cross-linking. Among such groups are cinnamate ($PhCH{=}CHCO_2R$), chalcone ($PhCH{=}CHCOPh$), and stilbene ($PhCH{=}CHPh$). The groups may be present as part of the backbone or, more commonly, as pendant groups. In most cases, cycloaddition occurs to give cyclobutane cross-links, as shown for cinnamate ester in Equation 1.41.

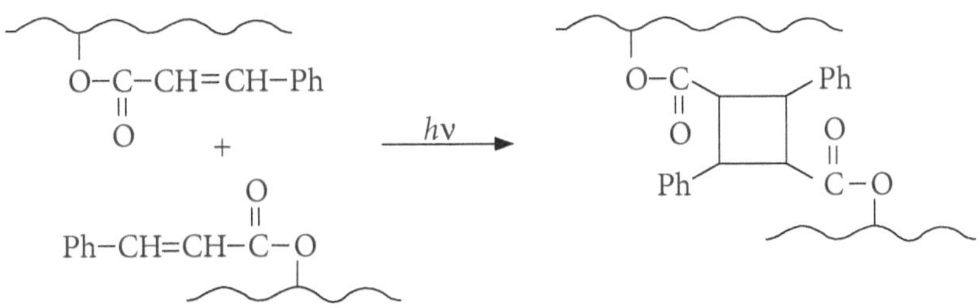

(1.41)

The reactive groups may be incorporated into polymer during the polymerization reaction; for example, β-vinyloxyethyl cinnamate undergoes cathionic polymerization through the vinyl ether (Equation 1.42) to yield a linear polymer containing pendant cinnamate ester. Alternatively, the group may be added to preformed polymer, as in the Schotten–Baumann reaction of poly(vinyl alcohol) with cinnamoyl chloride (Equation 1.43).

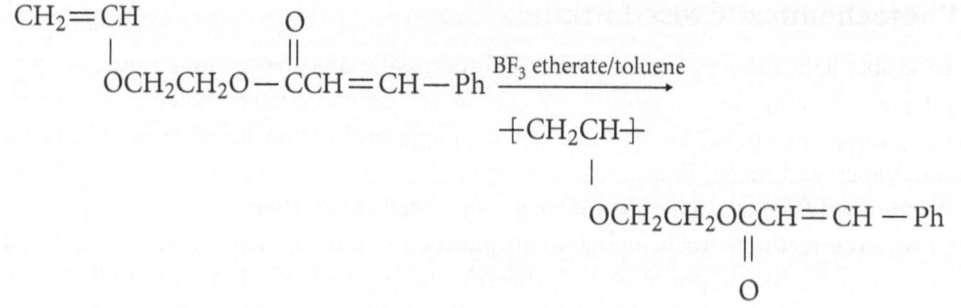

(1.42)

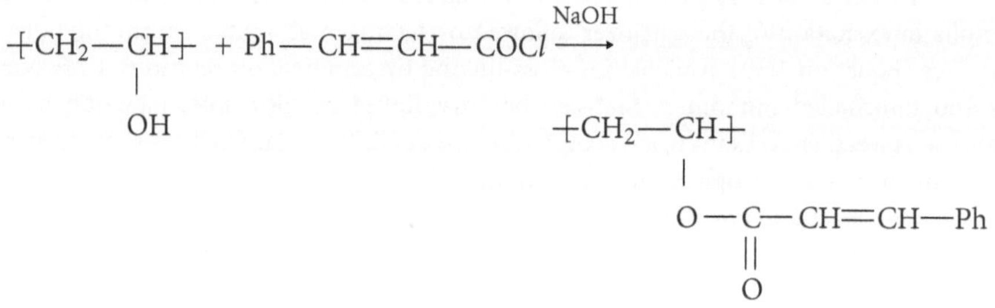

(1.43)

1.11.5 Ionic Cross-Linking

Ionic cross-links fall in the category of thermally labile cross-links, that is, chemical cross-links that break apart on heating and reform on cooling. Examples of ionic cross-linking are the hydrolysis of chlorosulfonated polyethylene with a divalent metal oxide and a source of water, and the neutralization of poly(ethylene-*co*-methacrylic acid) to salts of divalent metals. In a typical commercial product of chlorosulfonated polyethylene there is one chlorosulfonyl group for each 200 backbone carbon atoms. Magnesia (MgO) and a hydrogenated wood resin (as a source of H_2O) can be used as a cross-linking system to yield magnesium sulfonate cross-links in the polymer (Equation 1.44).

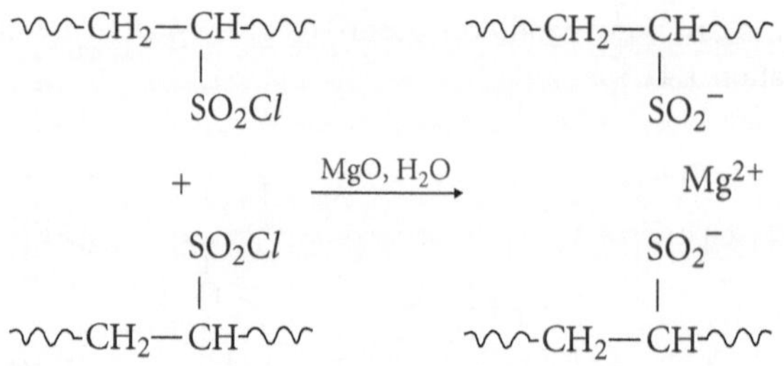

(1.44)

Neutralization of ethylene copolymers containing up to 5%–10% acrylic or methacrylic acid copolymer with a metal salt such as the acetate or oxide of zinc, magnesium, and barium yields products referred to as ionomers. (Commercial products may contain univalent as well as divalent metal salts.) Ionomers are marked by Du Pont under the trade name Surlyn. These have interesting properties compared with the nonionized copolymer. Introduction of ions causes disordering of the semicrystalline structure, which makes the polymer transparent. Ionomers act like reversibly cross-linked thermoplastics as a result of microphase separation between ionic metal carboxylate and nonpolar hydrocarbon segments. The behavior is similar to the physical cross-linking in thermoplastic elastomers (see Chapter 1 of *Industrial*

Polymers, Specialty Polymers, and Their Applications). Ionomers are discussed more fully in Chapter 2 of *Industrial Polymers, Specialty Polymers, and Their Applications.*

1.12 Solubility Behavior of Polymers

Knowledge of the solubility of various polymers in different solvents is important in assessing their chemical resistance and their application potentialities in the fields of paints, spinning fibers, and casting films. Important also is the knowledge of the solubility of various materials, such as plasticizers and extenders in the polymer, especially since this has an important bearing on plastics formulation.

Because of the size and shape of the polymer molecules and other factors, the solubility relations in polymer systems are complex in comparison to those among low-molecular-weight compounds. Some empirical solubility rules have, nevertheless, been derived for applying to polymer systems, and it is also possible to make certain predictions about solubility characteristics of such systems.

The underlying reason that one material can act as a solvent for another is the compatibility of the materials—i.e., the ability of the molecules of the two materials to coexist without tending to separate. If we denote the force of attraction between the molecules of one material A by F_{AA}, that between the molecules of another material B by F_{BB}, and represent that between one A and one B molecule as F_{AB}, then the system will be compatible and a solution will result if $F_{AB} > F_{BB}$ and $F_{AB} > F_{AA}$. On the other hand, if F_{AA} or $F_{BB} > F_{AB}$, the system will be incompatible and the molecules will separate, forming two phases. In the absence of any specific interaction (e.g., hydrogen bonding) between solvent and solute, we can reasonably assume the intermolecular attraction forces between the dissimilar molecules to be approximately given by the geometric mean of the attraction forces of the corresponding pairs of similar molecules; that is, $F_{AB} = (F_{AA}F_{BB})^{1/2}$. Consequently, if F_{AA} and equal, F_{AB} will also be similar and the materials should be soluble.

1.12.1 Solubility Parameter

A measure of the intermolecular attraction forces in a material is provided by the cohesive energy. Approximately, this equals the heat of vaporization (for liquids) or sublimation (for solids) per mol. The cohesive energy density in the liquid state is thus ($\Delta E_v/V$), in which ΔE_v is the molar energy of vaporization and V is the molar volume of the liquid. The square root of this cohesive energy density is known as the solubility parameter (δ), that is,

$$\delta = (\Delta E_v/V)^{1/2} \tag{1.45}$$

If the vapor behaves approximately like an ideal gas, Equation 1.45 can be written as

$$\delta = [(\Delta H_v - RT)/V]^{1/2} = [(\Delta H_v - RT)\rho/M]^{1/2} \tag{1.46}$$

where ΔH_v is the molar enthalpy of vaporization and ρ is the density of liquid with molecular weight M. For a volatile liquid cohesive energy density and, hence δ, can be determined experimentally by measuring ΔH_v and ρ.

Example 5

Calculate an estimate of the solubility parameter for water at 25°C from its heat of vaporization at the same temperature, given by

$$H_2O(l) = H_2O(g), \qquad \Delta H_{25^\circ C} = 10.514 \text{ kcal}$$

Answer. From Equation 1.46:

$$\delta^2 = [(10{,}514\ \text{cal mol}^{-1}) - (1.987\ \text{cal mol}^{-1}\ \text{K}^{-1})(298\ \text{K})](1\ \text{g cm}^{-3})/(18\ \text{g mol}^{-1})$$

$$= 551.2\ \text{cal cm}^{-3} = 2.3 \times 10^9\ \text{J m}^{-3}$$

$$\delta = 23.5\quad (\text{cal cm}^{-3})^{1/2} = 48.0 \times 10^3\quad (\text{J m}^{-3})^{1/2} = 48\ \text{MPa}^{1/2}$$

[Conversion factors: 1 cal cm^{-3} = 4.184×10^6 J m^{-3} = 4.184×10^6 Pa = 4.184 MPa. Hence, 1 $(\text{cal cm}^{-3})^{1/2}$ = 2.045 $\text{MPa}^{1/2}$]

Hildebrand [21] first used the solubility parameter approach for calculating estimates of the enthalpy of mixing, ΔH_{mix}, for mixtures of liquids. The equation employed is

$$\Delta H_{\text{mix}} = V_{\text{mix}} \phi_1 \phi_2 (\delta_1 - \delta_2)^2 \tag{1.47}$$

where V_{mix} is the molar volume of the mixture, and δ_1 and δ_2 are the solubility parameters of components 1 and 2, respectively.

A necessary requirement for solution and blending compatibility is a negative or zero Gibbs free energy change (ΔG_{mix}) when the solution or blend components are mixed, that is,

$$\Delta G_{\text{mix}} = \Delta H_{\text{mix}} - T\Delta S_{\text{mix}} \leq 0 \tag{1.48}$$

Since the ideal entropy of mixing (ΔS_{mix}) is always positive, the components of a mixture can be assumed to be miscible only if $\Delta H_{\text{mix}} \leq T\Delta S_{\text{mix}}$. Solubility therefore depends on the existence of a zero or small value of ΔH_{mix}, only positive (endothermic) heats of mixing being allowed, as in Equation 1.47. Miscibility or solubility will then be predicted if the absolute value of the $(\delta_1 - \delta_2)$ difference is zero or small [22]. Specific effects such as hydrogen bonding and charge transfer interactions can lead to negative ΔH_{mix} but these are not taken into account by Equation 1.47, and separate considerations must be applied in order to predict their effect on miscibility and solubility [23].

Solubility parameters of solvents can be correlated with the density, molecular weight, and structure of the solvent molecule. According to the additive method of Small [24], the solubility parameter is calculated from a set of additive constants, F, called molar attraction constants, by the relationship

$$\delta = \frac{\rho}{M} \sum F \tag{1.49}$$

where $\sum F$ is the molar attraction constants summed over the groups present in the compound; ρ and M are the density and the molar mass of the compound. The same procedure is applied to polymers and Equation 1.49 is used, wherein ρ is now the density of the amorphous polymer at the solution temperature, $\sum F$ is the sum of all the molar attraction constants for the repeat unit, and M is the molar mass of the repeat unit. Values of molar attraction constants for the most common groups in organic molecules were estimated by Small [24] from the vapor pressure and heat of vaporization data for a number of simple molecules. A modified version of a compilation of molar attraction constants [25,26] is reproduced in Table 1.7. An example of the use of the tabulated molar attraction constants is given in the problem worked out below.

Example 6

Calculate an estimate of the solubility parameter for the epoxy resin DGEBA (diglycidyl ether of bisphenol A) having the repeat unit structure as shown below and density 1.15 g/cm^3.

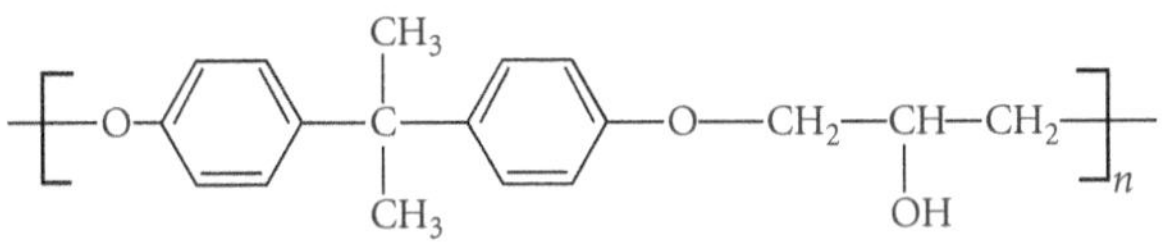

Answer. M (for repeating unit) = 284 g/mol.

Groups	ΣF (cal cm^3)$^{1/2}$/mol	No. of Groups	ΣF (cal cm^3)$^{1/2}$/mol
$-CH_3$	147.3	2	294.60
$-CH_2$	131.5	2	263.00
>CH–	85.99	1	85.99
>C<	32.03	1	32.03
–O– (ether)	114.98	2	229.96
–OH	225.84	1	225.84
–CH=(aromatic)	117.12	8	936.96
–C=(aromatic)	98.12	4	392.48
6-membered ring	−23.44	2	−46.88
Para substitution	40.33	2	80.66
			2494.64

$$\delta = \frac{(1.15\ \mathrm{g\ cm^{-3}})(2494.64\ \mathrm{cal^{1/2}\ cm^{3/2}\ mol^{-1}})}{(284\ \mathrm{g\ mol^{-1}})}$$
$$= 10.1(\mathrm{cal\ cm^{-3}})^{1/2} = 20.7\ \mathrm{MPa^{1/2}}$$

The units of the solubility parameter δ are in (energy/volume)$^{1/2}$. The δ values for some common solvents and polymers [27], listed in Table 1.8 and Table 1.9, have units of cal$^{1/2}$ cm$^{3/2}$, called *hildebrands.* The SI value in MPa$^{1/2}$ may be obtained by multiplying the δ value in hildebrand by 2.045. Most tabulated solubility parameters refer to 25°C. However, over the temperature range normally encountered in industrial practice, the temperature dependence of δ can be neglected.

While the solubility parameter of a homopolymer can be calculated from the molar attraction constants as illustrated in Example 6, the solubility parameter of random copolymers, δ_c, may be calculated from

$$\delta_c = \sum \delta_i w_i \tag{1.50}$$

where δ_i is the solubility parameter of the homopolymer that corresponds to the monomer i in the copolymer and w_i is the weight fraction of repeating unit i in the copolymer [28].

Solubility would be expected if the absolute value of $(\delta_1 - \delta_2)$ is less than about unity and there are no strong polar or hydrogen-bonding interactions in either the polymer or the solvent. To allow for the influence of hydrogen-bonding interactions, solvents have been characterized qualitatively as poorly, moderately, or strongly hydrogen bonded. The solvents listed in Table 1.8 are grouped according to this scheme. It is a useful practice to match both solubility parameter and hydrogen-bonding tendency for predicting mutual solubility.

Hansen [29] developed a three-dimensional solubility parameter system based on the assumption that the energy of evaporation, i.e., the total cohesive energy ΔE_t which holds a liquid together, can be divided into contribution from dispersion (London) forces ΔE_d, polar forces ΔE_p, and hydrogen-bonding forces ΔE_h. Thus,

$$\Delta E_t = \Delta E_d + \Delta E_p + \Delta E_h \tag{1.51}$$

TABLE 1.7 Group Molar Attraction Constants

Group	Molar Attraction, F $(\text{cal cm}^3)^{1/2}\ \text{mol}^{-1}$
$-CH_3$	147.3
$-CH_2-$	131.5
$>CH-$	85.99
$>C<$	32.03
$CH_2=$(olefin)	126.54
$-CH=$(olefin)	121.53
$>C=$(olefin)	84.51
$-CH=$(aromatic)	117.12
$-C=$(aromatic)	98.12
–O–(ether, acetal)	114.98
–O–(epoxide)	176.20
–COO–	326.58
$>C=O$	262.96
–CHO	292.64
$(CO)_2O$	567.29
–OH–	225.84
–OH aromatic	170.99
$-NH_2$	226.56
–NH–	180.03
–N–	61.08
$-C\equiv N$	354.56
$-N=C=O$	358.66
–S–	209.42
Cl_2	342.67
–Cl (primary)	205.06
–Cl (secondary)	208.27
–Cl (aromatic)	161.0
–Br	257.88
–Br (aromatic)	205.60
–F	41.33
Structure Feature	
Conjugation	23.26
Cis	−7.13
Trans	−13.50
5-membered ring	20.99
6-membered ring	−23.44
Ortho substitution	9.69
Meta substitution	6.6
Para substitution	40.33

Source: Hoy, K. L. 1970. *J. Paint Technol.*, 42, 76 and Brandrup J. and Immergut E. eds., 1975. *Polymer Handbook, 2nd Ed.* Wiley Interscience, New York.

Dividing this equation by the molar volume of a solvent, *V*, gives

$$\frac{\Delta E_t}{V} = \frac{\Delta E_d}{V} + \frac{\Delta E_p}{V} + \frac{\Delta E_h}{V} \tag{1.52}$$

or

$$\delta_t^2 = \delta_d^2 + \delta_p^2 + \delta_h^2 \tag{1.53}$$

where δ_d, δ_p, and δ_h are solubility parameters due to dispersion forces, dipole forces, and hydrogen-bonding (or, in general, due to donor–acceptor interactions), respectively. The three parameters, called Hansen parameters, were determined [29,30] empirically on the basis of many experimental observations

TABLE 1.8 Solubility Parameters for Some Common Solvents

Solvent	δ (cal/cm^3)$^{1/2}$
(i) Poorly H-bonded (Hydrocarbons and their halo-, nitro-, and cyano-Products)	
n-Hexane	7.3
Carbon tetrachloride	8.6
Toluene	8.9
Benzene	9.2
Chloroform	9.3
Methylene chloride	9.7
Nitrobenzene	10.0
Acetonitrile	11.9
(ii) Moderately H-bonded (esters, ethers, ketones)	
Isoamyl acetate	7.8
Dioctyl phthalate	7.9
Tetrahydrofuran	9.1
Methyl ethyl ketone	9.3
Acetone	9.9
1,4-Dioxane	10.0
Diethylene glycol	
Monoethyl ether	10.2
(iii) Strongly hydrogen-bonded (acids, alcohols, aldehydes, amides, amines)	
Piperidine	8.7
Acetic acid	10.1
Meta-cresol	10.2
t-Butanol	10.6
I-Butanol	11.4
Propylene glycol	12.6
Methanol	14.5
Ethylene glycol	14.6
Glycerol	16.5
Water	23.4

Source: Brandrup, J. and Immergut, E. eds. 1975. *Polymer Handbook, 2nd Ed.* Interscience, Wiley, New York.
SI value of δ in MPa$^{1/2}$ is obtained by multiplying the value in (cal/cm^3)$^{1/2}$ by 2.045.

for a large number of solvents (Table 1.10). Hansen's total cohesion parameter, δ_t, corresponds to the Hildebrand parameter δ, although the two quantities may not be identical because they are determined by different methods. Once the three component parameters for each solvent were evaluated, the set of parameters could then be obtained for each polymer (Table 1.11) from solubility ascertained by visual inspection of polymer–solvent mixtures (at concentrations of 10% w/v).

When plotted in three dimensions, the Hansen parameters provide an approximately spherical volume of solubility for each polymer in δ_d, δ_p, δ_h space. The scale on the dispersion axis is usually doubled to improve the spherical nature of this volume. The distance of the coordinates $(\delta_d^i, \delta_p^i, \delta_h^i)$ of any solvent i from the center point $(\delta_d^j, \delta_p^j, \delta_h^j)$ of the solubility sphere of polymer j is

$$d = [4(\delta_d^i - \delta_d^j)^2 + (\delta_p^i - \delta_p^j)^2 + (\delta_h^i - \delta_h^j)^2]^{1/2} \tag{1.54}$$

This distance can be compared with the radius R of the solubility sphere [31] of the polymer (Table 1.11), and if $d < R$, the likelihood of the solvent i dissolving the polymer j is high. This works well, despite the limited theoretical justification of the method.

Example 7

Using Hansen parameters (Table 1.10 and Table 1.11) determine if polystyrene is expected to dissolve in a solvent mixture of 60/40 v/v methyl ethyl ketone/*n*-hexane.

TABLE 1.9 Solubility Parameters for Some Common Polymers

Polymer	δ, $(cal/cm^3)^{1/2}$	H-Bonding Group[a]
Polytetrafluoroethylene	6.2	Poor
Polyethylene	8.0	Poor
Polypropylene	9.2	Poor
Polyisobutylene	8.0	Poor
Polybutadiene	8.4	Poor
Polyisoprene	8.1	Poor
Polystyrene	9.1	Poor
Poly(methyl methacrylate)	9.5	Medium
Poly(vinyl acetate)	9.4	Medium
Poly(vinyl chloride)	9.7	Medium
Cellulose diacetate	11.0	Strong
Poly(vinyl alcohol)	12.6	Strong
Polyacrylonitrile	12.7	Poor
Nylon-6,6	13.7	Strong

SI value of δ, in $MPa^{1/2}$ is obtained by multiplying the value in $(cal/cm^3)^{1/2}$ by 1.045.

Source: Burrell, H. 1975. *Polymer Handbook,* J. Brandrup and E. Immergut, eds., *2nd Ed.,* pp. IV-337–359. Wiley Interscience, New York.

[a] The hydrogen-bonding group of each polymer has been taken as equivalent to that of the parent monomer.

Answer. Denote the solubility parameter components of MEK, *n*-hexane, and polystyrene using superscripts *i, j,* and *k,* respectively. From Table 1.10:

$$\text{MEK:}\ \delta_d^i = 7.8, \quad \delta_p^i = 4.4, \quad \delta_h^i = 2.5 \text{ all in } (\text{cal cm}^{-3})^{1/2}$$

TABLE 1.10 Hansen Parameters for Solvents at 25°C

Liquid	δ_d $(cal/cm^3)^{1/2}$	δ_p $(cal/cm^3)^{1/2}$	δ_h $(cal/cm^3)^{1/2}$
Acetic acid	6.8	6.0	9.2
Acetone	6.3	4.8	5.4
Benzene	7.9	4.2	2.0
I-Butanol	7.8	2.8	7.7
Chloroform	5.4	6.7	3.1
Cyclohexane	8.0	1.5	0.0
1,4-Dioxane	8.0	4.9	3.9
Dioctyl phthalate	8.1	3.4	1.5
Ethyl acetate	6.5	4.2	4.3
Ethylene glycol	4.9	7.4	14.6
Glycerol	4.5	7.5	15.3
n-Hexane	7.3	0.0	0.0
Methyl ethyl ketone	7.8	4.4	2.5
Methanol	7.4	6.0	10.9
Nitrobenzene	8.6	6.8	0.0
Tetrahydrofuran	8.2	2.8	3.9
Toluene	8.0	3.9	0.8
Water	5.9	11.1	19.7
m-Xylene	8.1	3.5	1.2

SI value of a parameter in $MPa^{1/2}$ is obtained by multiplying the value in $(cal/cm^3)^{1/2}$ by 2.045.

Source: From "Tables of Solubility Parameters" *3rd Ed.* Chemicals and Plastics Research and Development Dept., Union Carbide Corporation, Tarrytown, NY, 1975.

TABLE 1.11 Hansen Parameters and Interactions Radius of Some Polymers and Resins

Polymer	δ_d (cal/cm^3)$^{1/2}$	δ_p (cal/cm^3)$^{1/2}$	δ_h (cal/cm^3)$^{1/2}$	R (cal/cm^3)$^{1/2}$
Acrylonitrile–butadiene elastomer	9.1	4.3	2.0	4.7
Cellulose acetate	9.1	6.2	5.4	3.7
Epoxy resin	10.0	5.9	5.6	6.2
Nitrocellulose	7.5	7.2	4.3	5.6
Polyamide, thermoplastic	8.5	−0.9	7.3	4.7
Polyisoprene	8.1	0.7	−0.4	4.7
Poly(methyl methacrylate)	9.1	5.1	3.7	4.2
Polystyrene	10.4	2.8	2.1	6.2
Poly(vinyl acetate)	10.2	5.5	4.7	6.7
Poly(vinyl chloride)	8.9	3.7	4.0	1.7
Styrene–butadiene rubber	8.6	1.7	1.3	3.2

SI value in MPa$^{1/2}$ is obtained by multiplying the value in (cal/cm^3)$^{1/2}$ by 2.045.

Source: Data from Hansen, C. M. and Beerbower, A. 1971. *Kirk Othmer Encyclopaedia of Chemical Technology*, A. Standen, ed., *2nd Ed.*, Suppl. 889, p. 910. Wiley Interscience, New York.

$$n-\text{hexane:}\ \delta_d^j = 7.3,\quad \delta_p^j = 0,\quad \delta_h^j = 0 \text{ all in } (\text{cal cm}^{-3})^{1/2}$$

The Hansen parameters are combined on a 60/40 volume fraction basis:

$$\delta_d^{ij} = 0.6 \times 7.8 + 0.4 \times 7.3 = 7.6\ (\text{cal cm}^{-3})^{1/2}$$

$$\delta_p^{ij} = 0.6 \times 4.4 + 0.4 \times 0.0 = 2.6\ (\text{cal cm}^{-3})^{1/2}$$

$$\delta_h^{ij} = 0.6 \times 2.5 + 0.4 \times 0.0 = 1.5\ (\text{cal cm}^{-3})^{1/2}$$

For polystyrene (Table 1.11):

$$\delta_d^k = 10.4,\quad \delta_p^k = 2.8,\quad \delta_h^k = 2.1,\quad R = 6.2 \text{ all in } (\text{cal cm}^{-3})^{1/2}$$

From Equation 1.54: $d = [4(10.4-7.6)^2 + (2.8-2.6)^2 + (2.1-1.5)^2]^{1/2} = 5.6\ (\text{cal cm}^{-3})^{1/2}$. As this value is less than the radius of the polymer solubility sphere (6.2 cal$^{1/2}$ cm$^{-3/2}$), the polymer is expected to be soluble. This is found to be the case for a 10% w/w solution.

Three-dimensional presentations of solubility parameters are not easy to use and it is more convenient to transform the Hansen parameters into fractional parameters as defined by [32]:

$$\begin{aligned} f_d &= \delta_d/(\delta_d + \delta_p + \delta_h) \\ f_p &= \delta_p/(\delta_d + \delta_p + \delta_h) \\ f_h &= \delta_h/(\delta_d + \delta_p + \delta_h) \end{aligned} \tag{1.55}$$

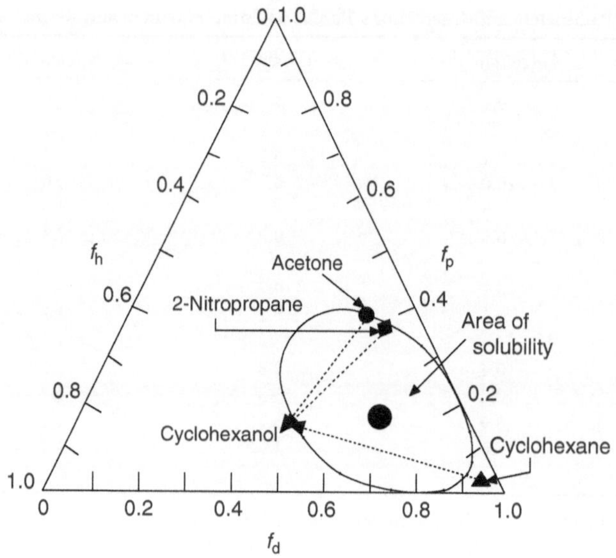

FIGURE 1.33 Limiting solubility boundary for chlorinated rubber. The solid circle represents the probable solubility parameter of the resin. (After Teas, J. P. 1968. *J. Paint Technol.*, 40, 519.)

The fractional parameters represent, in effect, the quantitative contribution of the three types of forces to the dissolving abilities for each solvent and can be represented more conveniently in a triangular diagram (Figure 1.33) to provide a visual presentation of the nature of the solvating powers of liquids, including in one picture such diverse liquids as water, alcohols, organic acids, and hydrocarbons [32]. The triangular solubility chart can be used conveniently for the prediction of solubility of polymers. A chart can be constructed for a given resin, identifying each solvent coordinate point to indicate either complete, partial or lack of solubility. In general, solvents which provide clear solutions are grouped in a reasonably well defined area of the chart thus forming a loop and solvents which tend to swell the resin lie near the border of such loops, as shown in Figure 1.33. The probable solubility parameters of the solute polymer will be at the heart of the solubility area. Any solvent possessing values of f_d, f_p, and f_h to place its plot inside the solubility area has a high probability of producing a clear solution. Conversely, liquids lying outside of the solubility area are very likely to be poor or nonsolvent.

Since δ_d values do not vary greatly, at least among common solvents, it has been suggested that a plot of δ_p versus δ_h should be sufficient for most practical purposes. Thus, another and probably more efficient two-parameter representation is made possible by defining [33]

$$\delta_v \equiv \left(\delta_d^2 + \delta_p^2\right)^{1/2} \tag{1.56}$$

The procedures outlined above have a practical use, but it should be realized that the parametric models are almost entirely empirical. Experimental uncertainties are also involved since solubility measurements are not very accurate. Solubility loops described by the models only indicate the limits of compatibility and always include doubtful observations.

1.13 Effects of Corrosives on Polymers

Polymers are resistant to electrochemical corrosion. When a polymer has a δ value similar to that of water, dissolution occurs. However, if the δ value of the polymer is lower, it is not attacked by water or other polar solvents. The carbon–carbon bonds in polymer backbones are not cleaved even by boiling water. When amide, ester, or urethane groups are present in the polymer backbone, they may be attacked

TABLE 1.12 Effects of Corrosive Environments on Plastics

Plastic	Environment															
	Aliphatic Solvents		Aromatic Solvents		Chlorinated Solvents		Esters and Ketones		Weak Bases		Strong Bases		Strong Acids		Strong Oxidants	
Acetal	A	B	A	B	A	B	A	B	A	A	A	B	E	E	E	E
Acrylic	B	C	E	E	E	E	E	E	A	C	B	E	D	E	E	E
Acrylonitrile–butadiene–styrene (ABS)	A	E	D	E	E	E	E	E	A	C	A	C	B	E	D	E
Cellulose acetate	A	B	A	C	A	D	E	E	A	C	C	E	C	E	C	E
Cellulose acetate butyrate	A	C	D	E	D	E	E	E	B	D	C	E	C	E	C	E
Cellulose acetate propionate	A	C	D	E	D	E	E	E	A	B	C	E	C	E	C	E
Epoxy (glass fiber filled)	A	B	A	B	A	C	B	C	A	A	B	C	B	C	D	D
Furan (asbestos filled)	A	A	A	A	A	A	A	A	B	B	B	B	A	A	E	E
Melamine	A	A	A	A	A	A	A	B	B	C	B	C	B	C	B	C
Phenolic (asbestos filled)	A	A	A	A	A	A	C	C	A	C	D	E	A	A	D	E
Polyamide	A	A	A	A	A	B	A	A	A	B	B	C	E	E	E	E
Polybenzimidazole	A	A	A	A	A	A	A	A	A	A	A	B	A	B	A	C
Polycarbonate	A	A	A	A	E	E	E	E	A	E	E	E	A	A	A	A
Poly(chlorotrifluoroethylene)	A	A	A	A	C	D	A	A	A	A	A	A	A	A	A	A
Polyester (glass fiber filled)	A	B	A	C	B	D	C	C	B	C	C	E	B	B	B	C
Polyethylene	C	E	C	E	D	E	D	E	A	A	A	A	A	A	A	A
Polypropylene	A	D	B	D	B	D	A	C	A	A	A	A	A	C	A	D
Polysulfone	A	A	D	D	E	E	C	D	A	A	A	A	A	A	A	A

(continued)

TABLE 1.12 (Continued)

Plastic	Environment															
	Aliphatic Solvents		Aromatic Solvents		Chlorinated Solvents		Esters and Ketones		Weak Bases		Strong Bases		Strong Acids		Strong Oxidants	
Polystyrene	D	E	D	E	E	E	D	E	A	E	A	E	D	E	D	E
Poly(tetra fluoroethylene)	A	A	A	A	A	A	A	A	A	A	A	A	A	A	A	A
Polyurethane	A	D	C	D	D	E	B	C	A	A	C	D	A	D	A	D
Poly(vinyl chloride)	A	E	D	E	E	E	D	E	A	E	A	E	A	E	B	E
Silicone	B	C	D	D	D	E	B	D	A	B	D	E	C	D	D	E
Urea	A	C	A	C	A	C	A	B	B	C	B	C	D	E	B	C

Code: A = no effect or inert; B = slight effect; C = mild effect; D = softening or swelling; E = severe deterioration. The effects were measured at 25°C (first letter in each column) and at 90°C (second letter in each column).

by hot acids or alkalies, and hydrolysis may occur. When these functional groups are present as pendant groups on the polymer chain, the reaction will be similar. However, the tendency for such attack is reduced when alkyl groups are present on the carbon atom attached to the functional group. Thus, poly (methyl methacrylate) but not poly(methyl acrylate) is resistant to acid or alkaline hydrolysis. Atoms or groups with strong carbon bonds, such as fluorine, chlorine, and ether groups, are resistant to attack by aqueous acids and alkalies. The chemical resistance of various plastics is summarized in Table 1.12.

1.14 Thermal Stability and Flame Retardation

Most organic polymers decompose when heated to moderate or high temperatures. It is for this reason that few synthetic polymers can be used for long periods of time above 150°C–200°C. This necessitates the use of metals and ceramics for many applications, even though synthetic polymers may be cheaper and, in some cases, stronger on a weight-for-weight basis.

Plastics can be grouped into eight temperature-time zones, depending on the temperature at which they retain 50% mechanical or physical properties when heated for different periods in air. These temperature–time zones are shown in Figure 1.34, and the materials falling in the different zones are listed in Table 1.13. The materials in zone 6 and above can compete with metals in high-performance applications because they perform in the same temperature range. Most polymers in zone 6 and above do not burn, but they may char and may be consumed very slowly in direct flames.

Rotation of chain segments is more difficult in polymers with cyclic rings in their chains. As a result, such polymers are stiffer and more resistant to deformation and have higher melting points and higher

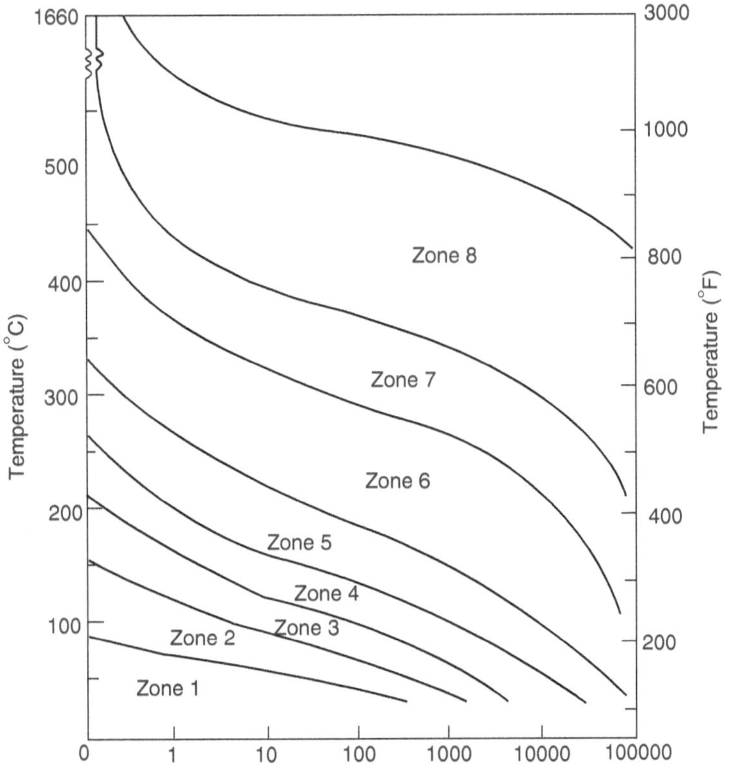

FIGURE 1.34 Time-temperature zones indicating thermal stability of plastics (see Table 1.13). (Adapted from Anon. 1968. *Plastics World*, 26(3), 30.)

TABLE 1.13 Plastics Retaining 50% Mechanical or Physical Properties at Temperatures in Air

Zone 1
Acrylic
Cellulose acetate
Cellulose acetate butyrate
Cellulose acetate propionate
Cellulose nitrate
Polyallomer
Polyethylene, low density
Polystyrene
Poly(vinyl acetate)
Poly(vinyl alcohol)
Poly(vinyl butyral)
Poly(vinyl chloride)
Styrene–acrylonitrile
Styrene–butadiene
Urea–formaldehyde
Zone 2
Acetal
Acrylonitrile–butadiene–styrene Copolymer
Ethyl cellulose
Ethylene–vinyl acetate copolymer
Furan
Lonomer
Polyamides
Polycarbonate
Polyethylene, high density
Polyethylene, cross-linked
Poly(ethylene terephthalate)
Polypropylene
Poly(vinylidene chloride)
Polyurethane
Zone 3
Poly(monochlorotrifluoroethylene)
Zone 4
Alkyd
Fluorinated ethylene propylene copolymer
Melamine formaldehyde
Phenol–furfural
Polyphenylene oxide
Polysulfone
Zone 5
Diallyl phthalate
Epoxy
Phenol–formaldehyde
Polyester
Poly(tetrafluoroethylene)
Zone 6
Polybenzimidazole
Polyphenylene
Silicone
Zone 7
Polyamide–imide
Polyimide
Zone 8
High-performance plastics being developed using intrinsically rigid linear macromolecules rather than the conventional crystallization and cross-linking

Source: Anon. 1968. *Plastics World*, 26(3), 30.

glass transition temperatures. Thermal stability is improved by the presence of aromatic and heterocyclic rings in polymer chains. This general approach has resulted in a number of new commercially available polymers with improved high-temperature properties. Polyimides and polybenzimidazoles are important examples of such high-temperature polymers. They can be exposed for a short time to temperatures as high as 1,112°F (600°C).

Polyimides are synthesized [35] by the reactions of dianhydrides with diamines in a two-stage process (Figure 1.35). In the first stage, the two materials form an intermediate poly(amic acid). Processing is accomplished after this stage, since the polymer at this point is still soluble and fusible. The poly(amic acid) is formed into the desired physical form of the final polymer product (e.g., film, fiber, laminate, coating, etc.) and then the second stage of the reaction is carried out, in which the poly(amic acid) is cyclized in the solid state to the polyimide by heating at moderately high temperatures (above 150°C). The polymer after the second stage of the process is insoluble and infusible. Instead of *m*-phenylenediamine, shown in Figure 1.35, other diamines have also been used to synthesize polyimides—for example, *p*-phenylenediamine, *p,p′*-diaminodiphenyl ether, and *p,p′*-diaminodiphenyl sulfide.

Polyimides have been used up to 100 h at 600°F (315.6°C). They have been used as a transparent head cover for fire fighters and as excellent flame- and high-temperature-resistant foams, which can be used as structurally stable insulation materials at high temperatures. Polyimide/glass fiber composites have many uses in the electronics and aerospace industries and in other high-performance applications.

Polybenzimidazoles are synthesized by the reactions of aromatic diacids and aromatic tetraamines (Figure 1.36). Polybenzimidazoles are mostly used as fibers in parachutes, for reentry vehicles, and so on. Composites of polybenzimidazoles with fibers have excellent basic strength and high-temperature performance. These composites have been extensively used in nose fairings, aircraft leading edges, reentry nose cones, radomes, and deicer ducts.

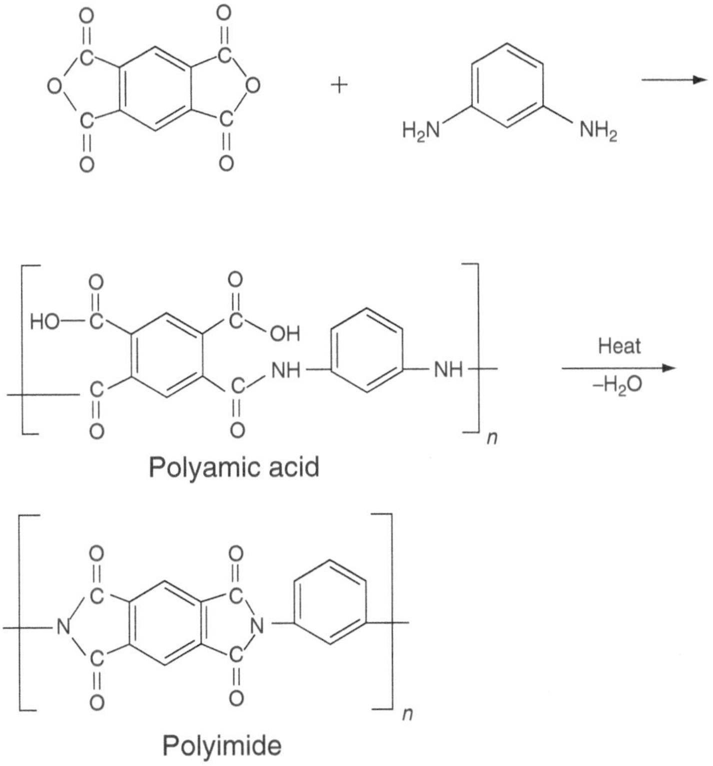

FIGURE 1.35 Synthesis of polyimide from pyromellitic anhydride and *m*-phenylenediamine.

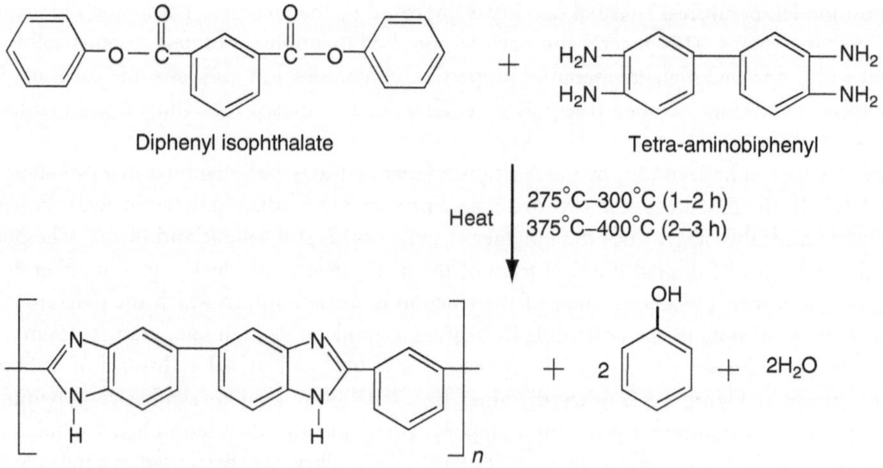

FIGURE 1.36 Synthesis of polybenzimidazole by condensation polymerization.

A weak link in the polybenzimidazole structure is the imino hydrogen. When this hydrogen is replaced by a phenyl group as in *N*-phenyl polybenzimidazole, a dramatic increase in high-temperature properties in oxidizing atmospheres is obtained.

In polymers such as polybhenylenes there are a few aliphatic linkages and many aromatic rings which account for their improved heat resistance. Several polyphenylenes are shown in Figure 1.37. Poly (phenylene oxide) has excellent dimensional stability at elevated temperatures. Repeated steam autoclaving does not deteriorate its properties. Poly(phenylene sulfide) is completely nonflammable. It is used in the form of composites with both asbestos and glass fibers.

Aliphatic linkage is completely eliminated in ladder polymers such as polybenzimidazopyrrolones, commonly called "pyrrones" (Figure 1.38). Such polymers are highly stable in air. They do not burn or melt when heated but form carbon char without much weight loss. They are potentially the ultimate in heat- and flame-resistant materials.

Another way to get good heat resistance is to use inorganic material as backbone chain as in silicone polymers. Here organic radicals are attached to silicone atoms on an inorganic silicon–oxygen structure. Presence of silicon–oxygen links gives such materials outstanding heat resistance. A silicone polymer has the structure

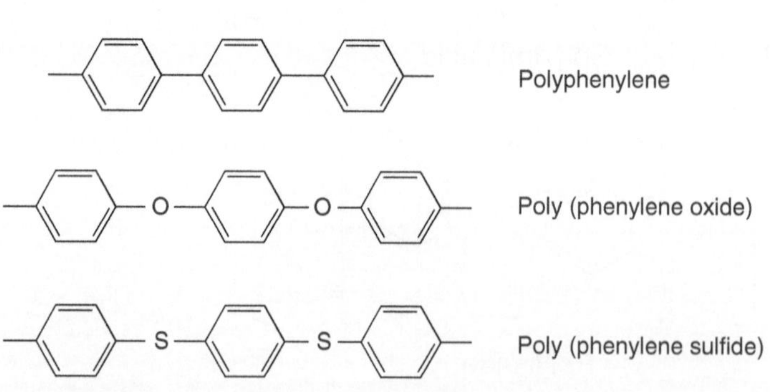

FIGURE 1.37 Structures of some polyphenylenes.

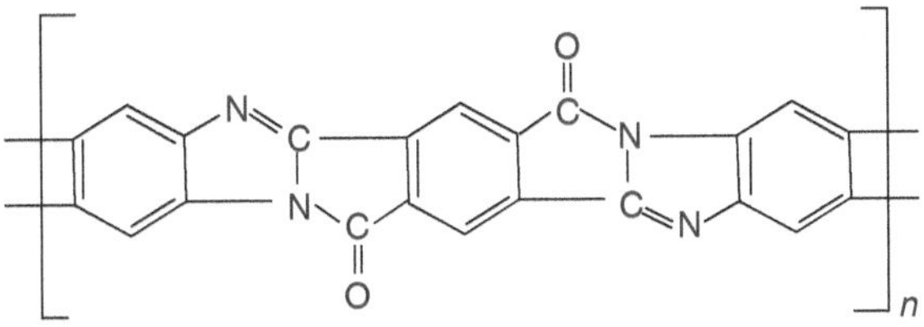

FIGURE 1.38 Typical structure of a polybenzimidazopyrrolone.

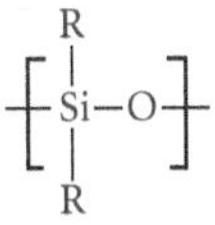

When *n* is a small number, the structure is that of a silicone oil, whereas silicon rubbers have high values of *n*. when the ratio R/Si is lower than 2, cross-linked polymers are obtained. Properties of silicone polymers are greatly affected by the type of organic radical present. For a given chain length, a methyl silicone can be an oily liquid, but a phenyl silicone is a hard and brittle resin.

1.14.1 Thermal Degradation

There are basically three types of thermal degradation reactions for vinyl polymers [36,37]: (1) nonchain scission; (2) random chain scission; and (2) depropagation. In practice, mechanisms 2 and 3 blend into one another, with many polymers showing evidence of both processes.

Nonchain scission refers to reactions involving pendant groups that do not break the backbone. Typical of such reactions are dehydrochlorination of poly(vinyl chloride) (Equation 1.57), elimination of acid from poly(vinyl esters)—for example, poly(vinyl acetate) (Equation 1.58)—and elimination of alkene from poly(alkyl acrylate)s (Equation 1.59).

$$\text{+}CH_2-\underset{\displaystyle Cl}{\underset{|}{CH}}\text{+} \longrightarrow \text{+}CH{=}CH\text{+} + HCl \tag{1.57}$$

$$\text{+}CH_2-\underset{\displaystyle \underset{\displaystyle O}{\underset{||}{OC}}-CH_3}{\underset{|}{CH}}\text{+} \longrightarrow \text{+}CH{=}CH\text{+} + HO-\underset{\displaystyle O}{\underset{||}{C}}-CH_3 \tag{1.58}$$

$$\text{+}CH_2-\underset{\displaystyle \underset{\displaystyle OCH_2CH_2R}{\underset{|}{C{=}O}}}{\underset{|}{CH}}\text{+} \longrightarrow \text{+}CH_2-\underset{\displaystyle \underset{\displaystyle OH}{\underset{|}{C{=}O}}}{\underset{|}{CH}}\text{+} + CH_2{=}CHR \tag{1.59}$$

The first two reactions lead to highly colored residues, indicating that the double bonds formed in the polymer backbone are primarily conjugated. Such elimination reactions are not satisfactory for synthesizing polyacetylene, however, since side reactions also occur. But nonchain scission has been used as one approach to solving problem of polyacetylene's intractability (see "Inherently Conducting Polymers" in Chapter 2 of *Industrial Polymers, Specialty Polymers, and Their Applications*).

Random chain scission results from homolytic bond-cleavage reactions at week points in the polymer chains. Complex mixtures of degradation products are formed, the origin of which may be explained in terms of radical transfer reactions such as

$$\sim CH_2CH_2CH_2CH_2 \sim \longrightarrow \sim CH_2\dot{C}H_2 + \cdot CH_2CH_2 \sim \longrightarrow \sim CH{=}CH_2 + CH_3CH_2 \sim$$

Random chain scission occurs with all vinyl polymers to varying degrees, but it occurs less with increasing substitution on the polymer backbone.

Depropagation, or depolymerization (unzipping), to give monomer occurs mainly with polymers prepared from 1,1-disubstituted monomers. The depolymerization is a free-radical process that may be initiated either at a chain end or at a random site on the backbone. For example, poly(methyl methyl methacrylate) appears to begin unzipping primarily at the chain ends, whereas poly(α-methylstyrene) does so at random sites along the chain. In both cases, tertiary radicals are formed with each depropagation step:

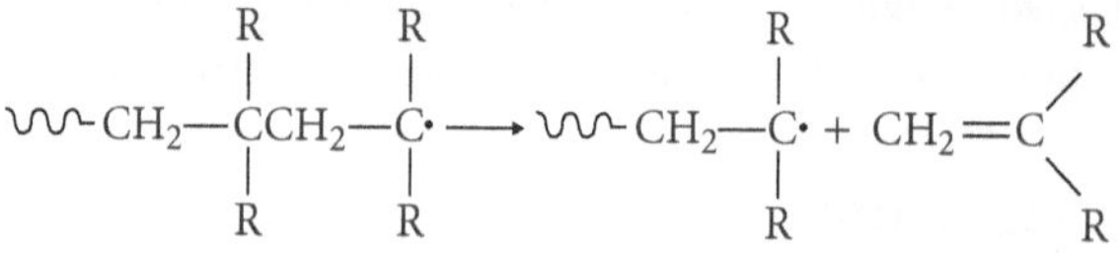

The nature of active chain ends in poly(methyl methacrylate) is a question for debate. At moderate temperatures (220°C) only half of the chains in this polymer unzip, and higher temperatures (350°C) may be needed to decompose the remaining polymer. Apparently, chains which unzip at 200°C are terminated by unsaturated groups, whereas those which depolymerize only at higher temperatures have saturated end groups.

Like poly(methyl methacrylate) and poly(α-methyl styrene), polytetrafluroethylene also undergoes 100% conversion to the monomer at elevated temperatures. However, it does so only at low pressures and high temperatures. At atmospheric pressure, the monomer molecules recombine to form dimer and other species. This polymer is one of the most thermally stable polyolefins known, but even so, it cannot withstand prolonged exposure to temperatures above about 350°C–400°C.

Polystyrene represents a case in which monomer is only one of several species formed by thermal degradation at 350°C as monomer, dimer, trimer, and tetramer are formed in the relative proportions of 40:10:8:1. The thermal breakdown process is believed to begin at unsaturated linkages which constitute the weak points along the chain. A cleavage at these sites initiates a free-radical mechanism leading to liberation of monomer and to an intramolecular back-biting process. The process liberates dimer, trimer, and so on, by a transfer mechanism such as the shown in Equation 1.60

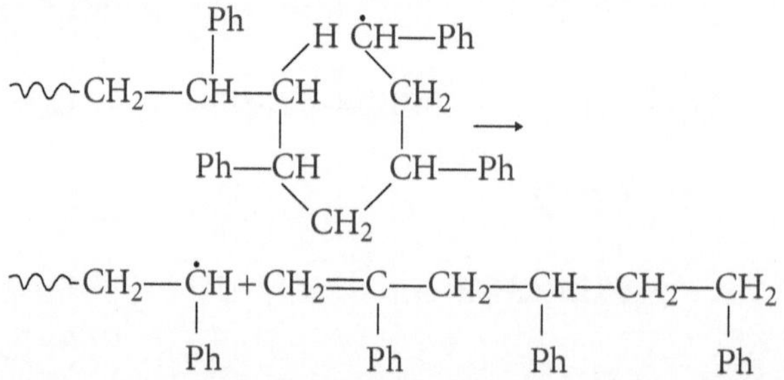

(1.60)

Polyethylene yields very little monomer by thermal degradation. Above about 300°C, the decomposition products form a continuous spectrum of unsaturated hydrocarbons which contain from 1 to at least 70 carbon atoms. The products are believed to be the results of random chain cleavage accompanied by inter- and intramolecular chain transfer. Oxygenated sites may act as weak links for cleavage and the existence of chain branch points may facilitate the transfer process. Polypropylene behaves in a similar manner to polyethylene.

The foregoing discussion applies specifically to olefin addition polymers. Some comments on the thermal behavior of condensation polymers are also appropriate. Polyamides can decompose during melt spinning or molding operations. Such decomposition, usually sight, is apparently initiated by free radicals formed by the hemolytic cleavage of $-NH-CH_2-$ skeletal bonds. Water and carbon dioxide are also liberated. The water serves to hydrolyze amide (–NH–CO–) linkages, further shortening the chains. Branches are also formed by reaction of terminal $-NH_2$ groups with carbonyl units (Equation 1.61), leading ultimately to gelation of the molten polymer.

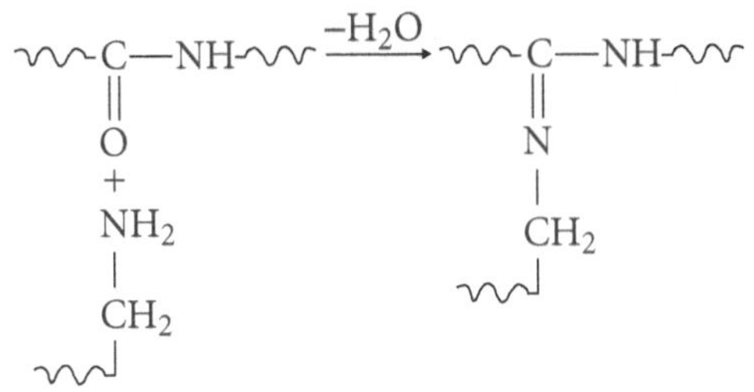

(1.61)

Polyesters, such as poly(ethylene terephthalate), are fairly stable at temperatures just above the melting point. However, at temperatures between 300 and 550°C, this polymer decomposes to yield carbon dioxide, acetaldehyde, and terephthalic acid together with smaller amounts of other decomposition products, such as water, methane, and acetylene.

1.14.2 Ablation

When subjected briefly to very high temperatures, some polymers, such as phenolics, can undergo rapid decomposition to gases and porous char, thereby dissipating the heat and leaving a protective thermal barrier on the substrate. This sacrificial loss of material accompanied by transfer of energy is known as *ablation*. Interaction of a high-energy environment (2,500°C–5,000°C) with the exposed ablative material results in sacrificial erosion of some amount of the surface material, and the attendant energy absorption controls the surface temperature and greatly restricts the flow of heat into the substrate.

Most notable applications of ablative materials are in protecting space vehicles during reentry into the earth's atmosphere, protecting missile nose cones subjected to aerodynamic heating during hypersonic fight in the atmosphere, insulating sections of rocket motors from hot propulsion gases, resisting the intense radiant heat of thermonuclear blasts, and providing thermal protection for structural materials exposed to very high temperatures.

Polymers have been used as ablative materials for a combination of reasons [38]. Some of their advantages are (1) high heat absorption and dissipation per unit mass expended, which may range from several hundred to several thousand calories per gram of ablative material; (2) automatic control of surface temperature by self-regulating ablative degradation: (3) excellent thermal insulation; (4) tailored performance by varying the individual material component and composition of ablative systems; (5) design simplicity and ease of fabrication; (6) light weight and low cost.

Polymer ablatives, however, can be used only for transitory periods of a few minutes or less at very high temperatures and heat load. Moreover, the sacrificial loss of surface material during ablation causes dimensional changes which must be predicted and incorporated into the design.

An ablative material should have a high heat of ablation, which measures the ability of the material to absorb and dissipate energy per unit mass. It should also possess good strength even after charring, since any sloughing off of chunk of material which have not decomposed or vaporized represent poor usage of the ablative system Figure 1.39 profiles the various stages of heating, charring, and melting in two phenolic composites, one reinforced with glass and the other with nylon. The glass-reinforced system appears to be mechanically superior because it produces a molten glass surface, whereas the nylon-reinforced system may have higher thermal efficiency.

1.14.3 Flame Retardation

All thermoset plastics are self-extinguishing. Among thermoplastics, nylon, polyphenylene oxide, polysulfone, polycarbonate, poly(vinyl chloride), chlorinated polyether, poly(chlorotrifluoroethylene) and fluorocarbon polymers have self-extinguishing properties [39]. The burning characteristics of some polymers are summarized in Table 1.14. Halogenation enhances the flame retardancy of polymers. Thus when the chlorine content of PVC, which usually has 56% chlorine, is increased, as in chlorinated PVC (61% chlorine), the oxygen index increases from 43 to 60. (The oxygen index is a minimum percent of oxygen in a N_2/O_2 gas mixture necessary to support combustion. A material with an oxygen index greater than 27 usually passes the vertical burn test. The burning characteristics of polymers are discussed more fully in Chapter 2 of *Industrial Polymers, Specialty Polymers, and Their Applications.*)

A discussion on flame-retardant additives is given in a later section on polymer compounding.

1.15 Deterioration of Polymers

Deterioration of polymers is manifested in loss of strength, loss of transparency, warpage, cracks, erosion, and so on [40–42]. Hydrophilic materials such as nylon or cellulose acetate can undergo swelling at high humidity or shrinkage due to low humidity. Degradation can occur due to imposition of energy in the form of heat, mechanical action, ultrasonic and sonic energy, radiation such as gamma rays, x-rays, visible light, ultraviolet light, infrared, and electrical action in the from of dielectric effects. Deterioration can occur by chemical effects such as oxidation, ozone attack, hydrolysis, attacks by solvents and detergents, cracking due to swelling of the plasticizer, hardening due to loss of the plasticizer, migration of plasticizers from layer to layer, crazing and cracking, delamination or debonding, void formation, etc.

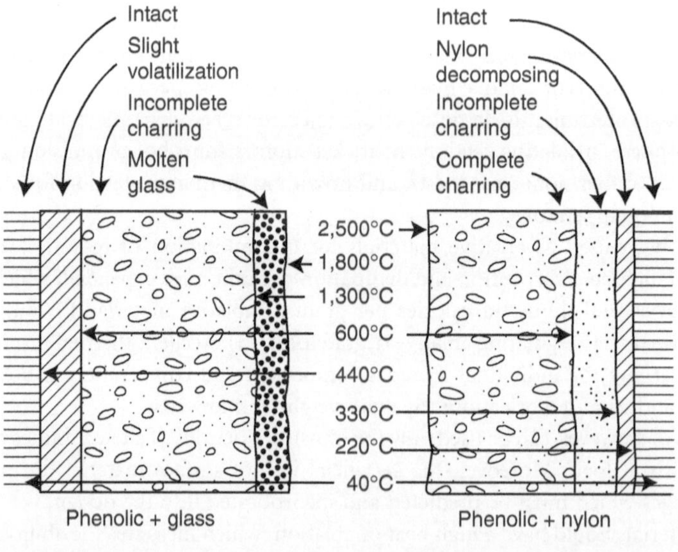

FIGURE 1.39 Temperature distribution in two plastics during steady-state ablation. (After Schmidt, D. L., *Mod. Plastics*, 37, 131 (Nov. 1960), 147 (Dec. 1960).)

TABLE 1.14 Burning Characteristics and Burn Rates of Some Polymers

Polymer	Burning Characteristics	Burn Rate[a] (cm/min)
Polyethylene	Melts, drips	0.8–3.0
Polypropylene	Melts, drips	1.8–4.0
Poly(vinyl chloride)	Difficult to ignite, while smoke	Self-extinguishing
Poly(tetrafluoroethylene)	Melts, chars, bubbles	Nonburning
Fluorinated ethylene propylene copolymer	Does not ignite	Nonburning
Polybutylene	Burns	2.5
Acetal	Burns, bluish flame	1.3–2.8
Cellulose acetate	Burns, yellow flame sooty smoke	1.3–7.6
Cellulose propionate	Burns, drips	1.3–3.0
Cellulose acetate butyrate	Burns, drips	0.8–4.3
Acrylonitrile–butadiene–styrene (general purpose)	Burns	2.5–5.1
Styrene–acrylonitrile	Melts, chars, bubbles	1.0–4.0
Polystyrene	Softens, bubbles, black smoke	1.3–6.3
Acrylic	Burns slowly, drips	1.4–4.0
Nylons	Burns slowly, froths	Self-extinguishing
Phenylene oxide		Self-extinguishing
Polysulfone		Self-extinguishing
Chlorinated polyether		Self-extinguishing
Polyimide		Nonburning

Source: Adapted from Kuryla, W. C. and Papa, A. J., eds. 1973. *Flame Retardancy of Polymeric Materials, Vol. 3.* Marcel Dekker, New York.

[a] ASTM D-635 test procedure.

1.15.1 Chemical Deterioration

Liner or thermoplastic polymers may deteriorate by scission. Ultraviolet light and neutrons can easily break a C–C bond of a vinyl-type polymer, producing smaller molecules (Figure 1.40).

Cross-linking is another way that linear or thermoplastic polymers may deteriorate. Two well-known examples are aging of polyethylene and natural rubber, with loss of flexibility due to cross-linking by oxygen under the catalytic action of sunlight (Figure 1.32). Vulcanized rubber has only 5%–20% of its possible positions anchored by sulfur cross-links. Over time there may be further cross-links of oxygen by the air, and the polymer may thus gradually lose its deformability and elasticity.

Because the oxygen molecule is a biradical, its reaction with a polymer usually results in a chin reaction involving free radicals. Thus, a polymer present in air or an atmosphere rich in oxygen can lose an H atom by reacting with O_2 to from a free radical, which then reacts with another oxygen molecule to form

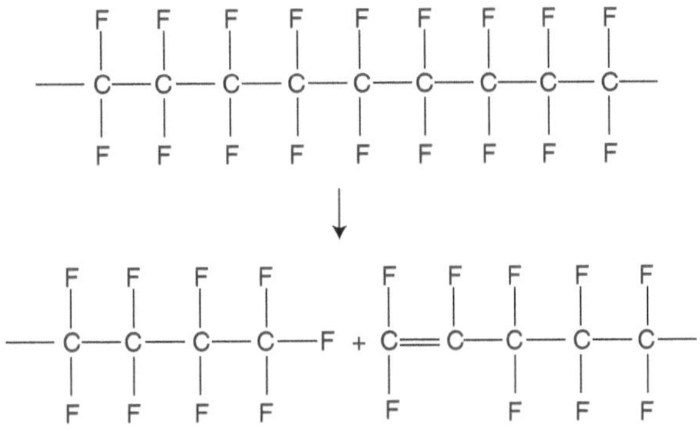

FIGURE 1.40 Scission of polytetrafluoroethylene by irradiation.

a peroxy free radical. The latter, in turn, reacts with another unit of a polymer chain to produce a hydroperoxide and another free radical.

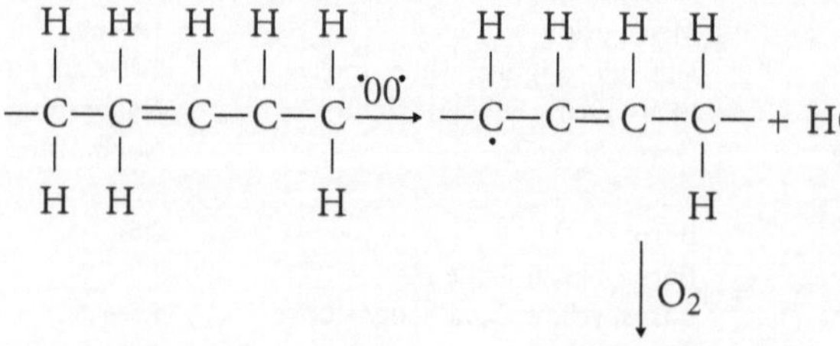

(1.62)

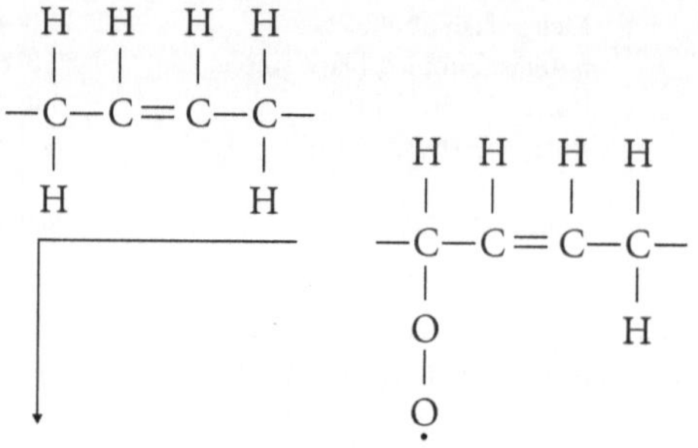

(1.63)

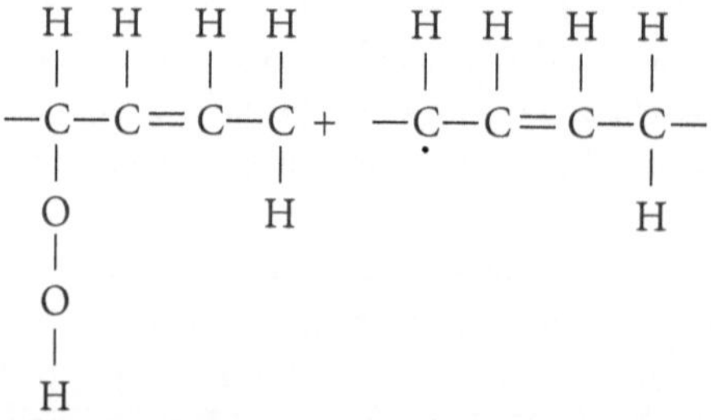

(1.64)

Free radicals can combine together and form a stable molecule.

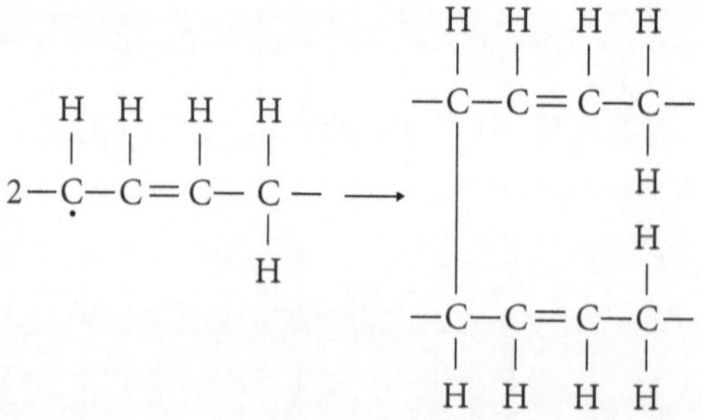

(1.65)

Termination of the chain reaction can also occur with the formation of a peroxide which may be transitory.

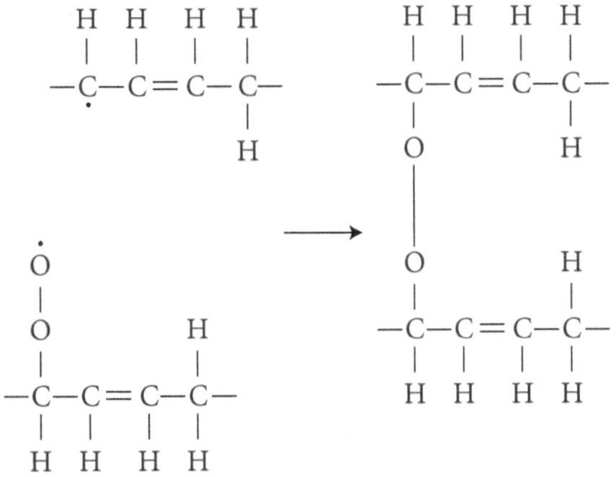

(1.66)

Inhibitors or antioxidants can also stop the process.

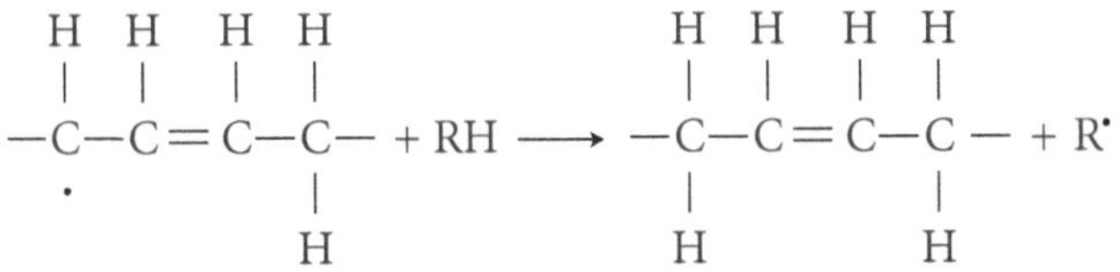

(1.67)

In this way, the chain is broken, and an inactive radical (R˙) is yielded. Aromatic amines or phenolic compounds can act as inhibitors.

However, new radicals can also be formed from the peroxy groups

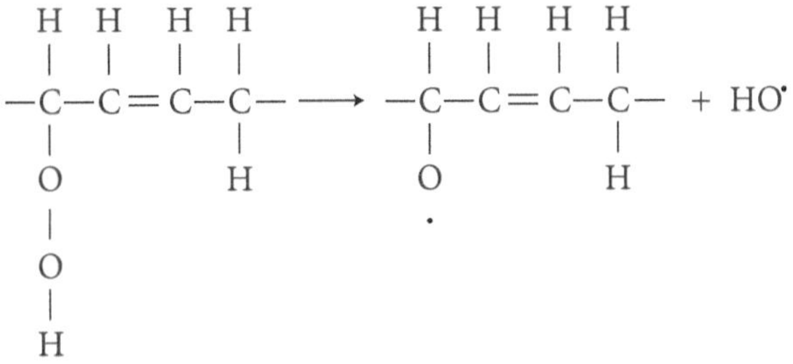

(1.68)

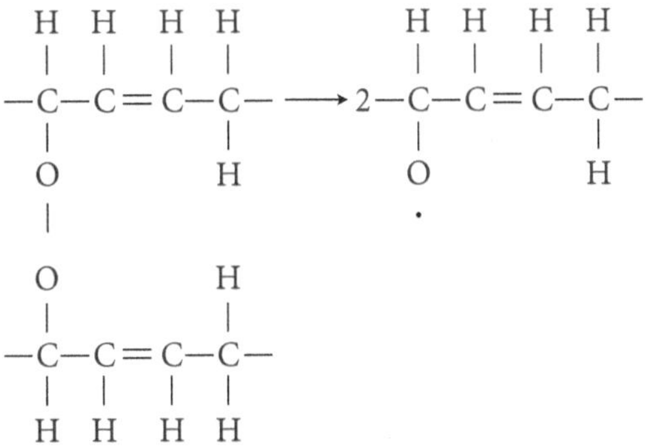

(1.69)

Thus for each original reaction with oxygen there can be numerous propagation reactions. The peroxides may also cleave to give aldehydes, ketones, acids, or alcohols. Due to this type of molecular cleavage, the product becomes soft with lower average molecular weight.

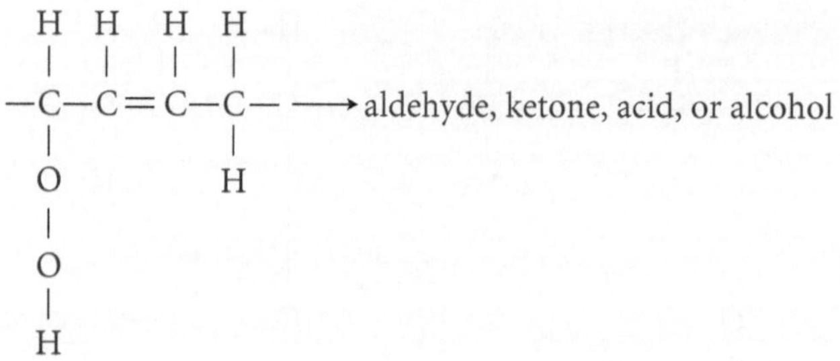

(1.70)

Ozone provides a very reactive source of oxygen because it breaks down to an oxygen molecule and a single reactive oxygen.

$$O_3 \rightarrow O_2 + O^{\cdot} \quad (1.71)$$

The temperature resistance of various plastics in air has been summarized in Figure 1.34 and Table 1.13.

Polymers containing hydrolysable groups or which have hydrolysable groups introduced by oxidation are susceptible to water attack. Hydrolyzable groups such as esters, amides, nitriles, acetals, and certain ketones can react with water and cause deterioration of the polymer. The dielectric constant, power factor, insulation resistance, and water absorption are most affected by hydrolysis. For polyesters, polyamides, cellulose, and cellulose either and esters, the hydrolysable groups are weak links in the chain, and hydrolysis of such polymers can cause serious loss of strength. A summary of water absorption characteristics of common plastic and rubbers is presented in Table 1.15.

1.15.2 Degradation by Radiation

It was mentioned earlier that radiation may cause both cross-linking or degradation. Which predominates depends on radiation dosage, polymer structure, and temperature, Ultraviolet (UV) light causes 1,1-disubstituted vinyl polymers to degrade almost exclusively to monomer at elevated temperatures, whereas cross-linking and chain scission reaction predominate at room temperature. Other vinyl polymers undergo cross-linking primarily, regardless of temperature.

Irradiation with a beam of x-rays, γ-rays, or electrons leads to much higher yields of monomer from 1,1-disubstituted polymers at room temperature. This phenomenon is used advantageously for microetching of resist coatings with electron beams in the preparation of integrated circuits. The process works on the principle that a silicon chip surface will be exposed if an electron beam depolymerizes the protective resin (resist) coating that covers the surface. The exposed silicon can then be doped. Because a beam of high energy radiation has a shorter wavelength than UV light, it is possible to reach finer details in the resist than via the alternative photocross-linking process using UV light. Poly(methyl methacrylate) has been studied extensively as an etch-type resist. This subject is discussed more fully in Chapter 2 of *Industrial Polymers, Specialty Polymers, and Their Applications.*

1.15.3 Microbiological Deterioration

One of the main advantages of synthetic polymers over naturally occurring polymeric materials such as cellulose or leather is their resistance to bacterial or fungal attack. Hence, the synthetic materials are, in general, more permanent. However, a few synthetic polymers are susceptible to biological breakdown and it is clearly important to know, from an applications point of view, which polymers are the most susceptible in a biological environment.

Polyurethanes in particular appear to be susceptible to microbial attack, though polyether polyurethanes (see Chapter 1 of *Industrial Polymers, Specialty Polymers, and Their Applications*) are more resistant to biological degradation than are polyester polyurethanes. The precise mechanisms of these degradations are not fully understood. Polyethylene, polypropylene, polyfluorocarbons, polyamides, polycarbonates, and many other polymer systems appear to be resistant to biological attack.

Table 1.16 provides a qualitative assessment of the resistance of plastic and rubbers to attack by microorganisms. The controversial assessment in certain cases may be due to discrepancies in reported data arising from differences in measurement technique, materials used, average molecular weight of materials, fillers, impurities in polymers, etc.

1.16 Stabilization of Polymers

We have seen that polymers tend to undergo degradation such as chain scission, depolymerization, cross-linking, oxidation, and so on [43–47]. These changes can be effected by various environmental factors such as heat, light, radiation, oxygen, and water. Various stabilizing ingredients are added to plastics to prevent or minimize the degradative effects. These ingredients act either by interfering with degradative processes or by minimizing the cause of degradation.

TABLE 1.15 Water Absorption Characteristics of Plastics and Rubbers

Material	Water abs. (%) 24 h on Sample 3.2 mm Thick
Phenol–formaldehyde resin cast (no filler)	0.3–0.4
Phenol–formaldehyde resin molded (wood-four filler)	0.3–1.0
Phenol–formaldehyde resin molded (mineral filler)	0.01–0.3
Phenol–furfural resin (wood-flour filler)	0.2–0.6
Phenol–furfural resin (mineral filler)	0.2–1.0
Urea–formaldehyde resin (cellulose filler)	0.4–0.8
Melamine–formaldehyde resin (cellulose filler)	0.1–0.6
Melamine–formaldehyde resin (asbestos filler)	0.08–0.14
Ethyl cellulose	0.8–1.8
Cellulose acetate (molding)	1.9–6.5
Cellulose acetate (high acetyl)	2.2–3.1
Cellulose acetate–butyrate	1.1–2.2
Cellulose nitrate	1.0–2.0
Casein plastics	7–14
Poly(vinyl chloride) (plasticized)	0.1–0.6
Poly(vinylidene chloride) (molding)	<0.1
Poly(vinyl chloride acetate) (rigid)	0.07–0.08
Poly(vinyl chloride acetate) (flexible)	0.40–0.65
Poly(vinyl formal)	0.6–1.3
Poly(vinyl butyral)	1.0–2.0
Allyl resins (cast)	0.3–0.44
Polyester resins (rigid)	0.15–0.60
Polyester resin (flexible)	0.1–2.4
Poly(methyl methacrylate)	0.3–0.4
Polyethylene	<0.01
Polypropylene	0.01–0.1
Polystyrene	0.03–0.05
Polytetrafluoroethylene	0.00
Nylon (molded)	1.5
Rubbers (extruded)	0.4
Chlorinated rubber	0.1–0.3

Source: Data mainly from Halim Hamid, S., Amin, M. B., and Maadhah, A. G., eds. 1992. *Handbook of Polymer Degradation.* Marcel Dekker, New York.

TABLE 1.16 Resistance of Plastics and Rubbers to Attack by Microorganisms

	Resistance
Plastics	
Poly(methyl methacrylate)	Good
Polyacrylonitrlle (orlon)	Good
Acrylonitrile-vinylchloride copolymer (Dynel)	Good
Cellulose acetate	Good, poor
Cellulose acetate butyrate	Good
Cellulose acetate propionate	Good
Cellulose nitrate	Poor
Ethyl cellulose	Good
Acetate rayon	Good
Viscose rayon	Poor
Phenol–formaldehyde	Good
Melamine–formaldehyde	Good, poor
Urea–formaldehyde	Good
Nylon	Good
Ethylene glycol terephthalate (Terylene)	Good
Polyethylene	Good, questionable
Polytetrafluoroethylene (Teflon)	Good
Polymonochlorotrifluoroethylene	Good
Polystyrene	Good
Poly(vinyl chloride)	Good, questionable
Poly(vinyl acetate)	Poor
Poly(vinyl butryral)	Good
Glyptal resins (alkyd resins)	Poor, moderate
Silicone resins	Good
Rubbers	
Pure natural rubber (caoutchouc)	Attacked
Natural rubber vulcanizate	Attacked
Crude sheet	Attacked
Pale crepe, not compounded	Attacked
Pale crepe, compounded	Resistant, attacked
Smoked sheet, not compounded	Attacked
Smoked sheet, compounded	Resistant, attacked
Reclaimed rubber	Attacked
Chlorinated rubber	Resistant
Neoprene, compounded	Resistant
GR-S, butadiene-styrene, compounded	Resistant, attacked
Hycar OR, butadiene-acrylonitrile compounded	Resistant, attacked
Buna N, butadiene-acrylonitrile, compounded	Attacked
GR-I (butyl), isobutylene-isoprene, compounded	Resistant, attacked
Thiokol, organic polysulfide, uncured	Attacked
Thiokol, organic polysulfide, vulcanized	Resistant
Thiokol, organic polysulfide, sheets for gasoline tank lining	Attacked
Silicone rubber	Resistant

Source: Adapted from Halim Hamid, S., Amin, M. B., and Maadhah, A. G., eds. 1992. *Handbook of Polymer Degradation.* Marcel Dekker, New York.

1.16.1 Antioxidants and Related Compounds

Oxidation, as we have noted, is a free-radical chain process. The most useful stabilizing agents will therefore be those which combine with free radicals, as shown by Equation 1.67, to give a stable species incapable of further reaction. These stabilizing agents are called antioxidants. They are the most frequently employed ingredients in plastics, fibers, rubbers, and adhesives. Stabilization is also achieved in some polymer systems by the use of additives which moderate the degradation reaction.

Two large, basic groups of antioxidants are normally distinguished: (1) chain terminating or primary antioxidants, and (2) hydroperoxide decomposers or secondary antioxidant, frequently called synergists.

The majority of primary antioxidants are sterically hindered phenols or secondary aromatic amines. They are capable of undergoing fat reaction with peroxy radicals and so are often called radical scavengers. For hindered phenols, for example, such reactions may be represented by the following scheme:

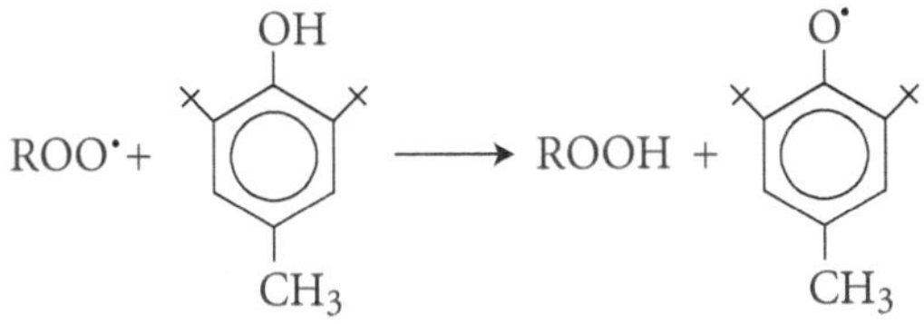

(1.72)

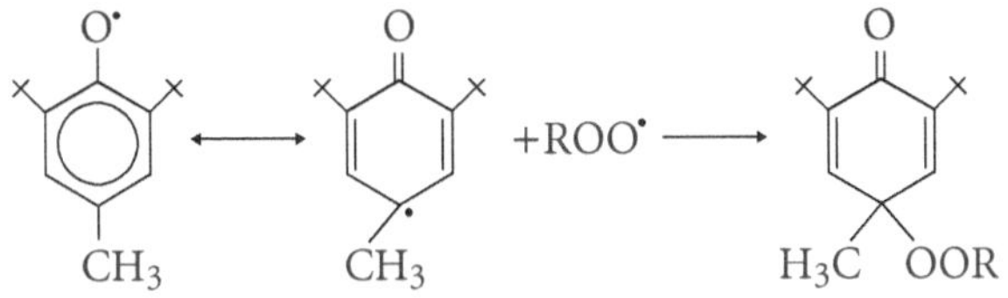

(1.73)

(+ = tertbutyl group)

Stabilization is achieved by the fact that Reaction 1.72 competes with peroxy radical reactions in polymer degradation, such as Reaction 1.64, transforming the reactive peroxy radical into a much less reactive phenoxy radical, which, in turn, is capable of reacting with a second peroxy radical according to Reaction 1.73.

The phenoxy radicals formed in Reaction 1.72 do not initiate new radical chains at the normal temperatures of use and testing, but such propagation reactions become possible at high temperatures. The thermal stability of peroxycyclohexadienones (formed by Reaction 1.73) is also limited, and their decomposition leads to new reaction chains. The effectiveness of sterically hindered phenols thus decreases with increasing temperature.

The secondary antioxidants are usually sulfur compounds (mostly thioethers and esters of thiodipropionic acid) or trimesters of phosphorous acid (phosphates). A remedy for many types of discoloration in plastics is often the use of a phosphite or a thioether. Both have the ability to react with hydroperoxides, as those formed in Reaction 1.72, to yield nonradical products, following heterolytic mechanisms. The reaction of phosphites to phosphates as an example:

$$P(OR')_3 + ROOH \rightarrow OP(OR')_3 + ROH \qquad (1.74)$$

Thioethers react with hydroperoxides in a first stage to yield sulfoxides and alcohols:

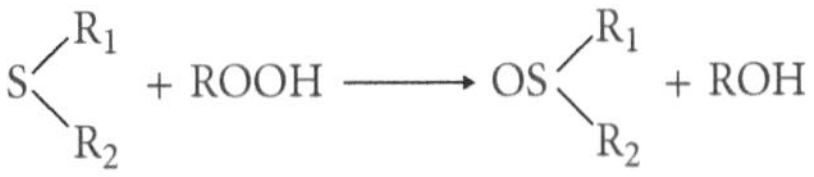

(1.75)

Sulfoxides themselves yield, on further oxidation, even more powerful hydroperoxide decomposers than the original sulfides, in that they are able to destroy several equivalents of hydroperoxides. This catalytic effect is explained by the intermediate occurrence of sulfenic acids and sulfur dioxide. The fact that the phenomenon of synergism, which is defined as a cooperative action such that the total effect is greater than the sum of two or more individual effects taken independently, is often observed when primary and secondary antioxidants are combined as has been explained with the concept of the simultaneous occurrence of the radical reactions (e.g., Equation 1.72) and the nonradical hydroperoxide decomposition (e.g., Equation 1.74 and Equation 1.75).

The preceding synergistic effect may be considered as an example of heterosynergism, which arises from the cooperating effect of two or more antioxidants acting by different mechanisms. Homosynergism, on the other hand, involves two compounds of unequal activity that are operating by the same mechanism, for example, a combination of two different chain-breaking antioxidants that normally function by donation of hydrogen to a peroxy free radical. The most likely mechanism of synergism in this case would involve the transfer of hydrogen from one inhibitor molecule to the radical formed in the reaction of the other inhibitor with a peroxy radical (see Equation 1.72), thus regenerating the latter inhibitor and prolonging its effectiveness.

1.16.2 Chemical Structures of Antioxidants

While sterically hindered phenols and secondary aromatic amines are the two main chemical classes of antioxidants for thermoplastics, the greatest diversity is, however, found in the class of the sterically hindered phenols. Most of these phenols are commercially available as relatively pure chemicals. Although aromatic amines are often more powerful antioxidants than phenols, their application is

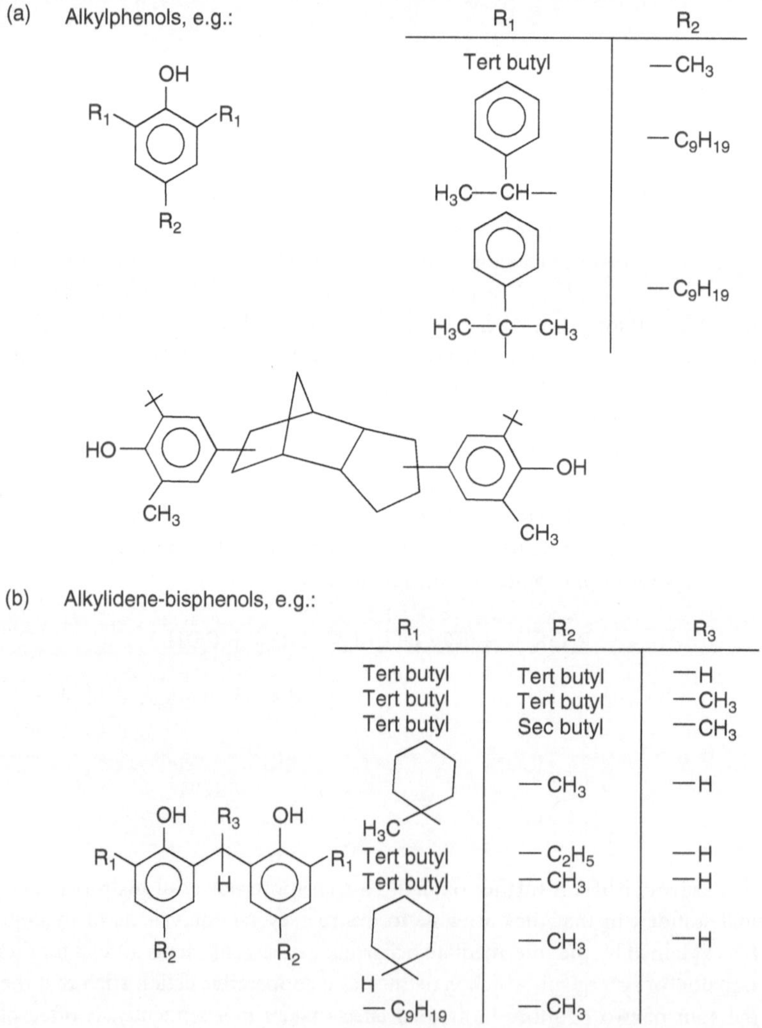

FIGURE 1.41 Structural formulas [45] of important sterically hindered phenols used as antioxidants for thermoplastics.

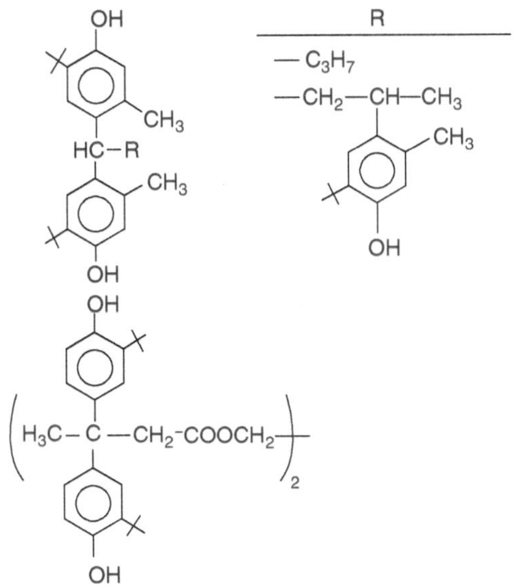

(c) Thiobisphenols, e.g.:

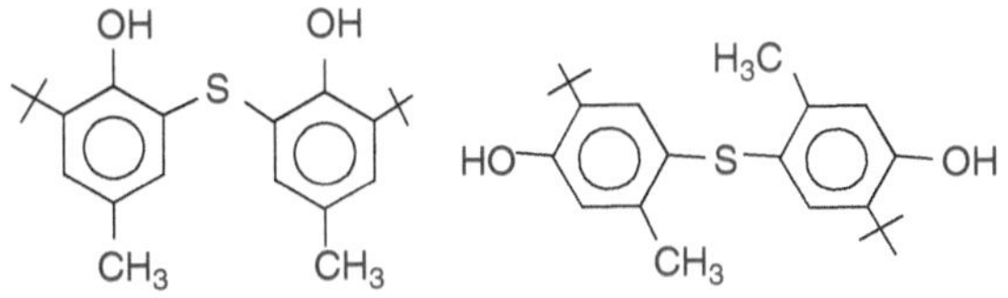

(d) Hydroxybenzyl compounds, e.g.:

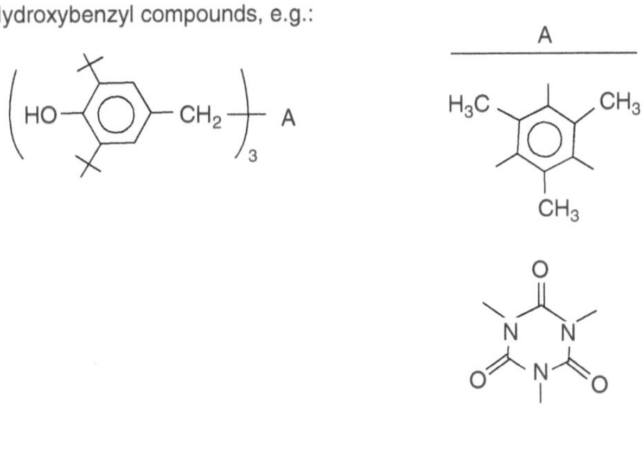

FIGURE 1.41 (*continued*)

limited to vulcanized elastomers since the staining properties of aromatic amines prohibit their use in most thermoplastics.

Sterically hindered phenols of commercial importance may be further classified, according to their structure, into (1) alkylphenols, (2) alkylidene-bisphenols, (3) thiobisphenols, (4) hydroxybenzyl compounds, (5) acylaminophenols, and (6) hydroxyphenyl propionates.

An important goal of antioxidant research has been to provide poly-phenols of high molecular weight and low volatility. Most of the commercial sterically hindered phenols have molecular weights in the range 300–600 and above 600. Figure 1.41 depicts structural formulas [45] of the important sterically hindered phenols used as antioxidants.

The majority of secondary antioxidants are esters of thiodipropionic acid with fatty alcohols and trimesters of phosphorus acid. Their chemical structures are shown in Figure 1.42. Phosphites are more

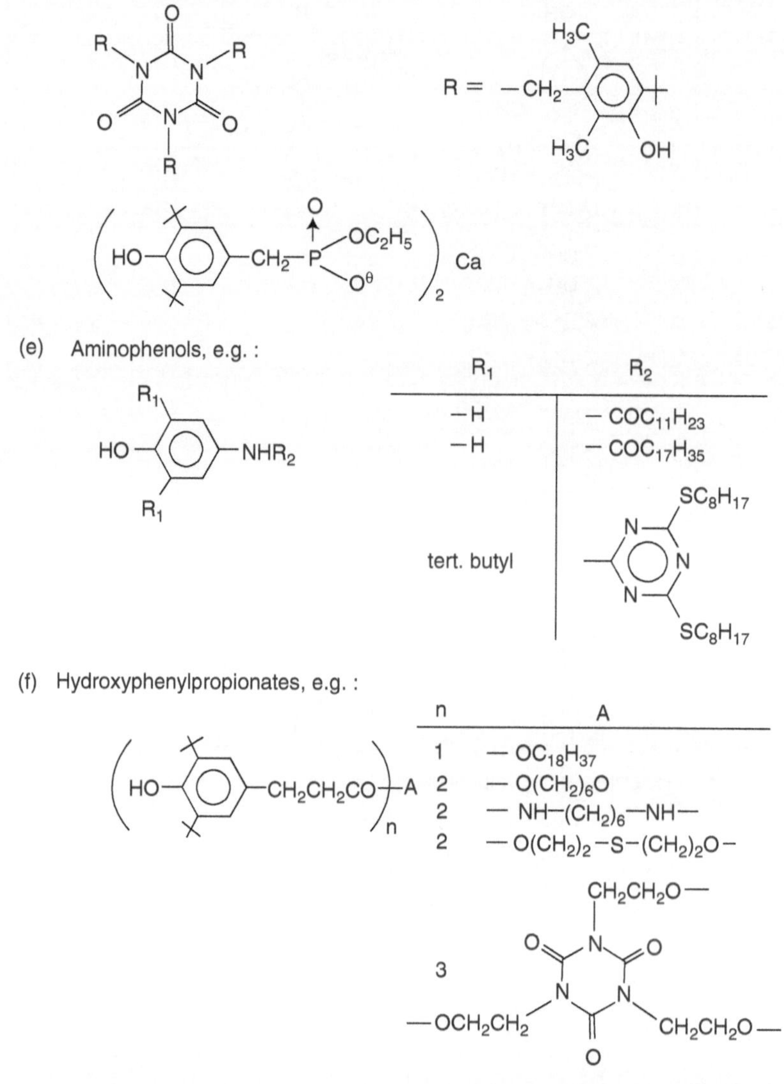

FIGURE 1.41 (*continued*)

sensitive towards hydrolysis than esters of carboxylic acids, and this point has to be taken into account when storing and using phosphates. Aromatic phosphates are often preferred since they are inherently more resistant to hydrolysis than aliphatic phosphates.

An index of trade names, manufacturers, and suppliers based on a choice of representative antioxidants for thermoplastics is given in Appendix A2. Detailed lists are found in Ref. [45,46].

1.16.3 Stabilization of Selected Polymers

Two important considerations in selecting antioxidants for polymers are toxicity and color formation. Thus, for use in food wrapping, any antioxidant or additive must be approved by the appropriate government agency. An antioxidant widely used in food products and food wrapping is butylated hydroxytoluene (BHT) (see Figure 1.41a: R_1 = *tert*-butyl and R_2 = $-CH_3$). It is also added in small amounts to unsaturated raw rubbers before shipping to protect them during storage.

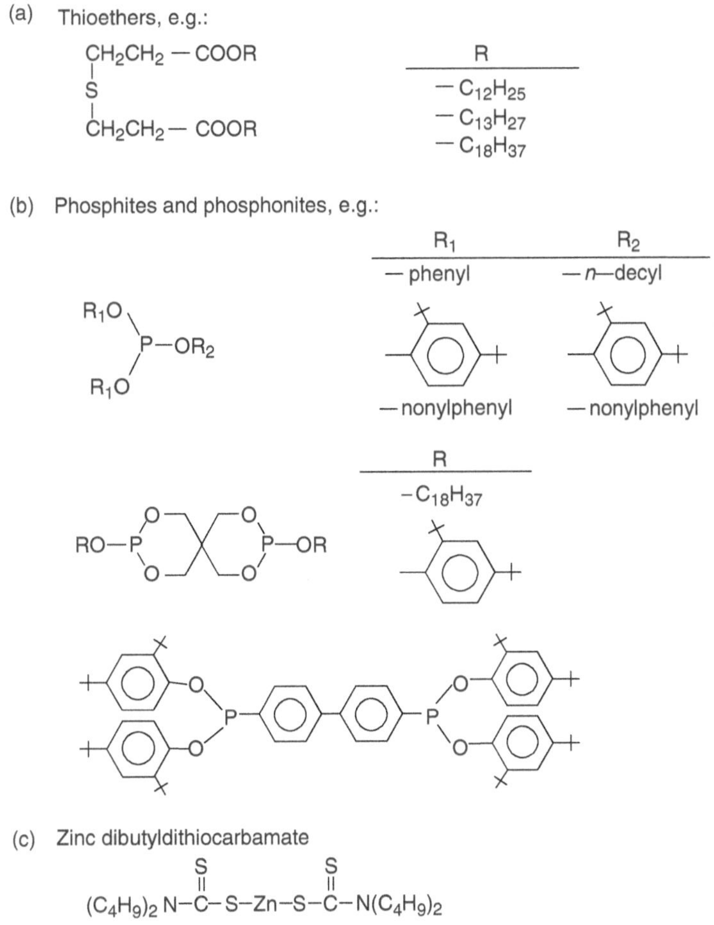

FIGURE 1.42 Structural formulas [45] of some commercial secondary antioxidants for thermoplastics.

Materials that are not effective inhibitors when used alone may nevertheless be able to function as synergists by reacting with an oxidized form of an antioxidant to regenerate it and thus prolong its effectiveness. For example, carbon black forms an effective combination with thiols, disulfides, and elemental sulfur, even though these substances may be almost completely ineffective alone under comparable conditions. Synergistic combinations from a variety of chain terminators and sulfur compounds have been reported. A widely used combination of stabilizers for polyolefins is 2,6-di-*tert*-butyl-4-methylphenol (BHT) (Figure 1.41a) with dilauryl thiodipropionate (DLTDP):

$$(H_{25}C_{12}\,O{-}\overset{\displaystyle O}{\overset{\|}{C}}{-}CH_2\,CH_2)_2\,S$$

This combination is of particular interest because both components are among the small group of stabilizers approved by the U.S. Food and Drug Administration for use in packaging materials for food products.

Other synergistic combinations have been reported involving free-radical chain terminators used with either ultraviolet absorbers or metal deactivators as the preventive antioxidants. The differences in mechanism of action of these different types of stabilizers permit them to act independently but cooperatively to provide greater protection than would be predicted by the sum of their separate effects.

1.16.3.1 Polypropylene

Because of the presence of tertiary carbon atoms occurring alternately on the chain backbone, propylene is particularly susceptible to oxidation at elevated temperatures. Since polypropylene is normally processed at temperatures between 220 and 280°C, it will degrade under these conditions (to form lower-molecular-weight products) unless it is sufficiently stabilized before it reaches the processor. The antioxidants are added at least partially during the manufacturing process and at the least during palletizing. Antioxidant systems in technical use are composed of processing stabilizers, long-term heat stabilizers, calcium- or zinc-stearate, and synergists if necessary.

Typical processing stabilizers for polypropylene and butylated hydroxy-toluene (BHT) as the primary antioxidant and phosphates and phosphonates as secondary antioxidants. Examples of the latter that are commonly used are: tetrakis-(2,4-di-*tert*-butyl-phenyl)-4-4′-bisphenylylenediphosphonite, distearyl-pentaerythrityl-diphosphonite, tris-(nonylphenyl)-phosphite, tris-(2,4-di-*tert*-butyl-phenyl)-phosphite and bis(2,4-di-*tert*-butyl-phenyl)- pentaerythrityl-diphosphite. In commercial polypropylenes, phosphorous compounds are always used together with a sterically hindered phenol. The compounds are commonly added in concentrations between 0.05 and 0.25%.

The most important long-term heat stabilizers for polypropylene are phenols of medium (300–600) and especially high (600–1,200) molecular weight, which are frequently used together with thioethers as synergists, e.g., dilauryl thiodipropionate (DLTDP), or distearyl thiodipropionate (DSTDP), or dioctadecyl disulfide.

1.16.3.2 Polyethylene

High-density polyethylenes (HDPE) are less sensitive to oxidation than polypropylene, so lower stabilizer concentrations are generally sufficient. As in polypropylene, antioxidants can be added during a suitable stage of manufacture or during palletizing. The antioxidants in technical use are the same as for polypropylene. Phenols of medium- and high-molecular weight are also active as long-term heat stabilizers. Concentrations between 0.03 and 0.15% are usual.

Low-density polyethylene (LDPE) is extensively used for the manufacture of films. During processing, which is carried out at temperatures of approximately 200°C, cross-linking, and thus formation of gel, can occur through oxidation if the polymer is not stabilized. Such gel particles are visible in the film as agglomerates, known as fish eyes or arrow heads. The processing stabilizers used in LDPE consist of systems commonly used for polypropylene, namely, combinations of a phosphite or phosphonite and a long-term heat stabilizer (hindered phenol) in overall concentrations up to 0.1%. Concentrations seldom exceed 0.1%, since the compatibility of any additive in LDPE is considerably lower than in any other polyolefins.

For primary cable insulation. Cable jackets and pipes manufactured from LDPE, medium to high-molecular-weight grades are used and are frequently cross-linked after processing. Long-term heat stabilizer is of primary importance in these applications, since lifetimes of up to 50 years are required (usually at elevated temperatures with short-time peaks up to 100°C for cables). The antioxidants have to be extremely compatible and resistant to extraction. Some typical antioxidants customary for insulation are as follows:

For power cable insulation. Pentaerythrityltetrakis-3-(3,5-di-*tert*-butyl-4-hydroxyphenyl)-propionate, polymeric 2,2,4-trimethyl-1,2-dihydroquino-line, 4,4′-thiobis-(3-methyl-6-*tert*-butyl-phenol), 2,2′-thiodiethyl-bis-3-(3,5-di-*tert*-butyl-4-hydroxyphenyl)-propionate, and distearyl thiodipropionate are used as synergists.

For communication cable insulation. 2,2′-Thiobis-(4-methyl-6-*tert*-butyl-phenol), 4,4′-thiobis-(3-methyl-6-*tert*-butyl-phenol), 2,2′-methylene-bis-(4-methyl-6-α-methylcyclohexyl-phenol), 1,1,3-tris-(5-*tert*-butyl-4-hydroxy-2-methylphenyl)-butane, 2,2′-thiodiethyl-bis-3-(3,5-di-*tert*-butyl-4-

hydroxy-phenyl)-propionate, pentaerythrityl-tetrakis-3-(3,5-di-*tert*-butyl-4-hydroxy-phenyl)-propionate, and dilauryl thiodipropionate are used as synergists.

The simultaneous use of metal deactivators has become increasingly important for cable insulation. This is discussed later.

1.16.3.3 Polystyrene

Unmodified crystal polystyrene is relatively stable under oxidative conditions, so that for many applications the addition of an antioxidant is not required. Nevertheless, repeated processing may lead to oxidative damage of the material, leading to an increase of melt flow index and to embrittlement of the material. Stabilization is effected by the addition of octadecyl-3-(3,5-di-*tert*-butyl-4-hydroxy-phenyl)-propionate at concentrations of up to 0.15%, if necessary in combination with phosphates or phosphonites to improve color.

Compared to unmodified crystal polystyrene, impact polystyrene consisting of copolymers of styrene and butadiene are more sensitive to oxidation. This sensitivity is a consequence of the double bonds in polybutadiene component and manifests itself in yellowing and the loss of mechanical properties of the polymer. In impact polystyrene, the following antioxidants or their mixture are used in total concentrations of 0.1%–0.25%: BHT, octadecyl-3-(3,5-di-*tert*-butyl-4-hydroxyphenyl)-propionate, 1,1,3-tris-(5-*tert*-butyl-4-hydroxy-2-methylphenyl)-butane, and dilauryl thiodipropionate.

1.16.3.4 Acrylonitrile–Butadiene–Styrene Copolymers

Like impact polystyrene, acrylonitrile–butadiene–styrene copolymers (ABS) are sensitive to oxidation caused by the unsaturation of the elastomeric component. The processes for the manufacture of ABS require the drying (at 100°C–150°C) of powdery polymers that are extremely sensitive to oxidation. Thus, antioxidants have to be added before the coagulation step, normally in emulsified form, although sometimes in solution. The primary antioxidants are frequently used together with a synergist. Primary anti-oxidants commonly used for ABS are BHT, 2,2′-methylenebis-(4-ethyl or methyl-6-*tert*-butyl-phenol), 2,2′-methylenebis-(4-methyl-6-cyclohexyl-phenol), 2,2′-methylenebis-(4-methyl-6-nonyl-phenol), octadecyl-3-(3,5-di-*tert*-butyl-4-hydroxyphenyl)-propionate, and 1,1,3-tris-(5-*tert*-butyl-4-hydroxy-2-methylphenyl)-butane. Important synergists are tris-(nonyl-phenyl)-phosphite and dilauryl thiodipropionate. These antioxidants are either liquids or show comparatively low melting points, which is an important prerequisite for the formation of stable emulsions.

1.16.3.5 Polycarbonate

Thermoxidative degradation of polycarbonate manifests itself in yellowing that is readily seen because of the transparency of polycarbonate. For this reason, stabilization against discoloration is considered to be important. Severe requirements are also imposed on the nonvolatility and thermostability of the stabilizers for polycarbonate, because the processing temperatures are extraordinarily high (about 320°C). The stabilizers are generally added during the palletizing step.

Yellowing of polycarbonate during processing is retarded by the addition of phosphates or phosphonites. They are used in concentrations of 0.05%–0.15%, possibly in combination with an epoxy compound as acid acceptor. The addition of these stabilizers not only decreases the yellowness index but also inhibits the rise of the melt flow index and negative influence on impact strength of processing.

The effective processing stabilizers are not suitable, however, for preventing the aging effect of long-term use. For long-term heat stabilization, a sterically hindered phenolic antioxidant is added. An effective antioxidant is octadecyl-3-(3,5-di-*tert*-butyl-4-hydroxyphenyl)-propionate.

1.16.3.6 Nylons

Degradation of nylons due to processing and aging causes discoloration and loss of mechanical properties, although not at the same rate. For example, yellowing is observed already after short periods of oven aging at 165°C, but tensile strength and elongation are hardly affected during the same time period. Discoloration of polyamides during processing can be suppressed to a certain extent by the addition of phosphates, e.g., tris-(2,4-di-*tert*-butyl-phenyl)-phosphite, in concentrations of 0.2%–0.4%.

The stabilization of nylons is mainly a matter of long-term stabilization. Three main groups of stabilizers have become known: (1) copper salts, especially in combination with halogen and/or phosphorus compounds (e.g., copper acetate ? potassium iodide/phosphoric acid), (2) aromatic amines (e.g., *N,N′*-dinaphthyl-*p*-phenylenediamine or *N*-phenyl-*N′*-cyclohexyl-*p*-phenylenediamine), and (3) hindered phenols.

Copper-halogen systems are effective in very low concentrations (10–50 ppm of copper; approximately 1000 ppm of halogen), although they are extracted quite easily with water and cause discoloration of the substrate. Next to the copper-halogen systems, aromatic amines are the most effective stabilizers. They are used in rather high concentrations, from 0.5 to 2%. However, since they have strongly discoloring properties they are used mainly for technical articles, which tolerate discoloration.

Hindered phenols do not show the above-mentioned disadvantages. They are the stabilizers of choice whenever good oxidation stability has to be coupled with good color stability and, possibly, food approval of the end article. The most important hindered phenols in use are the following: *N,N′*-hexamethylenebis-3-(3,5-di-*tert*-butyl-4-hydroxyphenyl)-propionamide, 1,1,3-tris-(5-*tert*-butyl-4-hydroxy-2-methylphenyl-butane, 1,3,5-tris-(3,5-di-*tert*-butyl-4-hydroxybenzyl)-mesitylene, and BHT. The antioxidants can be added already during polycondensation, the normal concentration ranging from 0.3 to 0.7%.

1.16.3.7 Thermoplastic Elastomers

Styrene-based thermoplastic elastomers (see Chapter 1 of *Industrial Polymers, Specialty Polymers, and Their Applications*) are sensitive to oxidation since they contain unsaturated soft segments. These elastomers are manufactured by solution polymerization process in aliphatic hydrocarbons. In order to prevent autoxidation during the finishing steps (stripping, drying), which manifests itself by a rise in melt flow index and discoloration of the raw polymer, antioxidant is added to the polymer solution before finishing. Hence the antioxidant has to be soluble in the polymerization solvent.

A number of hindered phenols are used in practice in a total concentration of approximately 0.5%. Examples of primary antioxidants used are BHT, 1,3,5-tris-(3,5-di-*tert*-butyl-4-hydroxybenzyl)-mesitylene, and octadecyl-3-(3,5-di-*tert*-butyl-4-hydroxyphenyl)-propionate. Tris-(nonylphenyl)-phosphite is used as a synergist. BHT may, however, be partially lost during finishing and drying. The phenols of higher molecular weight are sued as they have low volatility and have the further advantage of protecting the material also during processing and end use.

Thermoplastic polyester elastomers (see Chapter 1 of *Industrial Polymers, Specialty Polymers, and Their Applications*) contain readily oxidizable polyether soft segments that make stabilization necessary. Essentially two antioxidants, namely, 4,4′-di(α,α-dimethylbenzyl)-diphenylamine and *N,N′*-hexamethylenebis-3-(3,5-di-*tert*-butyl-4-hydroxyphenyl)-propionamide are in use in concentrations of up to 1%. The antioxidants can be added during pelletizing, or even better, already during polycondensation. When added during polycondensation, *N,N′*-hexamethylenebis-3-(3,5-di-*tert*-butyl-4-hydroxyphenyl)-propionamide is partly chemically bound to the polymer because of its amide structure. The antioxidant thus becomes highly stable to extraction.

The oxidative stability of thermoplastic polyurethane elastomers is determined by the length and the structure of the linear polyether or polyester soft segments. The stability of polyether urethanes against autoxidation is distinctly lower compared to that of polyester urethanes, but the latter are less stable to hydrolysis. Antioxidants may be used in polyurethanes for stabilization against loss of mechanical properties and against discoloration in injection molding grades and as gasfading inhibitors in elastomeric fibers.

The antioxidant used in thermoplastic polyurethane elastomers are hindered phenols, e.g., BHT, octadecyl-3-(3,5-di-*tert*-butyl-4-hydroxyphenyl)-propionate, and pentaerythrityl-tetrakis-3-(3,5-di-*tert*-butyl-4-hydroxyphenyl)-propionate, and aromatic amines, e.g., 4,4′-di-*tert*-octyl-diphenylamine, as well as their combinations. Aromatic amines can, however, be employed only in very limited concentrations (250–550 ppm at most) because of their discoloring properties.

1.16.3.8 Polyacetal

Polyoxymethylenes have a marked tendency to undergo thermal depolymerization with loss of formaldehyde. To prevent thermal depolymerization, polyoxymethylenes are structurally modified, the two possibilities being acetylation to block the reactivity of the end groups of co-polymerization with cyclic ethers, e.g., ethylene oxide. Polyacetals are also sensitive towards autoxidation, which invariably leads to depolymerization as a result of chain scission. The formaldehyde released by depolymerization is very likely to be oxidized to formic acid, which can catalyze further depolymerization.

The stabilizer systems for polyacetals are invariably composed of a hindered phenol with a costabilizer. The hindered phenols in use are 2,2′-methylenebis-(4-methyl-6-*tert*-butyl-phenol), 1,6-hexamethylenebis-3-(3,5-di-*tert*-butyl-4-hydroxyphenyl)-propionate, and pentaerythrityl-tetrakis-3-(3,5-di-*tert*-butyl-4-hydroxyphenyl)-propionate. A large number of nitrogen-containing organic compounds have been described as costabilizers for polyacetals, e.g., dicyandiamide, melamine, terpolyamides, urea, and hydrazine derivatives. The effectiveness of these compounds is based on their ability to react with formaldehyde and to neutralize acids, especially formic acid, formed by oxidation. In addition to nitrogen compounds, salts of long-chain fatty acids (e.g., calcium stearate, calcium ricinoleate, or calcium citrate) are also used as acid acceptors. The practical concentrations are 0.1–0.5% for the phenolic antioxidant and 0.1–1.0% for the costabilizer.

1.16.3.9 Poly(Vinyl Chloride)

Poly(vinyl chloride) (PVC) is relatively unstable under heat and light. The first physical manifestation of degradation is a change in the PVC color, which on heating, changes from the initial water-white to pale yellow, orange, brown, and finally black. Further degradation causes adverse changes in mechanical and electrical properties.

The most widely accepted mechanism for PVC degradation is one based on a free-radical chain. Thermal initiation probably involves loss of a chlorine atom adjacent to some structural abnormality, such as terminal unsaturation, which reduces the stability of the C–Cl bond [44]. The chlorine radical thus formed abstracts a hydrogen to form HCl, and the resulting chain radical then reacts to form a chain unsaturation with regeneration of another chlorine radical.

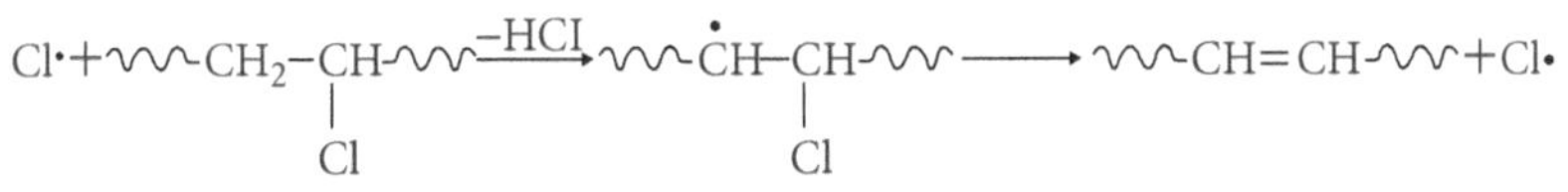

Thus, as hydrogen chloride is removed, polyene structures are formed:

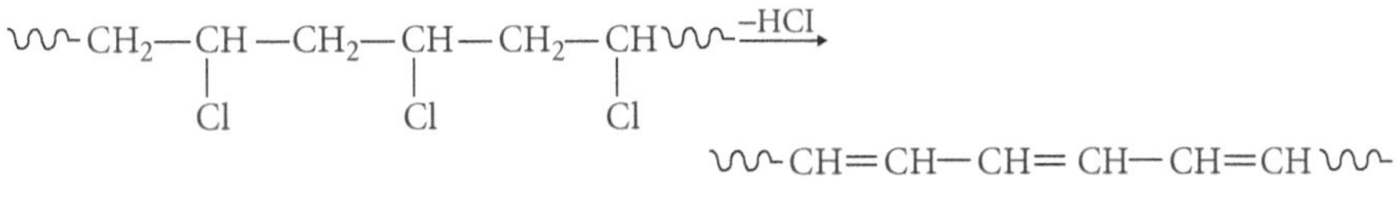

The reaction can also be initiated by ultraviolet light. In the presence of oxygen the reactions are accelerated (as evidenced by the acceleration of color formation), and ketonic structures are formed in the chain.

Stabilizers are almost invariably added to PVC to improve its heat and light stability. The species found effective in stabilizing PVC are those that are able to absorb or neutralize HCl, react with free radicals, react with double bonds, or neutralize other species that might accelerate degradation. Lead compounds, such as basic lead carbonate and tribasic lead sulfate, and metal soaps of barium, cadmium, lead, zinc, and calcium are used as stabilizers. Obviously, they can react with HCl. Epoxy plasticizers aid in stabilizing the resin. Another group of stabilizers are the organotin compounds, which find application because of their resistance to sulfur and because they can yield crystal-clear compounds.

1.16.3.10 Rubber

In stabilizing rubber, antioxidants are classified as staining or nonstaining, depending on whether they develop color in use. In carbon-black-loaded rubber tire a staining material is not harmful. For white rubber goods, however, it is important that the additives used be colorless and that they stay colorless while protecting against oxidation. Some antioxidants that find use in rubber are phenyl-β-naphthylamine (PBNA), which is staining, and 2,2′-thiobis (6-*tert*-butyl-para-cresol) (see Figure 1.41c), which is nonstaining.

It has been established that ozone formation is a general phenomenon that occurs when free radicals can combine with oxygen. Rubber products made form natural rubber, styrene-butadiene rubber (SBR), or nitrile rubber, when stretched moderately and exposed to low concentrations of ozone, crack rapidly and sometimes disastrously. The cracking is caused by cuts that appear on the surface and may penetrate deeply, causing serious damage. These cuts appear only when the rubber is under tension during exposure to ozone and if they cross the lines of tension at an angle of 90°. Since the cracking problem is associated only with elastomers containing some degree of residual unsaturation, the process is considered to be related to the attack of ozone on the unsaturation. Unlike molecular oxygen (O_2), ozone appears to add directly to the double bond, which is often followed by chain scission. Use of diene rubbers in an environment containing ozone, such as an automobile tire in some urban locations, therefore calls for stabilization against ozone.

The effect of ozone is both greatly delayed and reduced when various antiozonants, such as *N,N′*-di-alkyl-*p*-phenylene-diamines (X) or similar compounds, are incorporated in the rubber. They may react directly with ozone or with the ozone-olefin reaction products in such a way as to prevent chain scission. SBR is much more receptive to protection than either the nitrile or natural rubbers. Of the latter two the nitrile is more readily inhibited. If a wax is also added in small quantities with the antiozonant, the retardation of the attack of ozone is in certain instances enhanced several fold.

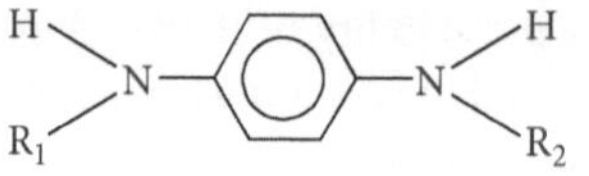

R_1,R_2= Alkyl

(X)

It is believed that waxes function by providing an unreactive layer on the rubber surface. Since, unlike oxygen, ozone reacts only at the surface and does not diffuse into the sample, a surface layer of relatively unreactive wax presents an impervious surface to ozone. The wax thus needs to bloom (i.e., exude to the surface) in order to afford protection. A great deal of attention has therefore been paid to the factors affecting migratory aptitude of wax. Waxes with good protective power have been found to have melting points between 65 and 72°C, refractive indices sin the range of 1.432–1.438, and branching to the extent of 30%–50% side chains [44].

Of the two general categories of waxes—paraffin and micro-crystalline—the latter are more strongly held to the surface. However, the use of waxes alone to provide protection against ozone attack is rather well restricted to static conditions of service. Whenever constant flexing is present, even the more strongly held microcrystalline waxes flake off and protection is lost. Combinations of waxes and chemical antiozonants are therefore used to provide protection under both static and dynamic conditions of service. In fact, it is felt that waxes can aid in the diffusion of chemical antiozonant to the rubber surface.

1.17 Metal Deactivators

As described previously, thermooxidative degradation of polyolefins proceeds by a typical free-radical chain mechanism in which hydroperoxides are key intermediates because of their thermally-induced hemolytic decomposition to free radicals, which in turn initiate new oxidation chains. However, since the monomolecular hemolytic decomposition of hydroperoxides into free radicals require relatively high activation energies, this process becomes effective only at temperatures in the range of 120°C and higher.

But in the presence of certain metal ions, hydroperoxides can undergo catalytic decomposition even at room temperature by a redox reaction to radical products. Two oxidation-reduction reactions can be involved depending on the metal and its state of oxidation:

$$ROOH + M^{n+} \rightarrow RO^{\cdot} + M^{(n+1)+} + OH^{-} \quad (1.76)$$

$$ROOH + M^{(n+1)+} \rightarrow ROO^{\cdot} + M^{n+} + H^{+} \quad (1.77)$$

$$2ROOH \xrightarrow{M^{n+}/M^{(n+1)+}} RO^{\cdot} + ROO^{\cdot} + H_2O \quad (1.78)$$

The electron transfer producing free radicals as shown above is preceded by the formation of unstable coordination complexes of the metal ions with alkyl hydroperoxides. The relative importance of Reaction 1.76 and Reaction 1.77 depends upon the relative strength of the metal ion as an oxidizing or reducing agent. When the metal ion has two valence states of comparable stability, both Reaction 1.76 and Reaction 1.77 will occur, and a trace amount of the metal can convert a large amount of peroxide to free radicals according to the sum of the two reactions (Reaction 1.78). This is true of compounds of metals such as Fe, Co, Mn, Cu, Ce, and V, commonly called transition metals.

The presence of the above-mentioned metal ions increases the decomposition rate of hydroperoxides and the overall oxidation rate in the autoxidation of a hydrocarbon to such an extent that even in the presence of antioxidants, the induction period of oxygen uptake is drastically shortened. In such a case, sterically hindered phenols or aromatic amines even at rather high concentrations, do not retard the oxidation rate satisfactorily. A much more efficient inhibition is then achieved by using metal deactivators, together with antioxidants. Metal deactivators are also known as copper inhibitors, because, in practice, the copper-catalyzed oxidation of polyolefins is by far of greatest importance. This is due to the fact that polyolefins are the preferred insulation material for communication wire and power cables, which generally contain copper conductors.

The function of a metal deactivator is to form an inactive complex with the catalytically active metal species. Specially suited for this purpose are chelating agents, which can form metal complexes of high thermal stability. The general feature of chelating agents is that they may contain several ligand atoms like N, O, S, P, alone or in combination with functional groups such as hydroxyl, carboxyl, or carbamide groups. The chelating agent *N,N'*-bis-(*o*-hydroxybenzal) oxalyldihydrazide is a good example for the above mentioned structural possibilities. It forms a soluble complex (XI) with the first mole of a copper salt by binding at the phenolic group, and a second mole of copper salt binds at amide nitrogens to form an insoluble complex.

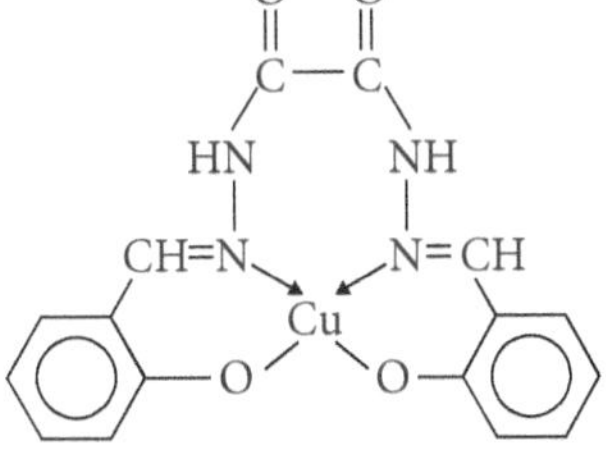

(XI)

Besides their main function to retard efficiently the metal-catalyzed oxidation of polyolefins, metal deactivators have to possess a number of ancillary properties to be useful in actual service. These include sufficient solubility or ease of dispersion, high extraction resistance, and low volatility and sufficient thermal stability under processing and service conditions.

Within the last two decades, a number of chemical structures have been proposed as metal deactivators for polyolefins. These include carboxylic acid amides of aromatic mono- and di-carboxylic acids and *N*-substituted derivatives such as *N,N′*-diphenyloxamide, cyclic amides such as barbituric acid, hydrazones and bishydrazones of aromatic aldehydes such as benzaldehyde and salicylaldehyde or of *o*-hydroxy-arylketones, hydrazides of aliphatic and aromatic mono- and di-carboxylic acids as well as *N*-acylated derivatives thereof, bisacylated hydrazine derivatives, polyhydrazides, and phosphorus acid ester of a thiobisphenol. An index of trade names and suppliers of a few commercial metal deactivators is given in Appendix A3.

Though there are metals other than copper (such as iron, manganese and cobalt) that can accelerate thermal oxidation of polyolefins and related polymers such as EPDM, in practice, however, the inhibition of copper-catalyzed degradation of polyolefins is of paramount importance because of the steadily increasing use of polyolefin insulation over copper conductors. Among polyolefins, polyethylene is still the most common primary insulation material for wire and cable. In the United States, high-density polyethylene and ethylenepropylene copolymers are used in substantial amounts for communications wire insulation.

For stabilization of polyolefins in contact with copper, it is often mandatory to combine a metal deactivator with an antioxidant. Metal deactivators in actual use are essentially *N,N′*-bis-[3-(3′,5′-di-*tert*-butyl-4′-hydroxy-phenyl)propionyl]-hydrazine and *N,N′*-dibenzaloxalyldihydrazide. The latter compound requires predispersion in a masterbatch because of its insolubility in polyolefins. This is not needed, however, for the former compound, which at commonly used concentrations is molecularly dispersed in polyolefins after processing. The required additive concentrations range from 0.05 to 0.5% depending on the polymer, the nature of the insulation (solid, cellular), whether the cable is petrolatum filled, and on service conditions.

Increasingly, communication cables are filled with petrolatum to improve waterproofing, whereas the insulation may be solid or cellular. Both trends, namely, petrolatum filling and cellular insulation, exert an influence on the oxidative stability of such cables. Thermal stability of high density polyethylene decreases by about 35% in changing from solid to cellular insulation. Moreover, in contact with petrolatum the stability of solid poly-ethylene decreases by 35% and that of cellular polyethylene decreases by 10–40%. Even in the most adverse instance, i.e., cellular insulation in contact with petrolatum, elevated concentrations of antioxidant and metal deactivator make it possible to achieve a high level of stability.

1.18 Light Stabilizers

Most polymers are affected by exposure to light, particularly the ultraviolet (UV) portion of sun's spectrum between 300 and 400 nm. (Fortunately, the earth's atmosphere filters out most of the light waves shorter than 300 nm.)Light contributes very actively to polymer degradation, especially when oxygen is present, which is a normal situation for most plastic materials. For example, when rubber is exposed to UV radiation at 45°C, it oxidizes three times as fast as in the dark at 70°C. Light and oxygen induce degradation reaction sin plastics that may not only discolor them but also exert a detrimental influence on numerous mechanical and physical properties. The inhibition of these degradation reactions is essential or else, the applications of many plastics would be drastically reduced. The inhibition can be achieved through addition of special chemicals, light stabilizers, or UV stabilizers, which are capable of interfering with the physical and chemical processes of light-induced degradation.

Carbon black is used as a stabilizer in a limited number offormulations where color is not a criterion. It not only absorbs light, but it can also react with free-radical species that might be formed. The weathering properties of polyethylene are improved by the incorporation of carbon blacks (at 2%–3% concentration). Weather-resistant wire and cable insulation, pipe for outdoor applications, films for mulching, and water conservation in ponds and canals may be made from polyethylene containing carbon black.

In spite of the fact that carbon black and other pigments may also protect the plastics from the effects of light, we shall consider in the following only those organic and organo-matallic compounds that impart only slight discoloration or no discoloration at all to the plastics to be stabilized.

The most important classes of light stabilizers from a practical point of view are 2-hydroxybenzophenones, 2-hydroxyphenylbenzotriazoles, hindered amines, and organic nickel compounds. In addition, salicylates, *p*-hydroxy-benzoates, resorcinol monobenzoates, cinnamate derivatives, and oxanilides are also used. Light stabilizers are generally used in concentrations between 0.05 and 2%, the upper limit of this range being employed only exceptionally. The sterically hindered amines represent the latest development in the field. They have outperformed the previously available light stabilizer classes in numerous synthetic resins.

1.18.1 Light Stabilizer Classes

The designation of the light stabilizer classes is based on mechanisms of UV stabilization. Free radicals are formed in polymers exposed to light, as a consequence of the excitation of absorbing functionalities in the polymer. This is a function of the energy of the light and of the structure of the polymer molecule. Since in the presence of oxygen, the polymer will simultaneously oxidize (photooxidation), it is often difficult to distinguish the pure photochemical processes from the thermal processes (oxidation), which are then superimposed. Some of the fundamental mechanisms involved have been evaluated [47]. The mechanism developed initially for the thermal oxidation of rubber can be applied to other substrates and also to photooxidation. The scheme shown below illustrates a possible reaction sequence to photooxidation:

Chain initiation

$$\left.\begin{array}{l}\text{Hydroperoxides ROOH}\\ \text{Carbonyl compounds} > C=O\\ \text{Catalyst residues (Ti, \ldots)}\\ \text{Charge transfer complexes (RH, } O_2)\end{array}\right\} \xrightarrow[M^{n+}/M^{(n+1)+}]{\Delta, h\nu} \begin{array}{l}\text{Free radicals}\\ (R\cdot, RO\cdot,\\ HO\cdot, HO_2^{\cdot}, \ldots)\end{array} \tag{1.79}$$

Chain propagation

$$R^{\cdot} + O_2 \rightarrow RO_2^{\cdot} \tag{1.80}$$

$$RO_2^{\cdot} + RH \rightarrow RO_2H + R^{\cdot} \tag{1.81}$$

Chain branching

$$ROOH \xrightarrow[M^{n+}/M^{(n+1)+}]{\Delta \text{ or } h\nu} RO\cdot + HO\cdot \tag{1.82}$$

$$2ROOH \xrightarrow[M^{n+}/M^{(n+1)+}]{\Delta \text{ or } h\nu} RO_2^{\cdot} + RO\cdot + H_2O \tag{1.83}$$

$$RO^{\cdot} + RH \rightarrow ROH + R^{\cdot} \tag{1.84}$$

$$HO^{\cdot} + RH \rightarrow H_2O + R^{\cdot} \tag{1.85}$$

Chain termination

$$\left.\begin{aligned} R^{\cdot} + R^{\cdot} &\rightarrow \\ R^{\cdot} + RO_2^{\cdot} &\rightarrow \\ RO_2^{\cdot} + RO_2^{\cdot} &\rightarrow \end{aligned}\right\} \text{nonradical products} \qquad (1.86)$$

In the light of the above photooxidation scheme there are four possibilities of protection against UV light. These are based on (1) prevention of UV light absorption or reduction of the amount of light absorbed by the chromophores; (2) reduction of the initiation rate through deactivation of the excited states of the chromophoric groups: (3) intervention in the photooxidative degradation process by transformation of hydroperoxides into more stable compounds, without generation of free radicals, before hydroperoxides undergo photolytic cleavage; and (4) scavenging of the free radicals as soon as possible after their formation, either as alkyl radicals or as peroxy radicals.

According to the four possibilities of UV protection mechanism described above, the light stabilizer classes can be designated as (1) UV absorbers, (2) quenchers of excited states, (3) hydroperoxide decomposers, and (4) free radical scavengers. It must be mentioned, however, that this classification is a simplification and that some compounds may be active in more than one way and often do so.

1.18.1.1 UV Absorbers

The protection mechanism of UV absorbers is based essentially on absorption of harmful UV radiation and its dissipation in a manner that does not lead to photosensitization, i.e., conversion to energy corresponding to high wavelengths or dissipation as heat. Besides having a very high absorption themselves, these compounds must be stable to light, i.e., capable of absorbing radiative energy without undergoing decomposition.

Hydroxybenzophenones and hydroxyphenyl benzotriazoles are the most extensively studied UV absorbers. Though the main absorptions of 2-hydroxybenzophenone are situated in the uninteresting wavelength domain around 260 nm, substituents such as hydroxy and alkoxy groups push this absorption towards longer wavelengths, between 300 and 400 nm, and at the same time total absorption in the UV absorbers (XII) are essentially derived from 2,4-dihyrdoxybenzophenone (X, R=H). Through choice of adequate alkyl group R in the alkoxy groups it is possible to optimize the protective power and the compatibility with the plastics to be stabilized.

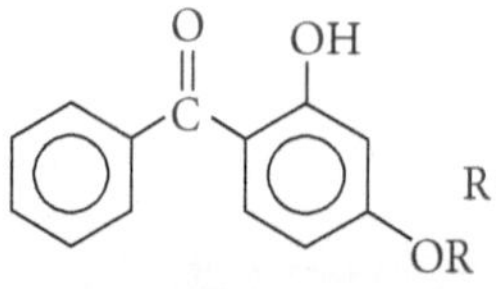

$R = H, CH_3$ to $C_{12}H_{25}$

(XII)

Derivatives of 2-hydroxybenzophenone have highly conjugated structures and a capacity to form intramolecular hydrogen bonds that exert a decisive influence on the spectroscopic and photochemical properties of these compounds. It has been shown with 2-hydroxybenzophenone (XIII) that on exposure to light (XIII) is transformed into enol (XIV), which turns back into its initial form (XIII) on losing thermal energy to the medium (Reaction 1.87):

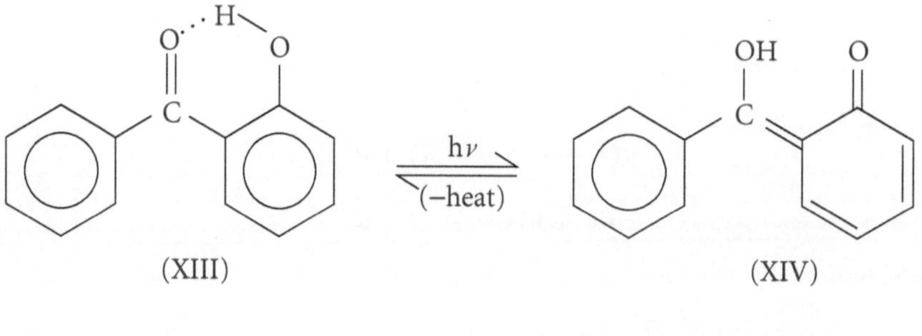

(1.87)

(XIII) (XIV)

The light energy consumed by the UV absorber corresponds to the quantity of energy needed to break the hydrogen bond. This explanation is supported by the fact that compounds that cannot lead to the formation of intramolecular hydrogen bonds (benzophenone or 2-methoxybenzophenone) do not absorb in the UV wavelength range.

Hydroxyphenyl benzotriazoles have the structure (XV) where X is H or Cl (chlorine shifts the absorption to longer wavelengths), R_1 is H or branched alkyl, and R_2 is CH_3 to C_8H_{17} linear and branched alkyl (R_1 and R_2 increase the affinity to polymers). Some technically important materials in this class are 2-(2′-hydroxy-5′-methyl-phenyl)-benzotriazole, 2-(2′-hydroxy-3′-5′-di-*tert*-butyl-phenyl)-benzotriazole, and 2-(2′-hydroxy-3′,5′-di-*tert*-butyl-phenyl)-5-chlorobenzotriazole. In comparison with 2-hydroxybenzophenones, the 2-(2′-hydroxyphenyl) benzotriazoles have higher molar extinction coefficients and steeper absorption edges towards 400 nm.

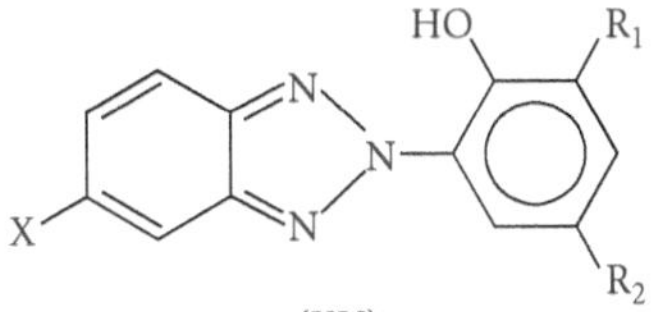

(XV)

The exact mechanism of light absorption by hydrobenzotriazoles is not known. However, the formation of intramolecular hydrogen bond, as in (XVI), and of zwitter ions having a quinoid structure, as in (XVII), may be responsible for the transformation of light radiation energy into chemical modifications:

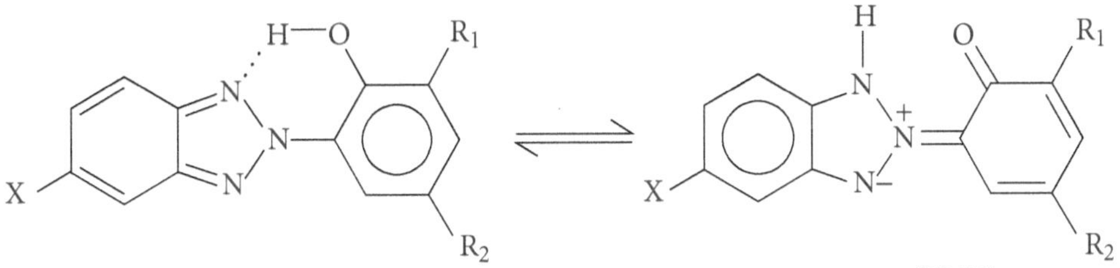

(XVI) (XVII) (1.88)

It may be noted that the tendency to form chelated rings by the creation of hydrogen bonds between hydroxide and carbonyl groups [as in (XIII)] or groups containing nitrogen [as in (XVI)] is a characteristic property of all UV absorbers.

A fundamental disadvantage of UV absorbers is the fact that they need a certain absorption depth (sample thickness) for good protection of a plastic. Therefore, the protection of thin section articles, such as films and fibers, with UV absorbers alone is only moderate.

1.18.1.2 Quenchers

Quenchers (Q) are light stabilizers that are able to take over energy absorbed from light radiation by the chromophores (K) present in a plastic material and thus prevent polymer degradation. The energy absorbed by quenchers can be dissipated either as heat (Reaction 1.91) or as fluorescent or phosphorescent radiation (Reaction 1.92):

$$K + h\nu \rightarrow K^* \quad (1.89)$$

$$K + Q \rightarrow K + Q^* \quad (1.90)$$

$$Q^* \rightarrow Q + \text{heat} \tag{1.91}$$

$$Q^* \rightarrow Q + h\nu' \tag{1.92}$$

For energy transfer to occur from the excited chromophore K* (donor) to the quencher Q (acceptor), the latter must have lower energy states than the donor. The transfer can take place by two processes: (1) long-range energy transfer or Föster mechanism and (2) contact, or collisional, or exchange energy transfer.

The Förster mechanism is based on a dipole-dipole interaction and is usually observed in the quenching of excited states. It has been considered as a possible stabilization mechanism for typical UV absorbers with extinction coefficients greater than 10,000. The distance between chromophore and quencher in this process may be as large as 5 or even 10 nm, provided there is a strong overlap between the emission spectrum of the chromophore and the absorption spectrum of the quencher.

However, for an efficient transfer to take place in the contact or exchange energy transfer process, the distance between quencher and chromophore must not exceed 1.5 nm. From calculations based on the assumption of random distribution of both stabilizer and sensitizer in the polymer, it is thus concluded that exchange energy transfer cannot contribute significantly to stabilization. This would not apply, however, if some kind of association between sensitizer and stabilizer takes place (for example, through hydrogen bonding).

Considering the dominant role of hydroperoxides in polyolefin photooxidation [cf. Equation 1.82 and Equation 1.83], quenching of excited –OOH groups would contribute significantly to stabilization. However, since the –OOH excited state is dissociative, i.e., its lifetime is limited to one vibration of the O–O bond, contact energy transfer during this very short time (about 10^{-3} s) appears highly unlikely if the –OOH group is not already associated with the quencher. The quenching action being thus independent of the thickness of the samples, quenchers are specifically useful for the stabilization of thin section articles such as films and fibers.

Metallic complexes that act as excited state quenchers are used to stabilize polymers, mainly polyolefins. They are nickel and cobalt compounds corresponding to the following structure:

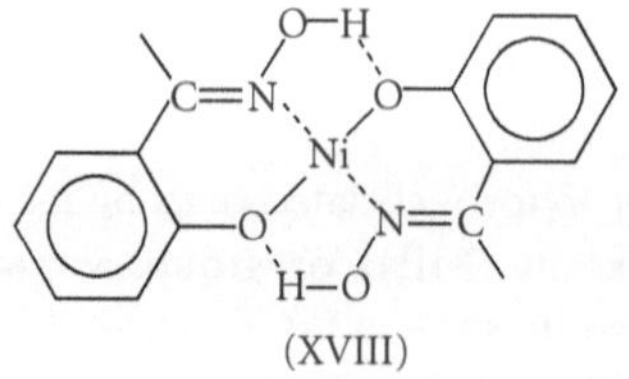

(XVIII)

Metallic complexes based on Ni, Co, and substituted phenols, thiophenols, dithiocarbamates, or phosphates are used. Typical representatives are nickel-di-butyldithiocarbamate, *n*-butylamin-nickel-2,2′-thio-bis-(4-*tert*-octyl-phenolate), nickel-bis-[2,2′-thio-bis-(4-*tert*-octyl-phenolate)] and nickel-(O-ethyl-3,5-di-*tert*-butyl-4-hydroxy-benzyl)-phosphonate. But their use is not as widespread as for other UV absorbers because they tend to be green.

1.18.1.3 Hydroperoxide Decomposers

Since hydroperoxides play a determining role in the photooxidative degradation of polymers, decomposition of hydroperoxides into more stable compounds, before the hydroperoxides undergo photolytic cleavage, would be expected to provide an effective means of UV protection. Metal complexes of sulfur-containing compounds such as dialkyldithiocarbamates (XIX), dialkyldithiophosphates (XX) and thiobisphenolates (XXI) are very efficient hydroperoxide decomposers even if used in almost catalytic quantities. Besides reducing the hydroperoxide content of pre-oxidized polymer films, they also can act as very efficient UV stabilizers. This explains the fact that an improvement in UV stability is often observed on combining UV absorbers with phosphite or nickel compounds.

1.18.1.4 Free-Radical Scavengers

Besides the absorption of harmful radiation by UV absorbers, the deactivation of excited states by quenchers, and the decomposition of hydroperoxides by some phosphorus and/or sulfur containing compounds, the scavenging of free-radical intermediates is another possibility of photostabilization, analogous to that used for stability against thermooxidative degradation. It has been shown that compounds (XXI), (XXII), (XXIII), and (XXIV) are effective radical scavengers. The radicals generated by hemolytic cleavage of hydroperoxide (Equation 1.93):

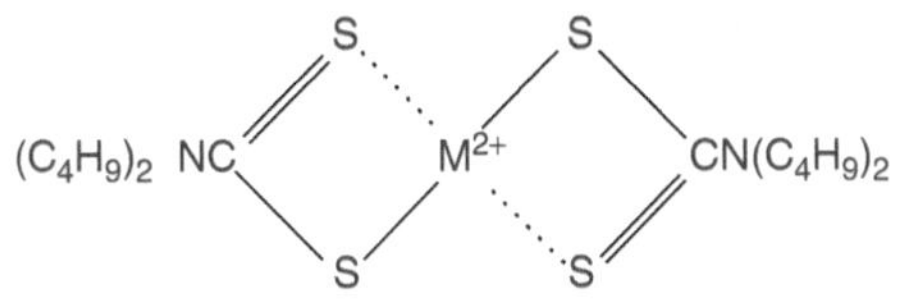

M = Zn, Ni

(XIX)

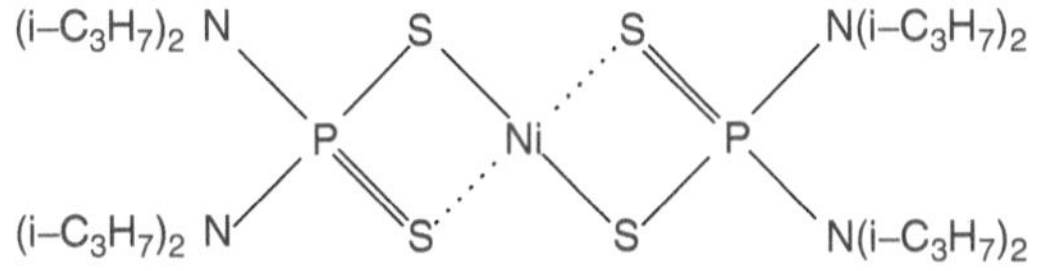

(XX)

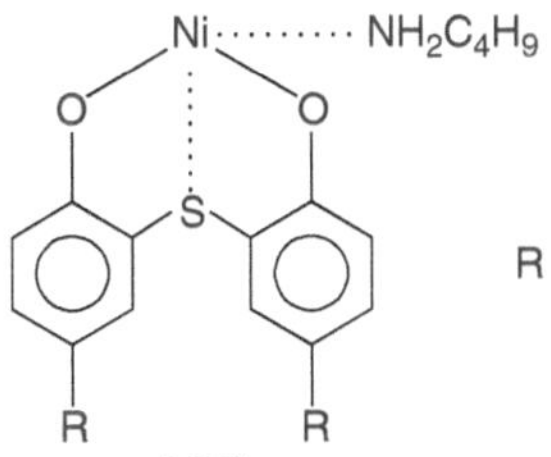

R = tert. octyl

(XXI)

$$PPOOH \rightarrow PPO^{\cdot} + {}^{\cdot}OH \tag{1.93}$$

may be removed by reactions, such as

$$HO^{\cdot} + InH \rightarrow [InH\ldots{}^{\cdot}OH] \tag{1.94}$$

$$PPO^{\cdot} + InH \rightarrow PPOH + In^{\cdot} \tag{1.95}$$

with a radical scavenger InH.

The latest development in the field of light stabilizers for plastics is represented by sterically hindered amine-type light stabilizers (HALS). A typical such compound is bis-(2,2,6,6-tetramethyl-4-piperidyl)-sebacate (XXV). Since it does not absorb any light above 250 nm, it cannot be considered a UV absorber or a quencher of excited states. This has been confirmed in polypropylene through luminescence measurements.

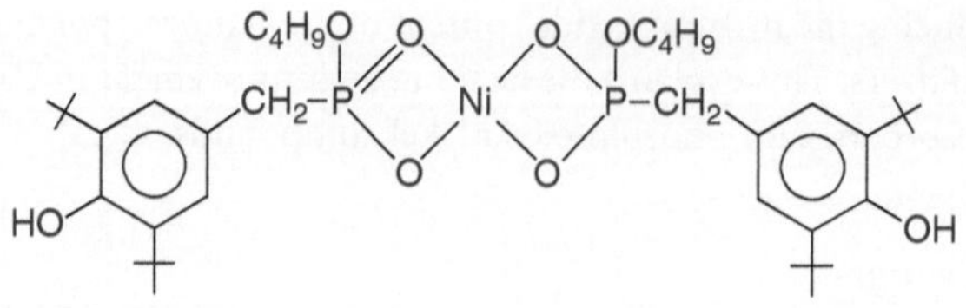

(—+ = tert butyl)

(XXII)

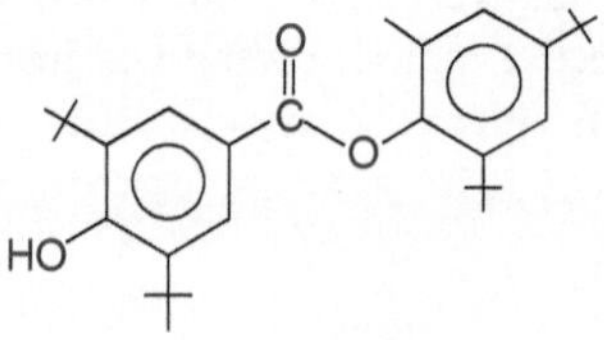

(XXIII)

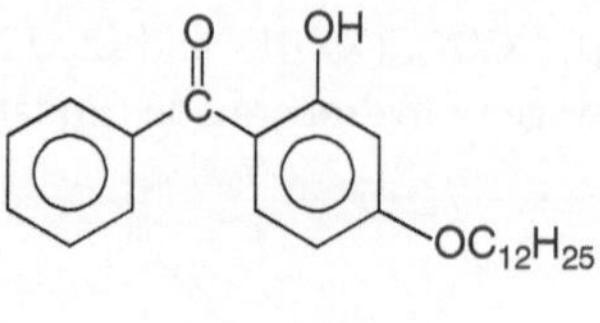

(XXIV)

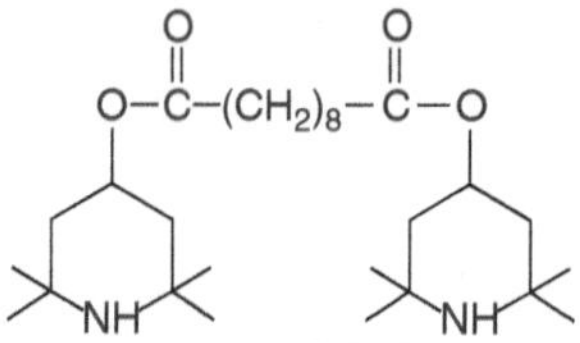

(XXV)

A low-molecular weight HALS such as (XXV), denoted henceforth as HALS-I, has the disadvantage of relative volatility and limited migration and extraction resistance, which are undesirable in special plastics applications (for example, in fine fibers and tapes). For such applications, it is advantageous to use polymeric sterically hindered amines such as poly-(*N*-β-hydroxyethyl-2,2,6,6-tetramethyl-4-hydroxypiperidyl succinate) represented by (XXVI) and a more complex polymeric hindered amine represented by (XXVII). In later discussions, they will be designated as HALS-II and HALS-III, respectively. Though they do not reach completely the performance of the low-molecular weight HALS-I, they are nevertheless superior to the other common light stabilizers used at several-fold higher concentrations.

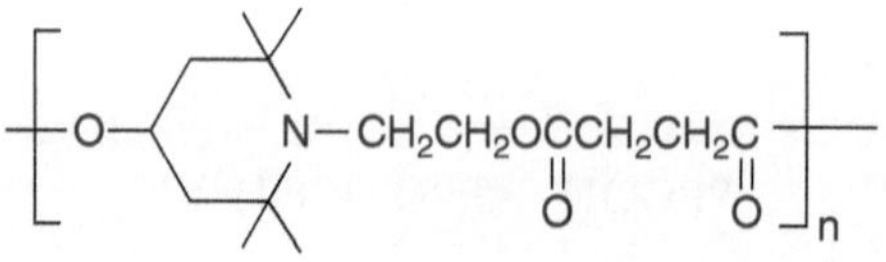

(XXVI)

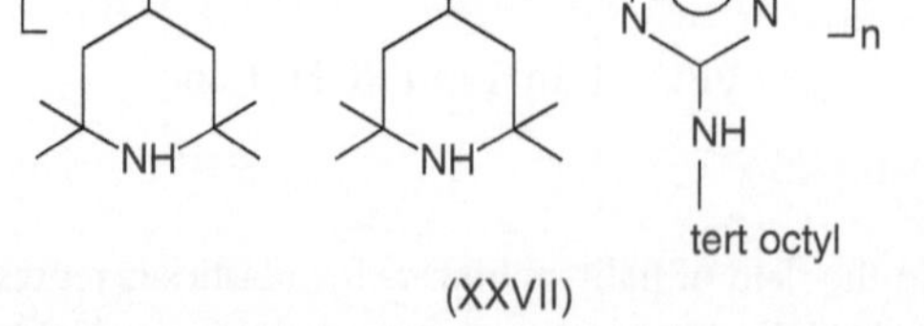

(XXVII)

The protection mechanisms of HALS, known so far mostly from studies with model systems, can be summarized as follows: From ESR measurements it is concluded that, under photooxidative conditions, HALS are converted, at least in part, to the corresponding nitroxyl radicals (XXVIII). The latter, through

Reaction 1.96, are thought to be the true radical trapping species.

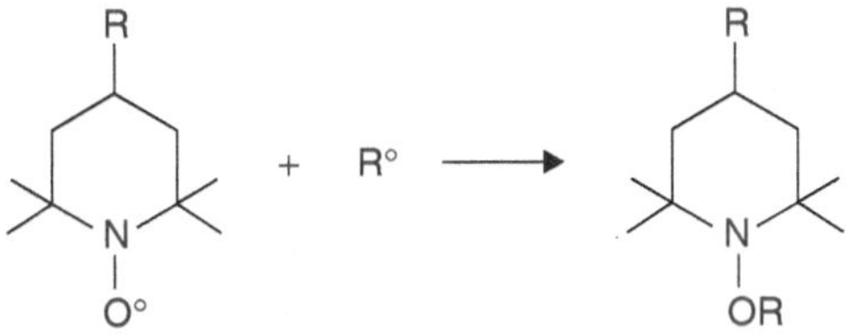

(XXVIII) (XXIX) (1.96)

Another explanation of the UV protection mechanism of HALS involves the hydroxylamine ethers (XXIX) formed in Reaction 1.96. There is indirect evidence that (XXIX) can react very quickly with peroxy radicals, thereby regenerating nitroxyl radicals (Reaction 1.97). Reaction 1.96 and Reaction 1.97, which constitute the "Denisov cycle," result in an overall slowdown of the usual chain oxidation Reaction 1.80 and Reaction 1.81.

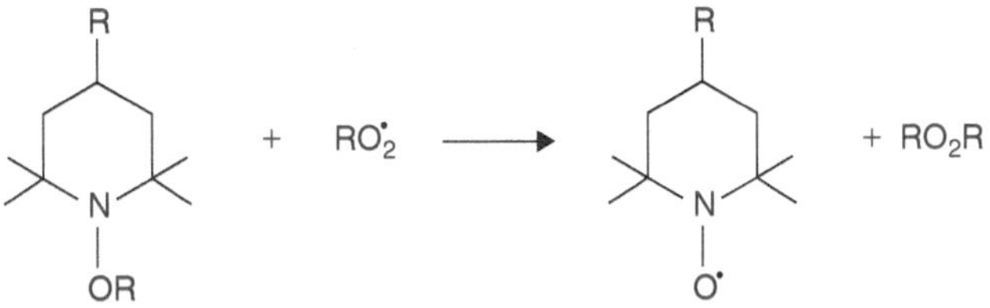

(XXIX) (XXVIII) (1.97)

Nitroxyl radicals may also react with polymer radical to form hydroxylamine (Reaction 1.98), and the latter can react with peroxy radicals and hydroperoxides according to Reaction 1.99 and Reaction 1.100:

$$>NO^{\bullet} + R^{\bullet} \longrightarrow\ >NOH + >C{=}\overset{|}{C}-R \tag{1.98}$$

$$>NOH + RO_2^{\bullet} \longrightarrow\ >NO^{\bullet} + RO_2H \tag{1.99}$$

$$>NOH + ROOH \longrightarrow\ >NO^{\bullet} + H_2O + RO^{\bullet} \tag{1.100}$$

The formation of associations between HALS and hydroperoxides followed by reaction of these associations with peroxy radicals (Reaction 1.101 and Reaction 1.102) represents another possibility of retarding photooxidation.

$$>NH + HOOR \rightleftharpoons [>NH\ldots HOOR] \tag{1.101}$$

$$[>NH\ldots HOOR] + RO_2^{\bullet} \longrightarrow ROOR + H_2O +\ >NO^{\bullet} \tag{1.102}$$

The hydroperoxides associated with HALS may undergo photolysis producing hydroxy and alkoxy radicals in close proximity to the amines. The radicals may then abstract a hydrogen atom from the amine and form hydroxylamine and hydroxylamine ethers (Reaction 1.103):

$$\rangle NH \ldots HOOR \rightarrow \left[\begin{matrix} & {}^{\bullet}OH \\ \rangle N{-}H & \\ & {}^{\bullet}OR \end{matrix} \right] \rightarrow \begin{cases} \rangle NOH + ROH \\ \rangle NOR + H_2O \end{cases} \tag{1.103}$$

Still other mechanisms have been postulated, e.g., the interaction of HALS with α,β,-unsaturated carbonyl compounds and the formation of charge-transfer complex between HALS and peroxy radicals. However, despite extensive publications in the field, a complete knowledge of the process occurring in polymer photooxidation in the presence of HALS is still not available.

1.19 Light Stabilizers for Selected Plastics

In choosing a light stabilizer for a given plastic, several factors, in addition to its protective power, play an important role. In this respect, one may cite physical form (liquid/solid, melting point), thermal stability, possible interaction with other additives and fillers that may eventually lead to discoloration of the substrate, volatility, toxicity (food packaging), and above all, compatibility with the plastics material considered. An additive can be considered as compatible if, during a long period of time, no blooming or turbidity is observed at room temperature and at the elevated temperatures that may be encountered during projected use of the plastic material.

Among the light stabilizer classes available commercially, only a few may be used in a broad range of plastics. Thus, nickel compounds are used almost exclusively in polyolefins, whereas poly(vinyl chloride) is stabilized with UV absorbers only.

A list of trade names and manufacturers of selected light stabilizers is given in Appendix A4.

1.19.1 Polypropylene

For the light stabilization of polypropylene, representatives of the following stabilizer classes are mainly used: 2-(2′-hydroxyphenyl)-benzotriazoles, 2-hydroxy-4-alkoxybenzophenones, nickel-containing light stabilizers, 3,5-di-*tert*-butyl-4-hydroxybenzoates, as well as sterically hindered amines (HALS). Nickel-containing light stabilizers are used exclusively in thin sections, such as films and tapes, whereas all other classes may be used in thin as well as thick sections, though UV absorbers have only limited effectiveness in thin section. Nickel-containing additives are also used as "dyesites" because they allow the dying and printing of polypropylene fibers with dyestuffs susceptible to complexation with metals.

1.19.2 Polyethylene

As all polyolefins, polyethylene is sensitive to UV radiation, although less than polypropylene. For outdoor use polyethylene needs special stabilization against UV light. The light stabilizers for polyethylene are in principle the same as for polypropylene. On accelerated weathering, HALS show much better performance in HDPE tapes than UV absorbers, despite the latter being used in much higher concentrations. The comparison between HALS is, however, in favor of the polymeric HALS-III, which has the same performance when added at a concentration of 0.05% as HALS-I and HALS-II at 0.1%.

Among the numerous commercial light stabilizers only a few are suitable for low density polyethylene (LDPE). This is mainly due to the fact that most light stabilizers are not sufficiently compatible at levels of concentration necessary for the required protection and so they bloom more or less rapidly. Initially, UV absorbers of the benzophenone and benzotriazole types were used to protect LDPE materials. With the development of nickel-quenchers, significant improvement in the light stability of LDPE films has been achieved. For reasons of economy, however, combinations of nickel-quenchers with UV absorbers are mainly used. The service life of LDPE films may be increased by raising the concentrations of light stabilizers. Additive contents close to 2% may be found in greenhouse films thought to last up to 3 years outdoors.

A further improvement in UV stability of LDPE was expected with the development of HALS. However, LDPE compatibility of HALS available in the early years was insufficient, resulting in relatively poor performance on outdoor weathering. It was only with the development of polymeric HALS that these difficulties were overcome.

Tests have shown that HALS-II is significantly superior to UV absorbers and Ni-stabilizers so that the same performance can be achieved with much smaller concentrations. However, use of combinations of the polymeric HALS-II with a UV absorber leads to a significant improvement of the efficiency in comparison with the HALS used alone at the same concentration as the combination. A further boost of the performance can be achieved through use of the polymeric HALS-III. For example, the performance of HALS-II can be reached by using HALS-III at about half the concentration. The superiority of HALS-III becomes even more pronounced in films of thickness below 200 μm.

In linear low density polyethylene (LLDPE) also, the polymeric HALS-II and HALS-III show much better performance than other commercial light stabilizers. Blooming is observed with low molecular weight HALS-I, similar to that found with LDPE.

In more polar substances such as ethylene-vinyl acetate copolymers (EVA), the low molecular weight HALS-I can be used. However, in this substrate too, the polymeric HALS-II and HALS-III are significantly superior to the low molecular weight HALS.

1.19.3 Styrenic Polymers

Double bonds are not regarded as the chromophores responsible for initiation of photooxidation in polystyrenes because they absorb below 300 nm. However, peroxide groups in the polymer chain, resulting from co-polymerization of oxygen with styrene, are definitely photolabile. Moreover, oxidation products such as aromatic ketones of acetophenone type, which have been detected by emission spectroscopy, are formed during processing of styrene polymers at high temperatures. Aromatic ketones (AK) in the triplet state are able to abstract hydrogen from polystyrene (PSH) (Reaction 1.104):

$$\begin{aligned} &\mathrm{AK} \rightarrow \mathrm{AK}^* \\ &\mathrm{AK}^* + \mathrm{PSH} \rightarrow \mathrm{AKH}^{\cdot} + \mathrm{PS}^{\cdot} \end{aligned} \qquad (1.104)$$

This reaction is considered the most important initiation mechanism for styrene photooxidation in the presence of aromatic ketones.

Styrenic plastics such as acrylonitrile/butadiene/styrene graft copolymers (ABS) and impact-resistant polystyrenes are very sensitive towards oxidation, mainly because of their butadiene content. Degradation on weathering starts at the surface and results in rapid loss of mechanical properties such as impact strength.

Because of the lack of efficient light stabilizers, ABS has not been used outdoors on a large scale. However, by combining two light stabilizers with different protection mechanisms, e.g., a UV absorber of the benzotriazole class and the sterically hindered amine HALS-I, it is possible to achieve good stabilization even in ABS. This is a case of synergism in which the UV absorber protects the deeper layers, while HALS-I assures surface protection. At the same time discoloration of the ABS polymer is also reduced significantly. The same holds for polystyrene and styrene–acrylonitrile copolymers (SAN), and the best protection is obtained with HALS/UV absorber combinations. Light stabilization is necessary for articles of these polymers for which UV exposure can be expected (e.g., covers for fluorescent lights).

1.19.4 Poly(Vinyl Chloride)

"Pure" poly(vinyl chloride) (PVC) does not absorb any light above 220 nm. Different functional groups and structural irregularities that may arise during polymerization and processing have thus been considered as possible initiating chromophores. They include irregularities in the polymer chain as well as hydroperoxides, carbonyl groups, and double bonds.

Thermal stabilizers used in PVC also confer some degree of light stability. Ba/Cd salts and organic tin carboxylates, for example, confer already some UV stability to PVC on outdoor exposure. However, for transparent and translucent PVC articles requiring high UV stability, the light stability conferred by thermal stabilizers is not sufficient. The addition of light stabilizers in such cases is therefore mandatory. So far, UV absorbers yield the best results in practical use. HALS-I has almost no effect.

1.19.5 Polycarbonate

Bisphenol-A polycarbonate (PC) absorbs UV light below 360 nm but its absorption is intense only below 300 nm. Insufficient light stability of PC on outdoor use is manifested by yellowing, which increases rapidly.

Studies indicate that on absorption of UV light, PC undergoes photo-Fries rearrangement, which gives first a phenyl salicylate, and after absorption of a second photon and subsequent rearrangement, it gives 2,2′-dihydroxybenzophenone groups (Figure 1.43). The absorption of these groups reaches into the visible region, and PC yellowing has been essentially attributed to them. In addition to the Fries reaction, the formation of O_2-charge transfer complexes in PC, similar to those found in polyolefins and leading to the formation of hydroperoxides, is considered to contribute significantly to photooxidation in the early stages.

Among the stabilizer classes, only UV absorbers are in use for stabilization of PC. In choosing a UV absorber, intrinsic performance, volatility, adequate thermal stability at the elevated processing temperatures (about 320°C), and effect on initial color of the PC articles should be considered. Benzotriazole, oxanilide, and cinnamate-type UV absorbers are effective photostabilizers for PC with benzotriazoles giving the best performance among the three types.

1.19.6 Polyacrylates

Poly(methyl methacrylate) (PMMA) is highly transparent in the UV region, and thus much more light stable than other thermoplastics. UV absorbers may therefore be used to confer a UV filter effect to PMMA articles. PMMA window panes for solar protection containing 0.05%–0.2% 2-(2′-hydroxy-5′-methylphenyl)-benzotriazole are well-known examples of this application. The rear lights of motor cars, electric signs, and covers for fluorescent lights are some applications for which PMMA is UV-stabilized. The excellent light-stabilizing performance of HALS is also found with PMMA.

1.19.7 Polyacetal

Polyacetal is markedly unstable towards lights because even UV radiation of wavelengths as high as 365 nm may initiate its degradation. Polyacetal cannot therefore be used outdoors if it does not contain any light stabilizers. Even after a short weathering, surface crazes and pronounced chalking are observed.

Carbon black (0.5%–3%) is a good stabilizer for polyacetal when sample color is not important. Other possibilities for stabilization are the use of 2-hydroxybenzophenone and, especially, hydroxyphenylbenzotriazole-type UV absorbers. Stabilization with HALS/UV absorber is superior to that with UV absorber alone.

1.19.8 Polyurethanes

Lights stability of polyurethanes depends to a large extent on their chemical structure, and both components (i.e., isocyanate and polyol) have an influence. Polyurethanes based on aliphatic isocyanates and polyester diols show the best light stability if yellowing is considered, whereas polyurethanes based on aromatic isocyanates and polyether diols are worst in this respect.

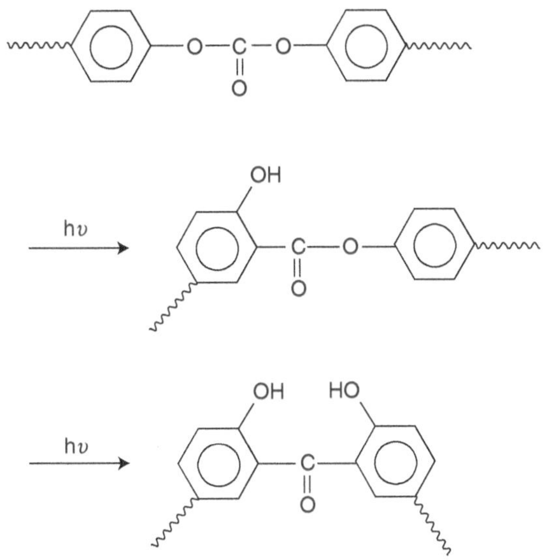

FIGURE 1.43 Photo-Fries rearrangement on polycarbonate.

Light stabilizers are used mainly in the coatings industry (textile coatings, synthetic leather). In addition to some UV absorbers of the 2-(2′-hydroxyphenyl)-benzotriazole type, HALS used alone or in combination with benzotriazoles are especially effective stabilizers.

1.19.9 Polyamides

The absorption of aliphatic polyamides in the short wavelength region of sunlight is attributed largely to the presence of impurities. Direct chain scission at wavelengths below 300 nm and photosensitized oxidation above 300 nm have been considered a long ago as responsible for photooxidation. Antioxidants used in polyamides often confer good light stability as well. However, enhanced performance is obtained with the combination of an antioxidant with a light stabilizer. The sterically hindered amines are significantly superior. For example, molded polyamide samples stabilized with HALS (0.5%) are found to exhibit approximately twice the light stability of the samples stabilized with a phenolic antioxidant (0.5%) or combination of the latter with a UV absorber (0.5%).

1.20 Diffusion and Permeability

There are many instances where diffusion and permeation of a gas, vapor, or liquid through a plastics material is of considerable importance in the processing and usage of the material. For example, dissolution of a polymer in a solvent occurs through diffusion of the solvent molecules into the polymer, which thus swells and eventually disintegrates.

In a plastisol process the gelation of PVC paste, which is a suspension of PVC particles in a plasticizer such as tritolyl phosphate, involves diffusion of the plasticizer into the polymer mass, resulting in a rise of the paste viscosity. Diffusion processes are involved in the production of cellulose acetate film by casting from solution, as casting requires removal of the solvent. Diffusion also plays a part in plastic molding. For example, lubricants in plastics compositions are required to diffuse out of the compound during processing to provide lubrication at the interface of the compound and the mold. Incompatibility with the base polymer is therefore an important criterion in the choice of lubricants in such cases.

Permeability of gases and vapors through a film is an important consideration in many applications of polymers. A high permeability is sometimes desirable. For example, in fruit-packaging applications of plastics film it is desirable to have high permeability of carbon dioxide. On the other hand, for making inner tube and tubeless tires, or in a child's balloon, the polymer used must have low air permeability.

1.20.1 Diffusion

Diffusion occurs as a result of natural processes that tend to equalize the concentration gradient of a given species in a given environment. A quantitative relation between the concentration gradient and the amount of one material transported through another is given by Fick's first low:

$$dm = -D\frac{dc}{dx}A\,dt \tag{1.105}$$

where dm is the number of grams of the diffusing material crossing area A (cm^2)of the other material in time dt (s). D is the diffusion coefficient (cm^2/s) whose value depends on the diffusing species and the material in which diffusion occurs, and dc/dx is the concentration gradient, where the units of x of and c are centimeters and grams per cubic centimeter.

Diffusion in polymers occurs by the molecules of the diffusing species passing through voids and other gaps between the polymer molecules. The diffusion rate will therefore heavily depend on the molecular size of the diffusing species and on the size of the gaps. Thus, if two solvents have similar solubility parameters, the one with smaller molecules with diffuse faster in a polymer. On the other hand, the size of the gaps in the polymer depends to a large extent on the physical state of the polymer—that is, whether it is crystalline, rubbery, or glassy.

Crystalline structures have an ordered arrangement of molecules and a high degree of molecular packing. The crystalline regions in a polymer can thus be considered as almost impermeable, and diffusion can occur only in amorphous regions or through region of imperfection: hence, the more crystalline the polymer, the greater will be its tendency to resist diffusion. Amorphous polymers, as noted earlier, exist in the rubbery state above the glass transition temperature and in the glassy state below this temperature. In the rubbery state there is an appreciable "free volume" in the polymer mass, and molecular segments also have considerable mobility, which makes it highly probable that a molecular segment will at some stage move out of the way of the diffusing molecule, the contributing to a faster diffusion rate. In the glassy state, however, the molecular segments cease to have mobility, and there is also a reduction in free volume or voids in the polymer mass, both of which lower the rate of diffusion. Thus, diffusion rates will be highest in rubbery polmers and lowest in crystalline polymers.

1.20.2 Permeability

Permeation of gas, vapor, or liquid through a polymer film consists of three steps: (1) a solution of permeating molecules in the polymer, (2) diffusion through the polymer due to concentration gradient, and (3) emergence of permeating molecules at the outer surface. Permeability is therefore the product of solubility and diffusion; so where the solubility obeys Henry's law one may write

$$P = DS \tag{1.106}$$

where P is the permeability, D is the diffusion coefficient, and S is the solubility coefficient [48].

Hence, factors which contribute to greater solubility and higher diffusivity will also contribute to greater permeability. Thus, a hydrocarbon polymer like polyethylene should be more permeable to hydrocarbons than to liquids of similar molecular size but of different solubility parameter, and smaller hydrocarbons should have higher permeability. The permeabilities of a number of polymers to the common atmospheric gases [49,50], including water vapor, are given in Table 1.17.

It appears that regardless of the film material involved, oxygen permeates about four times as fast as nitrogen, and carbon dioxide about 25 times as fast. The fact that the ratios of the permeabilities for all gases, apart from water vapor, are remarkably constant, provided there is no interaction between the film material and the diffusing gas, leads one to express the permeability as the product of three factors [49]: one determined by the nature of the polymer film, one determined by the nature of the gas, and one accounting for the interaction between the gas and the film; i.e.,

$$P_{i,k} = F_i G_k H_{i,k} \tag{1.107}$$

where $P_{i,k}$ is the permeability for the system polymer i and gas k; F_i, G_k and $H_{i,k}$ are the factor associated with the film, gas, and interaction respectively. When $H_{i,k} \approx 1$, there is little or no interaction, and as the degree of interaction increases, $H_{i,k}$ becomes larger. With little or no interaction, Equation 1.107 becomes

$$P_{i,k} = F_i G_k \tag{1.108}$$

It then appears that the ratio of the permeability of two gases (k,l) in the same polymer (i) will be the same as the ratio between the two G factors

$$\frac{P_{i,k}}{P_{i,l}} = \frac{G_k}{G_l} \tag{1.109}$$

If the G value of one of the gases, usually nitrogen, is taken as unity, G values of other gases and F values of different polymers can be calculated from Equation 1.109 and Equation 1.108. These values are reliable for gases but not for water vapor. Some F and G values for polymers are given in Table 1.18. Evidently, the F values correspond to the first column of Table 1.17, since G for N_2 is 1. The G values for O_2 and CO_2 represent the averages of P_{O2}/P_{N2} and P_{CO2}/P_{N2} in columns 5 and 6, respectively.

1.21 Polymer Compounding

In many commercial plastics, the polymer is only one of several constituents, and it is not necessarily the most important one. Such systems are made by polymer compounding—a term used for the mixing of polymer with other ingredients. These other ingredients or supplementary agents are collectively referred to as additives. They may include chemicals to act as plasticizing agents, various types of filling agents, stabilizing agents, antistatic agents, colorants, flame retardants, and other ingredients added to impart certain specific properties to the final product [45,51,52].

The properties of compounded plastics may often be vastly different from those of the base polymers used in them. A typical example is SBR, which is the largest-volume synthetic rubber today. The styrene-butadiene copolymer is a material that does not extrude smoothly, degrades rapidly on exposure to warm air, and has a tensile strength of only about 500 psi (3.4×10^6 N/m^2). However, proper compounding changes this polymer to a smooth-processing, heat-stable rubber with a tensile strength of over 3,000 psi 20×10^6 N/m^2). Since all properties cannot be optimized at once, compounders have developed thousands of specialized recipes to optimize one or more of the desirable properties for particular applications (tires, fan belts, girdles, tarpaulins, electrical insulation, etc.).

In SBR the compounding ingredients can be (1) reinforcing fillers, such as carbon black and silica, which improve tensile strength or tear strength; (2) inert fillers and pigments, such as clay, talc, and calcium carbonate, which make the polymer easier to mold or extrude and also lower the cost; (3) plasticizers and extenders, such as mineral oils, fatty acids, and esters; (4) antioxidants, basically amines or phenols, which stop the chain propagation in oxidation; and (5) curatives, such as sulfur for unsaturated polymers and peroxides for saturated polymers, which are essential to form the network of cross-links that ensure elasticity rather than flow.

TABLE 1.17 Permeability Data for Various Polymers

Polymer	Permeabiliety ($P \times 10^{10}$ $cm^3/cm^2/mm/s/cm$ Hg)				Ratios (to N_2 Permeability as 1.0)			
	N_2 (30°C)	O_2 (30°C)	CO_2 (30°C)	H_2O (25°C, 90% RH)	P_{O_2}/P_{N_2}	P_{CO_2}/P_{N_2}	P_{H_2O}/P_{N_2}	Nature of Polymer
Poly(vinylidene chloride) (Saran)	0.0094	0.053	0.29	14	5.6	31	1,400	Crystalline
Polychlorotrifluoro-ethylene	0.03	0.10	0.72	2.9	3.3	24	97	Crystalline
Poly(ethylene terephthalate) (Mylar A)	0.05	0.22	1.53	1,300	4.4	31	26,000	Crystalline
Rubber hydrochloride (Pliofilm ND)	0.08	0.30	1.7	240	3.8	21	3,000	Crystalline
Polyamide (Nylon 6)	0.10	0.38	1.6	7,000	3.8	16	70,000	Crystalline
Poly(vinyl chloride) (unplasticized)	0.40	1.20	10	1,560	3.0	25	3,900	Semicrystalline
Cellulose acetate	2.8	7.8	68	75,000	2.8	24	2,680	Glassy
Polyethylene (d=0.954, 0.960)	2.7	10.6	35	130	3.9	13	48	Crystalline
Polyethylene (d=0.922)	19	55	352	800	2.9	19	42	Semicrystalline
Polystyrene	2.9	11	88	12,000	3.8	30	4,100	Glassy
Polypropylene (d=0.910)	—	23	92	680	—	—	—	Crystalline
Butyl rubber	3.12	13.0	51.8	—	4.1	16.2	—	Rubbery
Polybutadiene	64.5	191	1,380	—	3.0	21.4	—	Rubbery
Natural rubber	80.8	233	1,310	—	2.9	16.2	—	Rubbery

Source: From Stannett, V. T. and Szwarc, M. 1955. *J. Polym. Sci.*, 16, 89 and Paine, F. A. 1962. *J. Roy. Inst. Chem.*, 86, 263.

TABLE 1.18 *F* and *G* Constants for Polymers and Gases

Polymer	F	Gas	G
Poly(vinylidene chloride) (Saran)	0.0094	N_2	1.0
Poly(chlorotrifluoroethylene)	0.03		
Poly(ethylene terephthalate)	0.05	O_2	3.8
Rubber hydrochloride (Pliofilm)	0.08	H_2S	21.9
Nylon 6	0.1	CO_2	24.2
Cellulose acetate (+15% plasticizer)	5		
Polyethylene ($d=0.922$)	19		
Ethyl cellulose (plasticized)	84		
Natural rubber	80.8		
Butyl rubber	3.12		
Nitrile rubber	2.35		
Polychloroprene	11.8		
Polybutadiene	64.5		

Source: From Stannett, V. T. and Szwarc, M. 1955. *J. Polym. Sci.*, 16, 89.

Polymer applications which generally involve extensive compounding are rubbers, thermosets, adhesives, and coatings. Fibers and thermoplastic polymers (with the exception of PVC) are generally not compounded to any significant extend. Fibers, however, involve complex after-treatment processes leading to the final product. PVC, which by itself is a rigid solid, owes much of its versatility in applications to compounding with plasticizers. The plasticize content varies widely with the end use of the product but is typically about 30% by weight.

Of the compounding ingredients, fillers and plasticizers are more important in terms of quantities used. Other additives used in smaller quantities are antioxidants, stabilizers, colorants, flame retardants, etc. The ingredients used as antioxidants and light stabilizers, and their effect have been discussed previously. Fillers, plasticizers and flame retardants are described next.

1.21.1 Fillers

Fillers play a crucial role in the manufacture of plastics. Alone many plastics are virtually useless, but they are converted into highly useful products by combining them with fillers. For example, phenolic and amine resins are almost always used in combination with substances like wood flour, pure cellulose, powdered mica, and asbestos. Glass fiber is used as a filler for fiber-reinforced composites with epoxy or polyester resins.

Another extremely important example is the use of carbon filler for rubber. Rubber would be of little value in modern industry were it not for the fact that the filler carbon greatly enhances its mechanical properties lie tensile strength, stiffness, tear resistance, and abrasion resistance. Enhancement of these properties is called reinforcement, and the fillers which produce the strengthening effect are known as reinforcing fillers. Other fillers may not appreciably increase strength, but they may improve other properties of the polymer, thus, making it easier to mold, which reduces cost.

Fillers used in plastics can be divided into two types; particulate and fibrous. Typical fillers in these two categories and the improvements they bring about are summarized in Table 1.19. In some instances they are added to perform one or more prime functions or to provide special properties. Asbestos, for example, provides high-temperature resistance and improves dimensional stability. Mica improves the electrical and thermal properties of all compounds. Glass fibers produce high strength. Carbon black is the only important reinforcing filler for most elastomers. It also imparts UV resistance and coloring.

Beryllium oxide-filled resins gain high conductivity without loss of electrical properties. Metal particles have been used as fillers for plastics to improve or impart certain properties. Thus, aluminum has been used for applications ranging from making a decorative finish to improving thermal conductivity. Copper particles are used in plastics to provide electrical conductivity. Lead is used because it dampens vibrations, acts as barrier to gamma-radiation, and has high density.

Of the new space-age products used as reinforcing fillers, carbon fibers and boron fibers have the greatest potential for use in high-strength advanced composites. Carbon fibers made by pyrolizing organic fibers such as rayon and polyacrylonitrile have tensile strengths approaching 3.3×10^5 psi (2.3 (10^9 N/m^2). Boron fibers made by depositing boron from a BCl_3–H_2 mixture onto tungsten wire have tensile strengths approaching 5×10^5 psi (3.5×10^9 N/m^2). The specific strengths and specific moduli of polymer composites made with these fibers are far above those attainable in monolithic structural materials such as high-strength aluminum, steel, and titanium. These composites can thus lead to significant weight savings in actual applications. A relatively recent addition to the high-performance fiber field is the organic polymeric fiber Kevlar-49, developed by Du Pont. It has a higher specific strength than glass, boron, or carbon. Furthermore, Kevlar, a polyamide, is cheap, having one-sixth the price of acrylic-based graphite fibers.

The extensive range of fillers and reinforcing agents used nowadays indicates the importance that these materials have attained. The main difference between inert and reinforcing fillers lies in the fact that modulus of elasticity and stiffness are increased to a greater or lesser extent by all fillers, including the spherical types such as chalk or glass spheres, whereas tensile strength can be appreciably improved only by a fiber reinforcement. Heat deflection temperature, i.e. stiffness at elevated temperatures, cannot be increased by spherical additives to the same extent as by fiber reinforcement. On the other hand, fillers in flake form, such as talc or mica, likewise produce a marked improvement in the heat deflection temperature but lead to a decrease in tensile strength and to elongation at break. The influences of common fillers on the properties of polyolefins are compared in Table 1.20.

1.22 Plasticizers

Plasticizers are organic substances of low volatility that are added to plastics compounds to improve their flexibility, extensibility, and processability. They increase flow and thermoplasticity of plastic materials by decreasing the viscosity of polymer melts, the glass transition temperature (T_g) the melting temperature (T_m), and the elasticity modulus of finished products [53].

Plasticizers are particularly used for polymers that are in a glassy state at room temperature. These rigid polymers become flexible by strong interactions between plasticizer molecules and chain units, which lower their brittle-tough transition or brittleness temperature (T_b) (the temperature at which a sample breaks when struck) and their T_g value, and extend the temperature range for their rubbery or viscoelastic state behavior (see Figure 1.19).

Mutual miscibility between plasticizers and polymers is an important criterion from a practical point of view. If a polymer is soluble in a plasticizer at a high concentration of the polymer, the plasticizer is said to be a primary plasticizer. Primary plasticizers should gel the polymer rapidly in the normal processing temperature range and should not exude from the plasticized material. Secondary plasticizers, on the other hand, have lower gelation capacity and limited compatibility with the polymer. In this case, two phases are present after plasticization process—one phase where the polymer is only slightly plasticized, and one phase where it is completely plasticized. Polymers plasticized with secondary plasticizers do not, therefore, deform homogeneously when stressed as compared to primary plasticizers. The deformation appears only in the plasticizer-rich phase and the mechanical properties of the system are poor. Unlike primary plasticizers, secondary plasticizers cannot be used alone and are usually employed in combination with a primary plasticizer.

Plasticizer properties are determined by their chemical structure because they are affected by the polarity and flexibility of molecules. The polarity and flexibility of plasticizer molecules determine their

TABLE 1.19 Some Fillers and Their Effects on Plastics

Properties Improved/Fillers	Chemical Resistance	Heat Resistance	Dimensional Stability	Tensile Strength	Stiffness	Impact Strength	Hardness	Lubricity	Electrical Insulation	Electrical Conductivity	Thermal Conductivity	Moisture Resistance	Processability	Recommend for Use in[a]
Aipha cellulose			+	+					+					S
Alumina	+	+	+											S/P
Aluminum powder										+	+			S
Asbestos	+	+	+		+	+	+		+					S/P
Calcium carbonate		+	+		+		+						+	S/P
Calcium silicate		+	+		+		+							S
Carbon black		+	+		+					+	+		+	S/P
Carbon fiber										+	+			S
Cellulose			+	+	+	+	+							S/P
Cotton (macerated/ chopped fibers)			+	+	+	+	+		+					S
Fibrous glass	+	+	+	+	+	+	+		+			+		S/P
Graphite	+		+	+	+		+	+		+	+			S/P
Jute					+	+								S
Kaolin	+	+	+		+		+	+				+	+	S/P
Kaolin (calcined)	+	+	+		+		+		+			+	+	S/P
Mica	+	+	+		+		+	+	+			+		S/P
Molybdenum disulfide					+		+	+				+	+	P
Nylon (macerated/ chopped fibers)	+	+	+	+	+	+	+	+	+				+	S/P
Acrylic fiber (Orion)	+	+	+	+	+	+	+		+			+	+	S/P
Rayon			+	+	+	+	+		+					S
Silica, amorphous									+			+	+	S/P
TFE-fluorocarbon			+		+		+	+						S/P
Talc	+	+	+		+		+	+	+			+	+	S/P
Wood flour			+	+					+					S

[a] P = in thermoplastics only; S = in thermosets only; S/P = in both thermoplastics and thermosets.

TABLE 1.20 Influence of Fillers on the Properties of Polyolefins

Material	Filler Content (% by wt)	Density (g/cm^3)	Elongation (%)	Tensile Strength (N/mm^2)	Modulus of Elasticity (N/mm^2)	Ball Indentation (N/mm^2)	Heat Deflection Temperature (1.85 N/mm^2) (°C)	Melt Index 190/5 (g/10 min)
Low density								
Polyethylene	—	0.92	500	10	210	16	35	1.7
Chalk	40	1.26	220		900	—	—	0.2
				16				
Glass fiber	30	1.11	65	24	1,200	33	—	1.6
Asbestos	30	1.13	20	20	670	22	—	2.3
Talc		1.14	40	16	600	30	—	1.1
	30							
Mica	30	1.16	46	13	440	27	—	1.2
Glass spheres (<50 μm)	30	1.11	73	10	290	19	—	2.6
Quartz powder	30	1.14	77	12	400	23	—	2.5
High density								
Polyethylene	—	0.95	>550	27	1,400	46	50	22
Glass fiber	30	1.16	2	60	7,800	68	108	6
Chalk	30	1.17	9	20	1,900	48	—	18
Wood flour	30	—	6	28	2,100	—	—	—
Kaolin	40	1.24	11	26	2,000	—	—	2.5
Asbestos	40	—	10	30	2,100	—	77	0.7
Polypropylene	—	0.90	620	31	1,600	75	62	1.0
Short glass fiber	30	1.20	5	49	7,500	100	120	6.0
Asbestos	40	1.28	7	45	5,000	84	96	4.0
Talc	40	1.21	11	30	3,100	83	95	3.0
Mica	30	—	—	30	6,900	—	—	—
Chalk	40	1.23	180	25	3,400	63	73	2.5
Wood flour	40	—	8	20	2,200	—	—	—

Source: From Gächter, R. and Müller, H., eds. 1987. *Plastics Additives Handbook.* Hanser Publishers, Munich/New York.

interaction with polymer segments. Plasticizers used in practice contain polar and nonpolar groups, and their ratio determines the miscibility of a plasticizer with a given polymer.

Plasticizers for PVC can be divided into two main groups [51] according to their nonpolar part. The first group consists of plasticizers having polar groups attached to aromatic rings and is termed the polar aromatic group. Plasticizers such as phthalic acid esters and tricresyl phosphate belong to this group. An important characteristic of these substances is the presence of the polarizable aromatic ring. It has been suggested that they behave like dipolar molecules and form a link between chlorine atoms belonging to two polymer chains or to two segments of the same chain, as shown in Figure 1.44a. Plasticizers belonging to this group are introduced easily into the polymer matrix. They are characterized by ability to produce gelation rapidly and have a temperature of polymerplasticizer miscibility that is low enough for practical use. These plasticizers are therefore called solvent-type plasticizers, and their kerosene extraction (bleeding) index is very low. They are, however, not recommended for cold-resistant materials.

The second group consists of plasticizers having polar groups attached to aliphatic chains and is called the polar aliphatic group. Examples are aliphatic alcohols and acid or alkyl esters of phosphoric acid (such as trioctyl phosphate). Their polar groups interact with polar sites on polymer molecules, but since their aliphatic part is rather bulky and flexible other polar sites on the polymer chain may be screened by plasticizer molecules. This reduces the extent of intermolecular interactions between neighboring polymer chains, as shown in Figure 1.44b.

Polar aliphatic plasticizers mix less well with polymers than do polar aromatics and, consequently, may exude (bloom) from the plasticized polymer more easily. Their polymer miscibility temperature is higher than that for the first group. These plasticizers are called oil-type plasticizers, and their kerosene extraction index is high. Their plasticization action is, however, more pronounced than that of polar aromatic plasticizers at the same molar concentration. Moreover, since the aliphatic portions of the molecules retain their flexibility over a large temperature range, these plasticizers give a better elasticity to finished products at low temperature, as compared to polar aromatic plasticizers, and allow the production of better cold-resistant materials. In PVC they also cause less coloration under heat exposure.

In practice plasticizers usually belong to an intermediate group. Mixtures of solvents belonging to the two groups discussed above are used as plasticizers to meet the requirements for applications of the plasticized material.

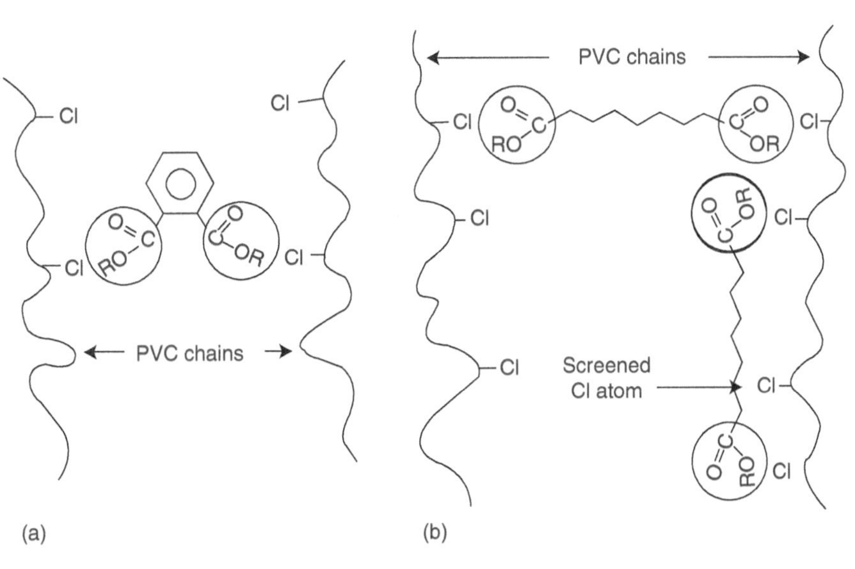

FIGURE 1.44 Action of (a) a polar aromatic plasticizer and (b) a polar aliphatic plasticizer on PVC chains.

Plasticizers can also be divided into groups according to their chemical structure to highlight their special characteristics. Several important plasticizers in each group (with their standard abbreviations) are cited below.

1.22.1 Phthalic Acid Esters

Di(2-ethyl hexyl) phthalate (DOP) and diisooctyl phthalate (DIOP) are largely used for PVC and copolymers of vinyl chloride and vinyl acetate as they have an affinity to these polymers, produce good solvation, and maintain good flexibility of finished products at low temperature. The use of *n*-octyl-*n*-decyl phthalate in the production of plastics materials also allows good flexibility and ductility at low temperature. Diisodecyl phthalate (DDP), octyl decyl phthalate (ODP), and dicapryl phthalate (DCP) have a lower solvency and are therefore used in stable PVC pastes. Butyl octyl phthalate (BOP), butyl decyl phthalate (BDP, and butyl benzyl phthalate (BBP) have a good solvency and are used to adjust melt viscosity and fusion time in the production of high-quality foams. They are highly valued for use in expandable plasticized PVC.

Dibutyl phthalate (DBP) is not convenient for PVC plasticization because of its relatively high volatility. It is a good gelling agent for PVC and vinyl chloride-vinyl acetate copolymer (PVCA) and so is sometimes used as a secondary plasticizer in plasticizer mixers to improve solvation. DBP is mainly used for cellulose-based varnishes and for adhesives. It has a high dissolving capacity for cellulose nitrate (CN).

Dimethyl phthalate (DMP) also has high dissolving capacity for CN. It has good compatibility with cellulose esters and are used in celluloid made from CN and plastic compounds or films made from other cellulosic polymers, cellulose acetate (CA), cellulose acetate-butyrate (CAB), cellulose acetate-propionate (CAP), and cellulose propionate (CP). It is light stable but highly volatile. Diethyl phthalate (DEP) possesses properties similar to DMP and is slightly less volatile.

1.22.2 Phosphoric Acid Esters

Tricresyl phosphate (TCP), trioctyl phosphate, diphenyl 2-ethylhexyl phosphate, and tri(2-ethylhexyl) phosphate (TOP) are used as plasticizers as well as flame retardants. They have a low volatility, resist oil extraction well, and are usually combined with other plasticizers. TCP is a good flame retardant plasticizer for technical PVC, PVAC, NC, and CAB. It is used for PVC articles especially in electrical insulation, but it is not recommended for elastic materials to be used at low temperature. Trioctyl phosphate is a better choice in low temperature applications but it offers a lower resistance to kerosene and oil extraction. Diphenyl 2-ethylhexyl phosphate is a good gelling agent for PVC. It also can be used for materials designed for low temperature application. TOP gels NC, PVCA, and PVC. It has markedly higher volatility than DOP, and it gives plastisols of low viscosity.

1.22.3 Fatty Acid Esters

The esters of aliphatic dicarboxylic acids, mainly adipic, azelaic, and sebacic acid, are used as plasticizers for PVC and PVCA. Di-2-ethylhexyl adipate (DOA), benzyl butyl adipate, di-2-ethylhexyl azelate (DOZ), and di-2-ethyl-hexyl sebacate (DOS) are good examples. They give the polymer outstanding low-temperature resistance and are distinguished by their high light stability. Another characteristic is their low viscosity, which is valuable in the manufacture of PVC plastisols. The solvating action of these esters on PVC at room temperature is weak. This also has a favorable effect on the initial viscosity and storage of plastisols, which is observed even when, for example, plasticizer mixtures of DOP and an aliphatic dicarboxylic ester with DOP contents of up to 80% are used. Such combinations of plasticizers help to improve gelation.

DOS exceeds the low-temperature resistance of all other products in the group. It has the least sensitivity to water and has a relatively low volatility. DOS is used most often because of these properties and because of its high plasticization efficiency.

Monoisopropyl citrate, stearyl citrate, triethyl citrate (TEC), butyl and octyl maleates, and fumarates are other important plasticizers for the preparation of stable PC pastes and of low-temperature-resistant PVC products. Triethyl citrate and triethyl acetylcitrate are among the few plasticizers to have good solvency for cellulose acetate. In comparison with diethyl phthalate, with which they are in competition in this application, there are slight but not serious differences in volatility and water sensitivity. The interest in citrate esters is due to a favorable assessment of their physiological properties. They are intended for plastic components used for packaging of food products.

Butyl and octyl maleates, being unsaturated, are used for copolymerization with vinyl chloride, vinyl acetae, and styrene to provide internal plasticization.

In the production of poly(vinyl butyral), which is used as an adhesive interlayer film between glass plates for safety glass, triethyleneglycol di(2-ethyl) butyrate is an extremely valuable plasticizer that has a proven record of success over many years. It has the required light stability and a plasticizing effect tailored to the requirements of poly(vinyl butyral) that gives the films suitable adhesion to glass ensuring good splinter adhesion over the temperature range from -40 to $+70$°C.

The ethylhexanoic esters of triethylene glycol and polyethylene glycol are good plasticizers for cellulose acetate butyrate (CAB). They, however, have only limited compatibility with poly(vinyl butyral).

Through the action of peracids on esters of unsaturated fatty acids, oxygen adds to the double bound forming an epoxide group. For example, epoxidized oleates have the following structure:

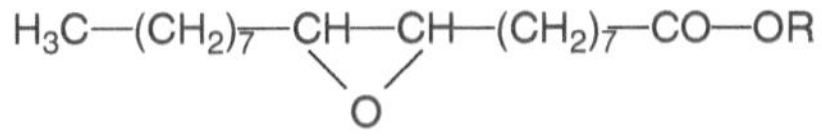

Since the epoxide group reacts with acids, epoxidized fatty acid esters have become popular as plasticizer for PVC with a good stabilizing effect. Such compounds are not only able to bind hydrogen chloride according to reaction:

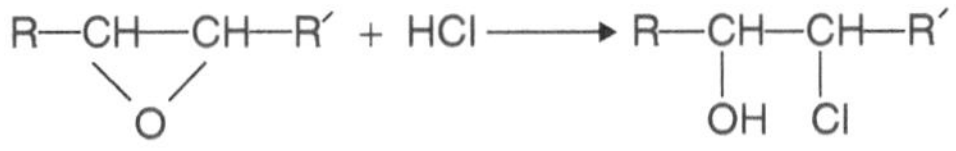

but they are also able to substitute labile chlorine atoms in PVC under the catalytic influence of metal ions;

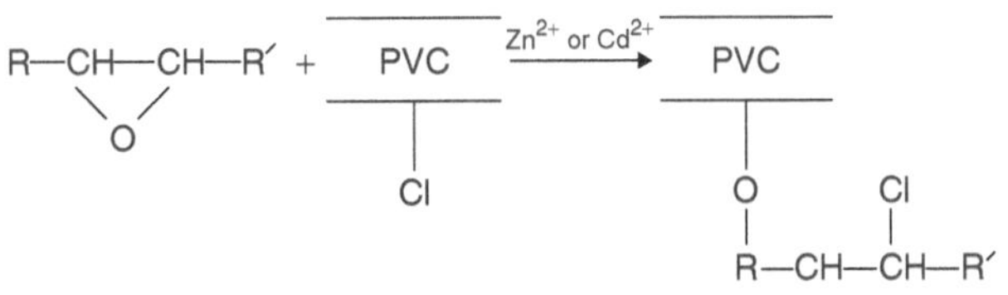

1.22.4 Polymeric Plasticizers

Polymeric plasticizers can be divided into two main types. Oligomers or polymers of molecular weight ranging from 600 to 8,000 belong to the first group. They include poly(ethylene glycol) and polyesters. High-molecular plasticizers comprise the second group.

Poly(ethylene glycol) is used to plasticize proteins, casein, gelatin, and poly(vinyl alcohol). Polyester plasticizers are condensation products of dicarboxylic acid with simple alcohols corresponding to the following two general formulas:

Ac–G–AcD–G–Ac

Al–AcD–G–AcD–Al

where Ac, monocarboxylic acid; G, glycol; AcD, dicarboxylic acid; Al, monofunctional alcohol

In practice, esters from adipic, sebacic, and phthalic acid are frequently used as polyester plasticizers. The value of n may vary from 3 to 40 for adipates and from 3 to 35 for sebacates. Polyester plasticizers are seldom used alone. They are used in combination with monomeric plasticizers to reduce the volatility of the mixed solvents. They offer a higher resistance to plasticizer migration and to extract by kerosene, oils, water, and surfactants. Polyester plasticizers are used specially in PVC-based blends and in nitrocellulose varnishes.

Ethylene-vinyl acetate (EVA) polymers (containing 65%–70% by weight of vinyl acetate) are of industrial interest as high-molecular weight plasticizers for PVC, mainly because of their low cost. A polymeric plasticizer PB-3041 available from Du Pont allows the preparation of a highly permanent plasticized PVC formulation. It is believed to be a terpolymer of ethylene, vinyl, acetate, and carbon monoxide. Also, butylene terephthalate-tetrahydrofuran block copolymers, with the trade name of Hytrel (Du Pont), are used as excellent permanent plasticizers of PVC.

1.22.5 Miscellaneous Plasticizers

Hydrocarbons and chlorinated hydrocarbons (chloroparaffins) belong to the secondary plasticizer type.

Aromatic and aliphatic hydrocarbons are used as extenders, particularly in the manufacture of PVC plastisols that must maintain as stable a viscosity as possible for relatively long periods of time (dip and rotational molding). The petroleum industry offers imprecisely defined products with an aliphatic-aromatic structure for use as extenders. They are added to plastisols in small amounts as viscosity regulators. Dibenzyl toluene also serves the same purpose. A disadvantage of this type of extenders is the risk of exudation if excessive quantities are used.

The normal liquid chlorinated paraffins used as plasticizers for PVC have viscosities ranging from 100 to 40,000 MPa.s at 20°C. Products with chlorine contents ranging from 30 to 70% are on the market. Compatibility with PVC increases with increasing chlorine content but the plasticizing effect is reduced. The low viscosity products (chlorine content 30%–40%) are used as secondary plasticizers for PVC. They have a stabilizing effect on viscosity in plastisols. Chlorinated paraffins can be used up to a maximum of 25% of the total plasticizer content of the PVC plastisol without the risk of exudation. As chlorine-containing substances, these plasticizers also have a flame-retarding effect.

At a chlorine content of 50%, chlorinated paraffins are sufficiently compatible with PVC to be used alone, i.e., as primary plasticizers, in semirigid compounds under certain circumstances. Products with extremely good compatibility, however, have low plasticizing action, requiring larger amounts to be used.

Among other plasticizers, *o*- and *p*-toluene-sulfonamides are used to improve the processibility of urea and melamine resins, and cellulose-based adhesives. For the same reason, *N*-ethyl-*o*-toluenesulfonamide and *N*-ethyl-*p*-toluenesulfonamide are used for polyamides, casein, and zein.

Diglycol benzoates are liquid plasticizers rather like benzyl butyl phthalate (BBP) in their plasticizing action in PVC. In the manufacture of PVC floor coverings, the benzoates offer the advantage of a highly soil resistant surface. In this respect, they are even superior to BBP, which itself performs well here.

Pentaerythritol esters can be used instead of phthalic acid esters when low volatility is one of the major factors in the choice of plasticizer for PVC. Pentaerythriol triacrylate can be used to plasticize PVC and to produce-cross-links under UV radiation. Such plasticized cross-linked PVC has some application in the cable industry.

Commonly used plasticizers for some polymers beside PVC are listed below:

Cellulosics: Dimethyl phthalate (DMP), diethyl phthalate (DEP), dibutyl phthalate (DBP), dimethyl-cyclohexyl phthalate, dimethyl glycol phthalate, trichloroethyl phosphate, cresyldiphenyl phosphate, some glycolates, some sulfonamides.

Poly(vinyl butyrals): Triethyleneglycol di(2-ethyl)butyrate, triethyleneglycol propionate, some adipates, some phosphates.

Polyurethanes: Dibutyl phthalate (DBP), diethyleneglycol dibenzoate (DEGB), and most of the plasticizers used for PVC are preferred for their good plasticization action.

1.23 Antistatic Agents

By virtue of their chemical constitution, most plastics are powerful insulators, a property that makes them useful for many electrical applications. A disadvantage of the insulation property, however, is the accumulation of static electricity, which is not discharged fast enough due to the low surface conductivity of most plastics—a difference between plastics and metals. On plastics and other nonmetallic materials, frictional contact is necessary to generate static charges. On immobile objects, for example, static electricity may build up simply by friction with the ambient air. Electrostatic charges can also be produced on the surfaces of polymeric materials in the course of processing operations such as extrusion, calendaring, and rolling up of plastic sheets or films.

The superficial electrical potential generated by friction may reach values up to a few tens of kilovolts, and this presents serious difficulties for practical applications and to users. Spark formation through heavy charging with subsequent ignition of dust or solvent/air mixtures have been a cause of many destructive explosions. In the service life of plastics products, static charging can give rise to many other troublesome effects, as for example, interference with the sound reproduction of records by dust particles picked up by electrostatic charges, production delays due to clinging of adjacent films or sheets, lump formation during pneumatic transportation, and static build-up on people passing over synthetic fiber carpeting or plastic floor-coverings, with a subsequent "shock" as the charge flows off, usually when the person touches a door handle.

There are many ways to eliminate surface electrostatic, for example, by increasing the humidity or the conductivity of the surrounding atmosphere, or by increasing the electric conductance of materials with the use of electroconducting carbon blacks, powdered metals, or antistatic agents.

Electroconducting carbon blacks are largely utilized to increase the electric conductivity of organic polymers. The electric conductivity of carbon blacks depends, inter alia, on the capacity to form branched structures in the polymer matrix, and on the size and size distribution of carbon black particles. The branched and tentacular structures of carbon in the polymer matrix are responsible for the electric conductivity, as is the case for lamp, acetylene, and furnace carbon blacks. The specific resistance of the carbon particles decreases with their size and then increases with further diminution of the size. A wide size distribution is believed to favor the formation of branched structures contributing to greater conductivity.

In spite of the effectiveness of some carbon blacks in reducing surface charges on plastics materials, the use of antistatic agents have increased steadily. The simplest antistatic agent is water. It is adsorbed on the surface of objects exposed to a humid atmosphere, and it forms a thin electroconducting layer with impurities adsorbed from the air. Such a layer is even formed on the surface of hydrophobic plastics, probably because of the existence of a thin layer of dirt.

Antistatic agents commonly used are substances that are added to plastics molding formulations or to the surface of molded articles in order to minimize the accumulation of static electricity. In general terms, antistatic agents can be divided according to the method of application into external and internal agents.

1.23.1 External Antistatic Agents

External antistatic agents are applied to the surface of the molded article from aqueous or alcoholic solution by wetting, spraying, or soaking the plastic object in solution, followed by drying at room temperature or under hot blown air. Their concentration varies between 0.1 and 2% by weight. Almost all surface active compounds are effective, as well as numerous hygroscopic substances such as glycerin, polyols, and polyglycols, which lack the surface activity feature. The most important external antistatic agents from the practical point of view are quaternary ammonium salts and phosphoric acid derivatives. The advantage of external agents lies in the fact that, in their performance, the properties of compatibility with the polymer and controlled migration in the polymer, which play an important part in the performance of internal antistatic agents described later, are not of any consideration.

It is generally assumed that surface-active molecules accumulate on the surface and are oriented such that the hydrophobic part containing the hydrocarbon chain extends into the plastic, and the hydrophilic part points outwards there it is able to absorb water on the surface. The phase boundary angle between water and plastic is reduced by the surface active antistatic agents thus allowing the absorbed water to be uniformly distributed on the plastic surface. A water film forms on the surface thus increasing the conductivity by means of an ion conduction process. This also explains why surface conductivity, and hence the antistatic action, are found to increase with increasing atmospheric humidity.

For the ion conduction in the surface film, a conductivity mechanism that is similar to protonic conductivity in water has been suggested:

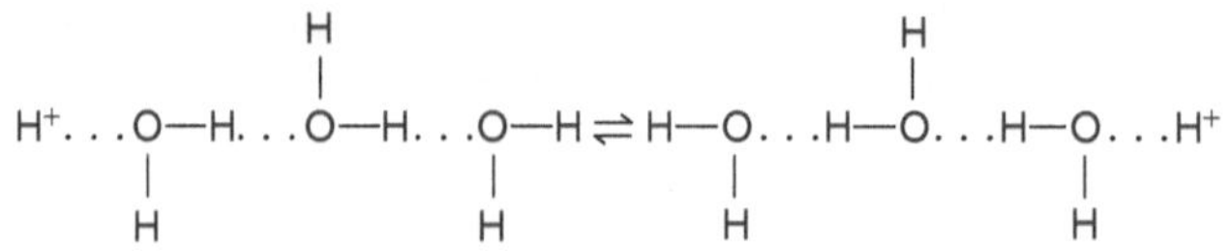

This mechanism is based on a comparison of the conductivities of substances of different chemical structure. For instance, primary amines are efficient as antistatic agents but secondary amines are not. The conductivity of tertiary amines, on the other hand, depends on the nature of *N*-hydroxyalkyl substituents. Among amides, only *N,N*-disubstituted derivatives and mainly those having two hydroxyalkyl substituents are effective. The presence of many OH groups in the molecule makes the efficiency of the antistatic agent less dependent on the humidity. Antistatic agents bearing OH or NH_2 groups are able to associate in chain form via hydrogen bonding and display antistatic activity even a low atmospheric humidity, unlike compounds that are able to form only intramolecular hydrogen bonds.

Many antistatic agents also show hygroscopic properties, thereby intensifying the attraction of water to the surface. At constant atmospheric humidity, a hygroscopic compound attracts more water to the surface and so increases antistatic effectiveness.

1.23.2 Internal Antistatic Agents

Internal antistatic agents are incorporated in the polymer mass as additives, either before or during the molding process. However, to be functional, they must be only partially miscible with the polymer so that they migrate slowly to the surface of the plastics material. Their action therefore appears after a few hours and even a few days after compounding, depending on the mutual miscibility of the agent and the polymer. The concentration of internal antistatic agents in plastics varies from 0.1 to 10% by weight.

Interfacial antistatic agents are all of interfacially active character, the molecule being composed of a hydrophilic and hydrophobic component. The hydrophobic part confers a certain compatibility with the particular polymer and is responsible for anchoring the molecule on the surface, while the hydrophilic part takes care of the binding and exchange of water on the surface.

1.23.3 Chemical Composition of Antistatic Agents

The selection of an antistatic agent for a given polymer is based on its chemical composition. Accordingly, antistatic agents can be divided in many groups, as presented below.

1.23.3.1 Antistatic Agents Containing Nitrogen

Antistatic agents containing nitrogen are mainly amines, amides, and their derivatives, such as amine salts, and addition compounds between oxiranes and aminoalcohols. Pyrrolidone, triazol, and polyamine derivatives also belong to this group. A few representative examples belonging to this group are:

Ethoxylated amines of the type 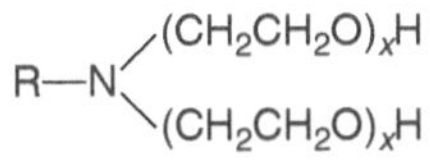

Amine oxides of the type 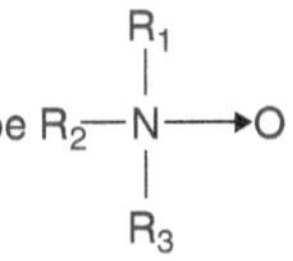

Fatty acid polyglycolamides of the type 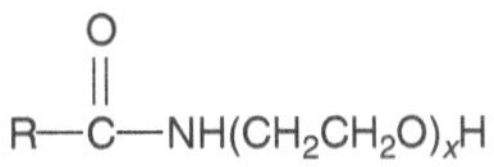

The following two commercial products are typical examples: (1) Alacstat C-2 (Alcolac Chemical Corp., U.S.A.). This is an *N,N*-bis(2-hydroxyethyl)-alkylamine used for polyolefins at 0.1% by weight. (2) Catanac 477 (American Cyanamid Co., U.S.A.). It is *N*-(3-dodecyloxy-2-hydroxypropyl)ethanolamine ($C_{12}H_{25}OCH_2CHOHCH_2NHCH_2CH_2OH$) used for linear polyethylene (0.15% by weight), for polystyrene (1.5% by weight), and for polypropylene (1% by weight). These compounds are not recommended for PVC.

Other compounds containing nitrogen and used as antistatic agents are quaternary ammonium salts, quaternized amines, quaternized heterocycles obtained from imidazoline and pyridine, and amides of quaternized fatty acids. Some of these compounds can be utilized for PVC for example:

Ammonium salts 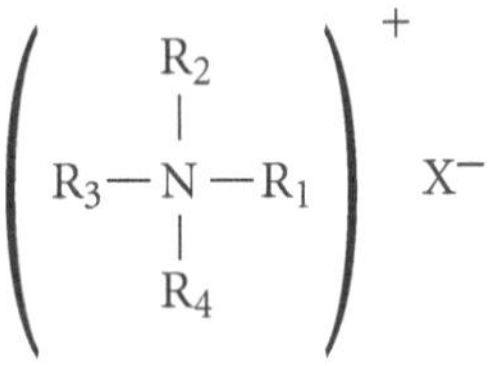

Quaternized ethoxylated amine 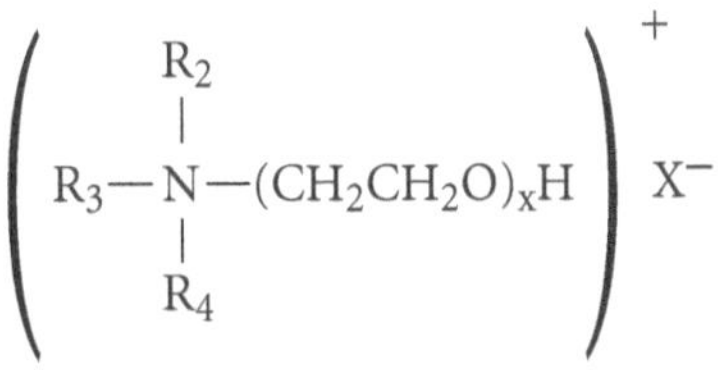

Amide of quaternized fatty acid 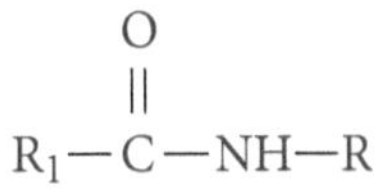

Some typical examples of commercial products are Catanac SN, Catanac 609, Catanac LS, and Catanac SP, all of American Cyanamid Co. Catanac SN is stearamidopropyl-dimethyl-β-hydroxyethylammonium nitrate:

$$\begin{array}{c}\qquad\qquad\qquad\qquad\qquad CH_3 \\ \qquad\qquad\qquad\qquad\qquad | \\ [C_{17}H_{35}CONHCH_2CH_2CH_2{-}N{-}CH_2CH_2OH]^+\ NO_3^- \\ \qquad\qquad\qquad\qquad\qquad | \\ \qquad\qquad\qquad\qquad\qquad CH_3\end{array}$$

It is available as a 50% by weight solution in a water-isopropanol mixture (pH 4–6). It begins to decompose at 180°C and decomposes very rapidly at 250°C. It is soluble in water, acetone, alcohols, and in a series of polar solvents. It is used as an internal and external antistatic agent for rigid PVC, impact-PS, polyacrylates, ABS terpolymers, paper, and textiles. For applying externally, the concentration of the solution is reduced to 1%–10% by weight. The solution can be applied on the surface of plastic products by rubbing them with a flannel cloth impregnated with the solution. For application as an internal antistatic agent, the commercial higher-concentration solution is directly added to the polymer mixture during the milling or mixing process.

Catanac 609 is *N,N*-bis(2-hydroxyethyl)-*N*-(3′-dodecyloxy-2′-hydroxypropyl) methylammonium methylsulfate:

$$\begin{array}{c}\qquad\qquad\qquad CH_2CH_2OH \\ | \\ [C_{12}H_{25}OCH_2CHOHCH_2{-}N{-}CH_3]^+\ CH_3SO_4^- \\ | \\ \qquad\qquad\qquad CH_2CH_2OH\end{array}$$

It is supplied mainly as a 50% by weight solution in a water-propanol mixture. It is applied as an external antistatic agent at 2% by weight concentration (pH 4–6) and is recommended for phonograph records and other products made of PVC and its copolymers.

Catanac LS is 3-lauramidopropyl-trimethylammonium methylsulfate:

$$\begin{array}{c}\qquad\qquad\qquad\qquad CH_3 \\ | \\ [C_{11}H_{23}CONHCH_2CH_2CH_2{-}N{-}CH_3]^+\ CH_3SO_4^- \\ | \\ \qquad\qquad\qquad\qquad CH_3\end{array}$$

It exists in a crystalline form with a melting point of 99°C–103°C.

Catanac SP is stearamidopropyl-dimethyl-β-hydroxyethyl-ammonium dihydrogenphosphate:

$$\begin{array}{c}\qquad\qquad\qquad\qquad CH_3 \\ | \\ [C_{17}H_{35}CONHCH_2CH_2CH_2{-}N{-}CH_2CH_2OH]^+\ H_2PO_4^- \\ | \\ \qquad\qquad\qquad\qquad CH_3\end{array}$$

supplied as a 35% by weight solution in a water-isopropanol mixture (pH 6–8). It begins to compose slightly at 200°C, and at 250°C its decomposition is rapid. It is soluble in water, acetone and alcohols, and is not corrosive to metals even during prolonged contact. Catanac SP can be used as an internal and external antistatic agent at 1%–3% by weight.

1.23.3.2 Antistatic Agents Containing Phosphorus

Antistatic agents that contain phosphorus can be used for all polymers, although they are recommended mainly for PVC with which they also function as plasticizers. They are phosphoric acid derivatives, phosphine oxide, triphosphoric acid derivatives, and substituted phosphoric amides. Typical examples are the following:

$$\text{Phosphoric acid esters } O = P(OR)_3$$

Ethoxylated alcohols and phosphoric acid esters $O = P[O(CH_2CH_2O)_x - R]_3$

Ammonium salts of phosphoric acid esters

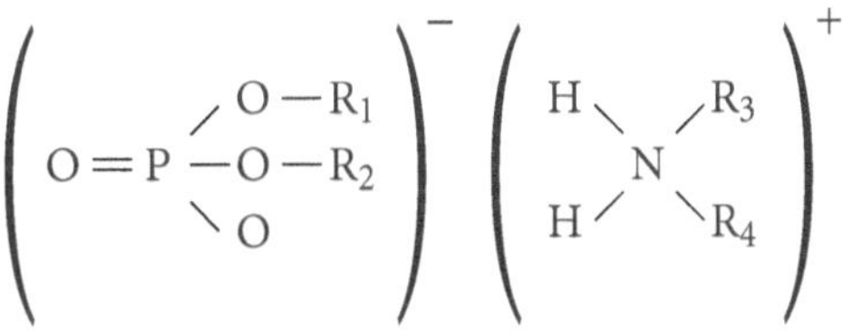

1.23.3.3 Antistatic Agents Containing Sulfur

Antistatic agents that contain sulfur include compounds such as sulfates, sulfonates, derivatives of aminosulfonic acids, condensation products of ethyleneoxide and sulfonamides, and sulfonates of alkylbenzimidazoles, dithiocarbamides, and sulfides.

Sulfur-containing antistatic agents are recommended mainly for PVC and PS because they do not interfere with heat stabilizers. They are not suitable for PMMA, polyolefins, and polyamides. As examples, there are alkylpolyglycolether sulfates, $RO(CH_2CH_2O)_xSO_3^-Me^+$, which are marketed under the trade name Statexan HA (Bayer).

1.23.3.4 Betaine-Type Antistatic Agents

Betaines are trialkyl derivatives of glycine, which exist as dipolar ions of the formula $R_3^+NCH_2CO_2^-$. Typical examples of betaine-type antistatic agents are stearylbetaine and dodecyldimethyl–ethanesulfobetaine. They are used mainly for polyolefins.

1.23.3.5 Nonionic Antistatic Agents

Nonionic antistatic agents are nonionizable, interfacially active compounds of low polarity. The hydrophilic portion of the molecule is usually represented by hydroxyl groups and the hydrophobic portion by organic groups. This group of antistatic agents includes the following: polyhydroxy derivatives of glycine, sugars, and fatty acids, sometimes modified by addition of oxiranes, most of the products being nontoxic and hygienically acceptable but having low antistatic effects; heavy alcohol derivatives such as alkylpolyglycol ethers, which are recommended for polyolefins and PVC; alkylphenolpolyglycol ethers having good heat stability, which are used mainly for polyolefins; polyglycol ethers ether obtained from the reaction of glycols with oxiranes, which are used for polyolefins; and fatty acid polyglycol esters $RCO_2(CH_2CH_2O)_xH$, which are used for many type of plastics.

Nonionic antistatic agents are supplied for the most parting liquid form or as waxes with a low softening region. The low polarity of this class makes its members ideal internal antistatic agents for polyethylene and polypropylene.

A number of substances that cannot be included in any of the groups above are also good antistatic agents. They are silicone copolymers, organotin derivatives, oxazoline derivatives, organoboron derivatives, and perfluorated surfactants. Some of them show excellent h eat stability.

1.24 Flame Retardants

Efforts to develop flame-retarding plastics materials have been focusing on the increasing use of thermoplastics. As a result, flame-retarding formulations are available today for all thermoplastics, which strongly reduce the probability of their burning in the initiating phase of a fire. The possibility of making plastics flame-retardant increases their scope and range of application. (It must be noted, however, that in the case of fire, the effectiveness of flame retardants depends on the period of time and intensity of the fire. Even a product containing the most effective flame-retardants cannot resist a strong and long-lasting fire) [54].

Flame retardants are defined as chemical compounds that modify pyrolysis reactions of polymers or oxidation reactions implied in combustion by slowing them down or by inhibiting them. The combustion of thermo-plastics (see "Fire Resistant Polymers" in Chapter 2 of *Industrial Polymers, Specialty Polymers, and Their Applications*) is initiated by heating the material up to its decomposition point through heat supplied by radiation, flame, or convection. During combustion, free radials are formed by pyrolysis and, besides noncombustible gases such as carbon dioxide and water, combustible gases are formed–mainly hydrocarbons, hydrogen, and carbon monoxide. The radicals combine with oxygen and combustible gases in a radical chain reaction, thereby causing heat release and further decomposition of the plastics material. For continued combustion it is necessary to have sufficient oxygen as well as combustible gaseous compounds.

The combustion will be slowed down or inhibited if free radicals, which are evolved by pyrolysis, are blocked. For example, it is presumed that the following reactions take place when thermoplastics containing organobromine compounds as flame retardants are used:

$$\text{Exothermic propagation} \quad HO^{\cdot} + CO^{\cdot} = CO_2 + H^{\cdot} \tag{1.110}$$

$$\text{Chain branching} \quad H^{\cdot} + O_2 = HO + O^{\cdot} \tag{1.111}$$

$$\text{Chain transfer} \quad O^{\cdot} + HBr = HO^{\cdot} + Br^{\cdot} \tag{1.112}$$

$$\text{Inhibition} \quad HO^{\cdot} + HBr = H_2O + Br^{\cdot} \tag{1.113}$$

$$\text{Regeneration} \quad RH + Br^{\cdot} = R^{\cdot} + HBr \tag{1.114}$$

The radical chain reaction is interrupted when the highly reactive $HO\cdot$ radical, which plays a key role in the combustion process, is replaced by the less reactive $Br\cdot$ radical (Reaction 1.113). While Equation 1.112 and Equation 1.113 show reactions with HBr (which is released during combustion), experiments have shown that halogen-containing organic compounds may also react as undecomposed molecules.

Flame retardants, in view of the above discussion, can be defined as chemical compounds that modify pyrolysis reactions of polymers or oxidation reactions implied in the combustion by slowing these reactions down or by inhibiting them. In practice, mixtures of flame retardants are often used to combine different types of retardancy effects. In spite of a few thousand references on flame retardants, only a small number of compounds are produced commercially. They are mainly chlorine, bromine, phosphorus, antimony, aluminum, and boron-containing compounds. Trade names and manufactures of some selected flame retardants are given in Appendix A6. Most flame retarding agents may be divided into three groups: (1) halogen compounds, (2) phosphorous compounds, and (3) halogen-antimony synergetic mixtures.

1.24.1 Halogen Compounds

As flame retardants, bromine compounds show superiority over chlorine compounds. For example, in case of combustion of n-heptane/air mixture, the minimum concentrations (in vol. %) found necessary for the extinction of flame are 5.2% for dibromomethane and 17.5% for trichloromethane. The efficiency of halogen-containing agents depends on the strength of the carbon-halogen bond. Thus the relative efficiency of aliphatic halides decreases in the following order:

$$\text{Aliphatic bromides} > \text{aliphatic chlorides} = \text{aromatic bromides} > \text{aromatic chlorides}$$

It is also known that aromatic bromides have a higher decomposition temperature (250°C–300°C) than aliphatic bromides (200°C–250°C).

Bromine-containing compounds on a weights basis are at least twice as effective as chlorine-containing ones. Because of the smaller quantities of bromine compounds needed, their used hardly influences the mechanical properties of the base resins and reduces markedly the hydrogen halide content of the combustion gases. Aliphatic, cycloaliphtic, as well as aromatic and aromatic-aliphatic bromine compounds are used as flame retardants.

Bromine compounds with exclusively aromatic-bound bromine are produced in large quantities. Occupying the first place is tetrabromobisphenol-A, which is employed as a reactive flame retardant in polycarbonates and epoxy resins. Similarly, tetrabromophthalic anhydride has been used in the production of flame-retarded unsaturated polyester resins.

Brominated diphenyls and brominated diphenylethers, for instance, octabromodiphenyl ether (melting point 200°C–290°C) and decabromodiphenyl ether (melting point 290°C–310°C), have gained great significance. Possessing excellent heat stability, they are excellent flame retardants for those thermoplastics that have to be processed at high temperatures, such as linear polyesters and ABS.

Aromatic-aliphatic bromine compounds, like the bis(dibromopropyl)-ether of tetrabromobisphenol A, or bromoethylated and aromatic ring-brominated compounds, such as 1,4-bis-(bromoethyl)-tetrabromobenzene, combine the high heat stability of aromatic-bound bromine with the outstanding flame retardancy of aliphatic-bound bromine. They are used mainly as flame retardants for polyolefins, including polyolefin fibers.

Polymeric aromatic bromine compounds, such as poly(tribromostyrene), poly(pentabromobenzyl acrylate) or poly(dibromophenylene oxide) are suitable for application in polyamides, ABS, and polyesters.

Oligomeric bromine-containing ethers on the basis of tetrabromoxylene dihalides and tetrabromobisphenol-A (molecular weight between 3,000 and 7,000) are useful flame retardants for polypropylene, polystyrene, and ABS.

The following cycloaliphatic compounds are suitable as flame retardants in polyolefins: hexabromocyclododecane and bromine-containing Diels-Alder reaction products based on hexachlorocyclopentadiene.

1.24.2 Phosphorus Compounds

Whereas halogen acids, which are released during the combustion of thermoplastics containing organic halogen compounds, interrupt the chain reaction (e.g., Reaction 1.94), the flame-retarding action of phosphorus compounds is not yet fully understood. It is supposed, however, that in case of fire, phosphorus compounds facilitate polymer decomposition, whereby the very stable poly(*meta*-phosphoric acid) formed creates an insulating and protecting surface layer between the polymer and the flame. The phosphoric acid also reacts with the polymer and produces a char layer protecting the surface.

Phosphorus-containing flame retardants may be divided according to the oxidation state of the element into phosphates [$(RO)_3PO$], phsophites [$(RO)_3P$], phosphonites [$(RO)_2PR'$], phosphinates [$(RO)R_2'PO$], phosphines [R_3P], phosphine oxides [R_3PO], and phosphonium salts [R_4PX]. The halogenated phosphorus-containing compounds represent a very important group of flame retardants with high efficiency A few examples are tris-(tribromoneopentyl)-phosphite, tris-(dibrompropyl)-phosphate, tris-(di-bromophenyl)-phosphate, and tris-(tribromophenyl)-phosphate. A similar flame retardancy may also be obtained by the combinations of phosphorus-containing compounds with various halogen derivatives. These combinations are relatively universal in their action.

Phosphorus-containing flame retardants are suitable for polar polymers such as PVC, but for polyolefins their action is not sufficient. In this case they are used together with Sb_2O_3 and with halogenated compounds. Alkyl-substituted aryl phosphates are incorporated into plasticized PVC and modified PPO (Noryl) to a great extent; for other plastics they are less important. Quaternary phosphonium compounds are recommended as flame retardants for ABS and polyolefins.

1.24.3 Halogen–Antimony Synergetic Mixtures

Antimony oxide is an important component of flame retarded thermoplastics containing halogen compounds. Used alone, the flame-retardant effect of antimony oxide is not sufficient, but the effect increases significantly as halogen is added to the system. The basis of this antimony-halogen synergism is probably the formation of volatile flame-retardant SbX_3 (see below), where X is Cl or Br. Antimony-halogen combinations have been widely used in polymer for flame retardation, and the interaction of antimony (mostly as antimony oxide) with halogenated polymers or polymers containing halogenated additives represents the classic case of flame-retardant synergism. Experimental results indicate that the best Sb:X ratio is 1 to 3. More effective are ternary synergetic mixtures such as Sb_2O_3- pentabromo-toluene-chloroparaffin.

The antimony oxide-alkyl (aryl) halide system functions on heating according to the following mechanism (shown for chloride):

$$RCl \rightarrow HCl + R'CH = CH_2 \quad (1.115)$$

$$Sb_2O_3 + 6HCl \rightarrow 2SbCl_3 + 3H_2O \quad (1.116)$$

$$Sb_2O_3 + 2HCl \rightarrow 2SbOCl + H_2O \quad (1.117)$$

Volatile $SbCl_3$ reduces the formation of radicals in the flame and also affects the oxidation process strongly, thus terminating the combustion. As with phosphorus compounds, it promotes carbonization of the polymer. It is believed that SbOCl and $SbCl_3$ already function as dehydrogenation agents in the solid phase of the polymer in flame. At 240°C–280°C, SbOCl is transformed to higher oxychlorides and $SbCl_3$, and at temperatures around 500°C the higher antimony oxychlorides are converted back into antimony oxide by releasing again $SbCl_3$:

$$5SbOCl \xrightarrow{240^\circ C\text{–}280^\circ C} Sb_4O_5Cl_2 + SbCl_3 \quad (1.118)$$

$$4Sb_4O_5Cl \xrightarrow{410^\circ C\text{–}475^\circ C} 5Sb_3O_4Cl + SbCl_3 \quad (1.119)$$

$$3Sb_3O_4Cl \xrightarrow{475^\circ C\text{–}565^\circ C} 4Sb_2O_3 + SbCl_3 \quad (1.120)$$

The flame retardation is also attributed to the consumption of thermal energy by these endothermic reactions, to the inhibition reactions in the flame, and to the release of hydrochloric acid directly in the flame. The pigmentation effect of mixtures containing Sb_2O_3 is a disadvantage that limits their use for light-colored and transparent products.

The demands on a flame retardant and thermoplastics formulated with such agents are manifold. The flame retardant should provide a durable flame-retarding effect by the addition of only small quantities of the additive; it should be as cheap as possible and the manner of incorporation should be easy; it should be nontoxic and should not produce fire effluents with increased toxicity; it should not decompose at the processing temperatures; it should not volatilize and smell; and the mechanical, optical, and physical properties of the thermoplastics should be affected as little as possible. It is understandable that these far-ranging demands cannot be satisfied by only one flame retardant and for all thermoplastics with their manifold applications. One is therefore forced to seek the optimum flame retarding formulation for each thermoplastic and the specific application. Examples of typical formulations of several flame-retarded thermoplastics are given in Appendix A7.

1.24.4 Intumescent Flame Retardants

While halogenated compounds are good fire-retardant additives for polyolefins, especially in synergistic combination with antimony trioxide, there are serious disadvantages, such as evolution of toxic gases and corrosive smoke. In the search for halogen-free flame retardants, intumescent flame-retardants (IFRs) have received considerable attention recently, and they have shown particularly good efficiency in the flame retardation of polypropylene (PP) [55]. IFRs that contain ammonium polyphosphate (APP) and pentaerythritol (petol) work mainly through a condensed phase mechanism [56]. APP is most often used as the acid source and petol as a carbon source. These APP-based products also have a low smoke density and do not emit corrosive gases during combustion.

Recently, the synergistic activity of IFRs with heavy metal ions has received wide attention as this can render an IFR more efficient at lower concentrations [57]. It has been found that divalent metal compounds and their derivatives have an excellent synergistic effect when used together with the polyphosphate and petol. Thus, using nickel formate, $Ni(HCOO)_2$ the limiting oxygen index (LOI) of PP/APP/petol (90: 6.75: 3.25 by wt.) has been found [58] to have the highest value of 30% at 2–3 wt.% Ni $(HCOO)_2$. It is proposed that during burning, the petol is first phosphorylated by APP, with the release of water and ammonia, while the phosphate ester is pyrolyzed with the formation of double bonds that subsequently cross-link, resulting in a three-dimensional structure.

1.25 Smoke Suppressants

In recent years, the hazards of smoke and toxic gas production by burning have received much greater recognition. Since burning of a large amount of polymer represents a fuel rich combustion, it produces large amounts of toxic carbon monoxide and carbon-rich black smoke, causing death and inhibiting escape from the burning environment. A comparison of smoke production from burning different polymers is given in Table 1.21.

The problem of smoke suppression has not yet been solved satisfactorily, because the demands on flame retardancy and reduced smoke development are principally of antagonistic nature. Most of the developmental work on smoke suppressants has been carried out in the field of PVC. For this polymer, highly dispersed aluminum trihydrate is widely used as a flame retardant and as a smoke suppressant. Antimony trioxide-metal borates (barium borate, calcium borate, and zinc borate) are good smoke reducers. Zinc borate of the formula $2ZnO{:}3B_2O_3{:}3.5H_2O$ has shown the best effect with regard to reduced flammability and reduced smoke development of PVC. It is possible to reduce smoke development of rigid PVC significantly by adding rather small quantities of molybdic oxide. For example, 2% of molybdic oxide reduces the smoke density to about 35%, but does not decrease the limiting oxygen index value of rigid PVC.

Several methods have been proposed for the determination of the smoke density. Reproducible values are obtained with the laboratory method worked out by the National Bureau of Standards, whereby a specimen of 7.6×7.6 cm is heated by radiation. The smoke density is then measured optically.

1.26 Colorants

A wide variety of inorganic acid organic materials are added to polymers to impart color. For transparent colored plastics materials, oil-soluble dyes or organic pigments (such as phthalocyanines) having small particle size and refractive index near that of the plastic are used. Others, including inorganic pigments, impart opaque color to the plastic. Some of the common colorants for plastics, among many others, are barium sulfate and titanium dioxide (white), ultramarine blues, phthalocyanine blues and greens, chrome green, molybdate organs, chrome yellow, cadmium reds and yellows, quinacridone reds and

TABLE 1.21 Smoke Emission on Burning (NBS Smoke Chamber, Flaming Condition)

Polymer	Maximum Smoke Density (D_{max})
ABS	800
Poly(vinyl chloride)	520
Polystyrene	475
Polycarbonate	215
Polyamide-imide	169
Polyarylate	109
Polytetrafluoroethylene	95
Polyethersulfone	50
Polyether ether ketone	35

magentas, and carbon black. Flake aluminum is added for a silver metallic appearance, and lead carbonate or mica for pearlescence.

The principal requirements of a colorant are that it have a high covering power-cost ratio and that it withstand processing and service conditions of the plastic. The colorant must not catalyze oxidation of the polymer nor adversely affect its desirable properties. Colorants are normally added to the powered plastic and mixed by tumbling and compounding on a hot roll or in an extruder.

1.27 Antimicrobials

Antimicrobials impart protection against mold, mildew, fungi, and bacterial growth to materials. Without antimicrobials, polymers can experience surface growths, producing allergic reactions, unpleasant odor, staining, embrittlement, and even permanent product failure. The effectiveness of an antimicrobial depends on its ability to migrate to the surface of the product where microbial attack first occurs. Antimicrobials are thus usually carried in plasticizers which are mobile. High mobility, however, can result in leaching of the additives due to weather exposure. The durability of protection is, therefore, determined by a proper balance between migration and leaching.

Of the hundreds of chemicals that are effective as antimicrobials, only a few are used in plastics applications. The most common group of active ingredients consists of 10,16-oxybisphenoxy arsine (OBPA, tradenames: Intercide, Akzo Chemical; Vinyzene, Morton International), 2-*n*-octyl-4-isothia-zolin-3-one (tradenames: Micro-Check 11, Ferro Corporation; Vinyzene IT, Morton International), *N*-trichloro-methylthio-4-cyclohexene-1,2-dicarboximide (Captan, tradenames: Vancide 89, R.T. Vanderbilt; Fungitrol C, Hüls America), and *N*-(trichloromethylthio) phthalimide (Flopet, trade name; Fungitrol 11, Hüls America). Other antimicrobial active ingredients include amine-neutralized phosphate and zinc 2-pyridinethiol-1-oxide.

OBPA is the most toxic and requires approximately 0.04% active ingredient in the final product. Less toxic active ingredients such as Flopet require a loading of 1.0% to achieve a similar level of protection. Antimicrobials are generally formulated with a carrier into concentrations of 2–5% active ingredient and are available to plastic processors in powder, liquid, or solid-pellet form. The carrier is usually a plasticizer, such as epoxidized soybean oil or diisodecyl phthalate.

Among the antimicrobials, OBPA has broad-spectrum effectiveness against gram-positive and gram-negative bacteria and fungi. Another additive, 2-*n*-octyl-4-isothiazolin-3-one, is a fungicide only and is commonly used to prevent molds in paints. Captan is an agricultural fungicide that has important plastics applications.

The majority of all antimicrobial additives are used in flexible PVC. The remaining portion is used in polyurethane foams and other resins. PVC applications using antimicrobials include flooring, garden hose, pool liners, and wall coverings.

1.28 Toxicity of Plastics

With the expanding use of plastics in all walks of life, including clothing, food packaging, and medical and paramedical applications, attention has been focused on the potential toxic liability of these man-made materials.

Toxic chemicals can enter the body in various ways, particularly by skin absorption, inhalation, and swallowing. Although some chemicals may have an almost universal effect on humans, others may attack few persons. The monomers used in the synthesis of many of the polymers are unsaturated compounds with reactive groups such as vinyl, styrene, acrylic, epoxy, and ethylene imine groups. Such compounds can irritate the mucous membranes of the eyes and respiratory tract and sensitize the skin. They are suspected of including chemical lesions and carcinogenic and mutagenic effects. It was long ago that carcinogenic properties of the monomers vinylchloride and chloroprene were reported, and there may be more trouble ahead in this respect. However, the monomers used in the production of polyester or polyamide resins are usually less reactive and may be expected to be less harmful.

Although many monomers are harmful chemicals, the polymers synthesized from them are usually harmless macromolecules. But then one has to take into account that the polymers may still contain small amounts of residual monomer and catalyst used in the polymerization process. Moreover, polymers are rarely used as such but are compounded, as we saw earlier, with various additives such as plasticizers, stabilizers, curing agents, etc., for processing into plastic good. Being relatively smaller in molecular size, the residual monomer, catalyst residues, and the additives can migrate from the plastic body into the environment. They may eventually migrate into food products packed in plastic containers, or they may interact with the biological substrate when plastics materials are used as parts of tissue and organs implanted into humans. The toxic potential of thermodegradation and combustion products of plastics when these materials are burned, either deliberately or by accident, is also an important consideration in view of the widening use of plastics as structural and decorative lining material in buildings, vehicles, and aircraft.

1.28.1 Plastic Devices in Pharmacy and Medicine

Within the past two decades there has been an increase in the use of a variety of plastic materials in the pharmaceutical, medical, dental, and biochemical engineering fields. Such plastic devices can be classified according to use into six basic groups: (1) collection and administrative devices—e.g., catheters, blood transfusion sets, dialyzing units, injection devices; (2) storage devices—e.g., bags for blood and blood products, containers for drug products, nutritional products, diagnostic agents; (3) protective and supportive devices—e.g., protective clothes, braces, films; (4) implants having contact with mucosal tissue—e.g., dentures, contact lenses, intrauterine devices; (5) permanent implants—e.g., orthopedic implants, heart valves, vascular grafts, artificial organs, and (6) nanomedicines and drug delivery.

The toxic potential of plastic devices becomes more relevant in those applications which involve long periods of contact with the substrate. For short periods of contact, however, a great number of plastic materials manufactured today as medical and paramedical devices will produce little or no irritant response.

With polyethylene, polypropylene, Teflon, Dacron, polycarbonate, and certain types of silicone rubbers, the migration of additives from the material is so small that a biological response cannot be detected. Besides, many of these materials contain additives only in extremely low concentrations, some perhaps having only a stabilizing agent in concentrations of less than 0.1 wt.%. Incidence of tissue response increases when the plastic materials require greater concentrations of various types of additives. It should be stressed, however, that the toxic effects of a material will depend on the intrinsic toxicity of the additives and their ability to migrate from the material to the tissue in contact with it. It is important to know precisely the toxic potential of each of the ingredients in the final polymerized and formulated material to be used in an implantable or storage device.

1.28.1.1 Packing

Many pharmaceutical manufacturers have now taken to packaging their drug product in plastic containers. The plastics used most in these applications are those manufactured from polyethylene and polypropylene, though other materials have also been used. Since both polyethylene and polypropylene contain extremely small amounts of additives (mostly as antioxidants and antistatic agents), the possibility of their release in sufficient concentrations to endanger the patient is negligible.

1.28.1.2 Tubings and Blood Bag Assemblies

Plastic tubings are used for many medical applications, such as catheters, parts of dialysis and administration devices, and other items requiring clear, flexible tubings. The most successful tubing is PVC, which has been plasticized to give it the desired flexibility. Plasticized PVC is also the chief material used in America for making bags for blood and products. In Europe polyethylene and polypropylene have been mainly used. The plasticizers used most for flexible PVC are the esters of phthalic acid; one that is most employed from this group is di-2-ethylhexyl phthalate. Long-term feeding studies have demonstrated the nontoxic nature of the plasticizer under the experimental conditions used. Organotin compounds, which are one of the best groups of stabilizers for vinyl polymers, have been used to stabilize PVC, but, in general, their toxicity has decreased their used in medical applications.

1.28.1.3 Implants

Man-made materials have been used for making implants to save human life. It is possible that these materials, depending on their specific nature and the site of implant, will degrade with long periods of contact and release polymer fragments in the body. These in turn may elicit one or more biological responses, including carcinogenesis. However, no well-substantiated evidence has been reported showing that man-made materials have caused cancer in humans, although the real answer will be available only when these materials have been implanted for periods of time exceeding 20 to 30 years.

Perhaps the most commonly used implantable material is silicone rubber. This material, if properly prepared by the manufacturer, does not cause local toxic response. Various types of epoxy polymers and polyurethane materials also have found one or more medical applications.

1.28.1.4 Adhesives

In most medical and paramedical applications of plastics, the materials used are those already produced by the manufacturers in a polymerized or formulated form. Certain surgical and dental applications, however, require that the material be polymerized or formulated just prior to use. Surgical cements and adhesives, a host of dental filling materials, materials for dentures, cavity liners, and protective coatings for tooth surfaces are in this category.

Cyanoacrylates have become very useful as tissue adhesives in surgical applications, because they polymerize rapidly in contact with moisture and create an extremely tenacious film. Methyl cyanoacrylate, used initially, has now fallen out of favor because of its toxic properties. The butyl and heptyl analogs are, however, quite satisfactory and do not produce objectionable tissue response. They also degrade at a much slower rate than the methyl compound in a biological environment.

1.28.1.5 Dental Materials

Sometimes known as white filling or synthetic porcelain, polymeric dental composites are commonly used as a tooth-colored restorative material, for example, in the fabrication of fillings and veneers, and the cementation of crowns. Dental composites are complex mixtures generally containing an organic resin matrix, reinforcing inorganic filler (such as Sr-glass, 0.7 μm), a silane coupling agent, which connects the filler and the resin matrix, and stabilizers to maximize storage [59]. Composites without the filler and coupling agent are commonly used as sealants, which effectively isolate pits and fissures to help prevent caries.

The dental resins commonly based on the highly viscous bisphenol A glycidyl methacrylate (abbreviated as bis-GMA, also known as Bowen's monomer after the inventor) have been used for over four decades. Because of its high viscosity (1,369 Pa.s), bis-GMA is blended with diluent, lower molecular weight monomers, to provide a workable matrix resin for composites, for example, bisphenol-A dimethacrylate (bis-DMA), ethylene glycol dimethacrylate (EGDMA), and triethylene glycol dimethacrylate (TEGDMA). Camphorquinone (CQ) is traditionally used as photosensitizer for dental composites. It undergoes a redox reaction with a tertiary amine to produce radicals for free-radical polymerization of the acrylate resin. In a typical procedure, a mixture of bis-GMA and TEGDMA with 1 wt% of CQ and tertiary amine is photopolymerized using visible light device with an intensity of 900 mW/cm^2. The photopolymerization of the dental resin is fast with most of the reaction taking place within 40 s, causing gelation and vitrification accompanied by shrinkage of about 8%. Shrinkage continues, as also stiffening, during post-polymerization at a much slower rate (see Table 1.22).

Based on data reported in several studies involving application of sealant to teeth, it appears that low levels of bisphenol A (BPA) may be released from certain sealants, although only during a short time period immediately after application of the sealant. Further, no detectable levels of BPA have been found in blood after application of a sealant that releases low levels of BPA into saliva and the sealant possesses no risk to human health [60].

1.28.1.6 Nanomedicines and Drug Delivery

The term "nano" refers to all molecules and devices/technologies in the size range 1–1,000 nm (a nanometer is 10^{-9} m). Nanomedicine has been defined as "the science and technology of diagnosing, treating, and preventing disease, relieving pain, and preserving and improving human health by using molecular tools and molecular knowledge of the human body." New drugs and new delivery systems both thus come under the umbrella of nanomedicines. Innovative devices and drug delivery systems are needed to exploit many of the drugs developed from advances in molecular biology, to guide the therapeutic to its correct location of action and ensure that pharmacological activity is maintained for an adequate duration once it is there. In many cases, water soluble polymers are used to carry the active drug into the body without affecting the sites other than the target sites, where the local environment breaks down the polymer releasing the active drug.

"Nanoparticles" are solid colloidal polymeric carriers (less than 1 μm in size) that have received much attention over the recent years due to their ability to control drug release and distribution and due to their biodegradability [62]. Furthermore, these systems have proven their potential to administer peptides or other drugs either by intravenous or oral routes, increasing their bioavailability and protection of the drug against degradation, and reducing the associated adverse effects [63,64].

Most commonly used methods for preparing polymer-based nanoparticles include emulsion evaporation [65], in situ monomer polymerization [66], a method based on the salting out effect [67] and nanoprecipitation [68]. The last named method represents an easy and reproducible technique, which has been widely explored for producing both vesicle and matrix type nanoparticles, also termed nanocapsule and nanosphere, respectively [69]. For the last decade, surfactant-free nanoparticles formation has been investigated by many researchers. Fessi et al. [68] developed surfactant-free nanocapsules of poly(D,L-lactide) (PLA) by the nanoprecipitation technique, using a novel and simple procedure which involved interfacial deposition of a preformed, well-defined and biodegradable polymer following displacement of a semi-polar solvent miscible with water from a lipophilic solution. Nanocapsules of PLA containing indomethacin as a drug model were thus prepared [68]. More recently, poly(lactide-*co*-glycolide) nanoparticles were prepared by using poly(ethylene glycol)-based block copolymers as substitutes for conventional surfactants [70].

Progress in the development of nano-sized hybrid therapeutics and nano-sized drug delivery systems over the last decade has been remarkable. A growing number of products have already secured regulatory authority approval and, in turn, are supported by a healthy clinical development pipeline. They include products used to treat cancer, hepatitis, muscular sclerosis, and growth hormone deficiency. (See Therapeutics Packaging and Nanomedicines in Chapter 3 of *Industrial Polymers, Specialty Polymers, and Their Applications.*)

TABLE 1.22 Young's Modulus, Density, and Shrinkage of Dental Resin Matrix after Photopolymerization

Time after Photopolymerization (s)	Young Modulus (GPa)	Density (g/cm^3)	Shrinkage (%)
0	2.22	1.203	8.02
24	3.27	1.220	9.44
48	3.31	1.222	9.67

Source: From Truffier-Boutry, D., Demonstier-Champagne, S., Devaux, J., Brebuyck, J.-J., Larbanois, P., Mestdagh, M., and Leloup, G. 2005. *Eur. Cell Mater.*, 9, Suppl. 1, 60.

1.28.2 Biodegradable Plastics

The ability to undergo biodegradation producing nontoxic by-products is a useful property for some medical applications. Biodegradable polymers [71] have been formulated for uses such as sutures, vascular grafts, drug delivery devices, and scaffolds for tissue regeneration, artificial skin, orthopedic implants, and others. The polymers commonly known in the medical field for such applications include poly(α-hydroxy esters), poly(ε-caprolactone), poly(ortho esters), polyanhydrides, poly(3--hydroxybutyrate), polyphosphazenes, polydioxanones, and polyoxalates (see Chapter 2 of *Industrial Polymers, Specialty Polymers, and Their Applications*).

The homopolymers poly(*L*-lactic acid) (PLLA) and poly(glycolic acid) (PGA) as well as poly (*DL*-lactic-*co*-glycolic acid) (PLGA) copolymers are poly(α-hydroxy esters). They are biocompatible, biodegradable and are among the few biodegradable polymers with Food and Drug Administration (FDA) approval for human clinical use. PLLA, PLGA copolymers, and PGA are produced by a catalyzed ring—opening polymerization of lactide and/or glycolide. Their degradation in vivo is brought about by simple, non-enzymatic hydrolysis of the ester bond leading, eventually, to lactic acid and/or glycolic acid. These acids are processed through normal metabolic pathways and eliminated from the body ultimately through the respiratory system as carbon dioxide. PGA is a highly crystalline, hydrophilic poly(α-hydroxy ester) and was one of the first synthetic degradable polymers to find application as a biomaterial. One of the first applications of PGA was as a biodegradable suture material. PLLA is one of the strongest known biodegradable polymers and has therefore found many applications in areas, such as orthopedics.

Although organ transplantation has saved many lives, the harsh reality remains that the need for donor organs far outweighs the supply. It is recognized that tissue engineering may provide an alternative to organ transplantation. Tissue engineering involves the creation of natural tissue with the ability to restore missing organ function. This may be achieved either by transplanting cells seeded into a porous material or, in some cases, by relying on ingrowth of tissue and cells into such a material. PGA, PLLA and PLGA copolymers have been used as an artificial scaffold [72] in cell transplantation and organ regeneration (see Chapter 2 of *Industrial Polymers, Specialty Polymers, and Their Applications*).

1.28.3 Oxo-Biodegradable Plastics

Discarded conventional plastics remain in the environment for decades. They block sewers and drains, disfigure the streets, beaches and countryside, and kill wildlife on land, in rivers and oceans. To overcome these problems increasing attention has been paid to the development of degradable plastics: (1) starch-based, biodegradable, (2) aliphatic polyesters, biodegradable, (3) photodegradable, and (4) oxo-biodegradable.

The starch-based plastics do not degrade totally, since only the starch constituent is consumed by microbial activity, and the plastic residues can be harmful to the soil and to birds and insects.

Aliphatic polyesters, described above, are relatively expensive. In the same way as starch, they rely on microbial activity in compost or soil to degrade. Both these products degrade by a process of hydro-degradation.

Photodegradable plastics degrade after prolonged exposure to sunlight, so will not degrade if buried in a landfill, a compost heap, or other dark environment, or if heavily overprinted. Oxo-biodegradable plastics (Symphony Plastic Technologies, Borehamwood, Herts, U.K.) are a new development. The plastics degrade by a process of oxo-degradation. The technology is based on a small amount (typically 3%) of pro-degradent additive (proprietary) being introduced into the conventional manufacturing process, thereby making the plastic degradable. The degradation, which does not rely on microbes, starts immediately after manufacture and accelerate when exposed to heat, light or stress. The process is irreversible and continues until the whole material reduces to CO_2 and water, leaving no fragments in the soil. The oxo-biodegradable plastics (OBPs) are also consumed by bacteria and fungi after the additive has broken down the molecular structure sufficiently, allowing access to living micro-organisms. So organic wastes in homes, restaurants, hospitals etc. can be put into oxo-biodegradable plastic sacks for disposing straight into the composting plant without emptying the sacks. Even if OBPs were eaten by cow, deer, turtle or other animal whilst still intact, they would degrade even faster due to the temperature and bacteria present in the gut without causing blockage, unlike conventional plastic bags which could kill the animal.

OBPs have an advantage over plastics produced from starch or other agricultural products in that they biodegrade and can be composed but do not need to be buried in a compost heap or landfill in order to degrade. The fact that they can degrade in a normal environment is a significant factor in relation to litter, because a large amount of plastic waste on land and at sea cannot be collected and buried. OBPs can be made transparent and can be used for direct food contact. The length of time OBP takes to degrade totally can be "programmed" at the time of manufacture by varying the amount of additive and can be as little as a few months or as much as a few years.

1.28.4 Toxicity of Plastic Combustion Products

Fires involving plastics produce not only smoke but also other pyrolysis and combustion products. Since most of the polymeric materials contain carbon, carbon monoxide is one of the products generated from the heating and burning of these materials. Depending on the material, the temperature, and the presence or absence of oxygen, other harmful gases may also evolve. These include HCl, HCN, NO_2, SO_2, and fluorinated gases. Presently, various flame-retarding agents are added to plastics to reduce their combustion properties. When heated or subjected to flame, these agents can change the composition of the degradation products and may produce toxic responses not originally anticipated [73].

1.28.5 Toxicity Testing

Ideally, materials for medical and paramedical applications should be tested or evaluated at three level: (1) on the ingredients used to make the basic resin, (2) on the final plastic or elastomeric material, and (3) on the final device. Organizations such as the Association for the Advancement of Medical Instrumentation, the U.S.A. Standard Institute, and the American Society for Testing and Materials (F4 Committee) have developed toxicity testing programs for materials used in medical applications. The American Dental Association has recommended standard procedures for biological evaluation of dental materials [74].

References

1. Seymour, R. B. and Carraher, C. E. 1992. *Polymer Chemistry, An Introduction.* Marcel Dekker, New York.
2. Billmeyer, F. W. 1969. *Appl. Polymer Symposia*, 10, 1.
3. Debye, P. 1947. *J. Phys. Coll. Chem.*, 51, 18.
4. Price, B. A., Halwer, M., and Speiser, R. 1950. *J. Opt. Soc. Am.*, 40, 768.
5. Onyon, P. F. 1951. Viscometry, Chap. 6. In *Techniques of Polymer Characterization*, P. W. Allen, ed. Butterworths, London, UK.
6. Mayerhoff, G. 1961. *Fortschr. Hochpolym. Forsch.*, 3, 59.
7. Patat, F. and Sinn, H. 1958. *Angew. Chem.*, 70, 496.

8. Cossee, P. 1967. The mechanism of Ziegler-Natta polymerization II. Quantum-chemical and crystal-chemical aspects, Chap. 3. In *The Stereochemistry of Macromolecules, Vol. 6*, A. D. Ketley, ed., Marcel Dekker, New York.
9. Brunsveld, L., Folmer, B. J. B., Meijer, E. W., and Sijbesma, R. P. 2001. *Chem. Rev.*, 101, 4071.
10. Sijbesma, R. P., Beijer, F. H., Brunsveld, L., Folmer, B. J. B., Hirschberg, J. H. K., Lange, R. F. M., Lowe, J. K. L., and Meijer, E. W. 1997. *Science*, 278, 1601.
11. Folmer, B. J. B., Sijbesma, R. P., Versteegen, R. M., van der Rijt, J. A. J., and Meijer, E. W. 2000. *Adv. Mater.*, 12, 12, 874.
12. Lenz, R. W. 1970. *Ind. Eng. Chem.*, 62, 2, 54.
13. Wall, F. T. 1944. *J. Am. Chem. Soc.*, 66, 2050.
14. Flory, P. J. 1944. *Chem. Rev.*, 35, 51.
15. Nielsen, L. C. 1969. *J. Macromol. Sci. Rev. Macromol. Chem. C369.*
16. Boyer, R. F. 1963. *Rubber Chem. Tech.*, 36, 1303.
17. Boyer, R. F. 1954. *J. Appl. Phys.*, 25, 825.
18. Beamen, R. G. 1952. *J. Polym. Sci.*, 9, 470.
19. Mark, H. F. 1967. *Sci. Am.*, 217(3), Sept. 19, 148, see also p.156.
20. Overberger, C. G. and Moore, J. A. 1970. *Fortschr. Hochpolym. Forsch.*, 7, 113.
21. Hildebrand, J. H., Prausnitz, J. M., and Scott, R. L. 1970. *Regular and Related Solutions*, Van Nostrand Reinhold, New York.
22. Burrell, H. 1955. *Official Dig.*, 27, 726.
23. Nelson, R. C., Hemwall, R. W., and Edwards, G. D. 1970. *J. Paint Technol.*, 42, 636.
24. Small, P. A. 1953. *J. Appl. Chem.*, 3, 71.
25. Hoy, K. L. 1970. *J. Paint Technol.*, 42, 76.
26. Brandrup, J. and Immergut, E. eds. 1975. *Polymer Handbook*, p. 251. Wiley Interscience, New York.
27. Burrell, H. 1975. Solubility parameter values, In *Polymer Handbook, 2nd Ed.*, J. Brandrup and E. Immergut, eds., p. 251. Wiley Interscience, New York, pp. IV-337–359.
28. Krause, S. 1972. *J. Macromol. Sci. Macromol. Rev.*, C7, 251.
29. Hansen, C. M. 1967. *J. Paint Technol.*, 39, 104 see also p. 511.
30. Barton, A. F. M. 1983. *Handbook of Solubility Parameters and Other Cohesion Parameters.* CRC Press, Boca Raton, FL.
31. Hansen, C. M. and Beerbower, A. 1971. *Kirk Othmer Encycl. Chem. Technol., 2nd Ed.*, A. Standen, ed., Suppl. 889, 910, Wiley Interscience, New York.
32. Teas, J. P. 1968. *J. Paint Technol.*, 40, 519.
33. Bagley, E. B., Nielsen, T. P., and Scigliano, J. M. 1971. *J. Paint Technol.*, 43, 35.
34. Anon. 1968. *Plastics World*, 26, 3, 30.
35. Lee, H., Stoffey, D., and Neville, K. 1967. *New Linear Polymers*, McGraw-Hill, New York.
36. Pielichowski, K. and Njuguna, J. 2005. . *Thermal Degradation of Polymeric Materials. Rapra. Technology, Schrewsbury, Shoropshire, UK.*
37. Bhuiyan, A. L. 1982. *Adv. Polym. Sci.*, 47, 43.
38. Schmidt, D. L. *Mod. Plastics*, 37, 131 (Nov.1960), 147 (Dec. 1960).
39. Kuryla, W. C. and Papa, A. J. eds. 1973. *Flame Retardancy of Polymeric Materials*, 3, p. 129. Marcel Dekker, New York.
40. Halim Hamid, S., Amin, M. B., and Maadhah, A. G. eds. 1992. *Handbook of Polymer Degradation*, p. 129. Marcel Dekker, New York.
41. Rabek, J. F. 1996. *Photodegradation of Polymers*, Springer, Berlin-Heidelberg, NY.
42. Rivaton, A., Gardette, J-L. , Mailhot, B., and Morlat-Therlas, S. 2005. *Macromol. Symp.*, 225, 129.
43. Grassie, N. and Scott, G. 1985. *Polymer Degradation and Stabilization.* Cambridge Univ. Press, Cambridge.
44. Hawkins, W. L. 1972. *Polymer Stabilization.* Wiley, New York.
45. Gächter, R. and Müller, H. eds. 1987. *Plastics Additives Handbook*, p. 89. Hanser Publishers, Munich/New York.

46. Henman, T. J. 1982. *World Index of Polyolefin Stabilizers*, Kogan Page Ltd., London, UK.
47. Ranby, B. and Rabek, J. F. 1975. *Photodegradation, Photooxidation, and Photostabilization of Polymers*, Interscience, New York.
48. Massey, L. 2002. *Permeability Properties of Plastics and Elastomers, 2nd Ed.* ChemTec Pullishing, Toronto-Scarborough.
49. Stannett, V. T. and Szwarc, M. 1955. *J. Polym. Sci.*, 16, 89.
50. Paine, F. A. 1962. *J. Roy. Inst. Chem.*, 86, 263.
51. Seymour, R. B. 1978. *Additives for Plastics., Vol. 2.*, Academic Press, New York
52. Stepak, J. and Daoust, H. 1982. *Additives for Plastics*, Springer, New York.
53. Wypych, G. 2004. *Handbook of Plasticizers*, ChemTec Publishing, Toronto-Scarborough.
54. Dufton, P. W. 2003. *Flame Retardants for Plastics*, ChemTec Publishing, Toronto-Scarborough.
55. Camino, G. and Delobel, R. 2000. *Fire Retardancy of Polymeric Materials*, A. F. Grand and C. A. Wilke, eds., p. 217. Marcel Dekker, New York.
56. Baljinder, K. K. and Horrocks, A. R. 1996. *Polym. Degrad. Stab.*, 54, 289.
57. Lewin, M. and Endo, M. 2003. *Polym. Adv. Technol.*, 14, 3.
58. Chen, X., Ding, Y., and Tang, T. 2005. *Polym. Int.*, 54, 904.
59. Soderholm, K. and Mariotti, A. 1999. *J. Am. Dent. Assoc.*, 130, 201.
60. Fung, E. Y. K., Ewoldsen, N. O., St. Germain, H. S., Marx, D. B., Miaw, C., Siew, C., Chou, H., Grunniger, S. E., and Meyer, D. M. 2000. *J. Am. Dent. Assoc.*, 131, 51.
61. Truffier-Boutry, D., Demonstier-Champagne, S., Devaux, J., Brebuyck, J-J. , Larbanois, P., Mestdagh, M., and Leloup, G. 2005. *Eur. Cell Mater.*, 9, Suppl. 1, 60.
62. Kreuter, J. 1991. *J. Controlled Release*, 16, 169.
63. Maincent, P., Le Verge, R., Sado, P., Couvreur, P., and Devissaguet, J. P. 1986. *J. Pharma. Sci.*, 75, 955.
64. Couvreur, P., Grislain, L., Lenaerts, V., Brasseur, P., Guiot, P., and Biernacki, A. 1986. *Polymeric Nanoparticles and Microspheres*, P. Guiot and P. Couvreir, eds., pp. 27–93. CRC Press Inc., Boca Raton.
65. Zambaux, M. F., Bonneaux, F., Gref, R., Maincent, P., Dellacherie, E., Alonso, M. J., Labrude, P., and Vigneron, C. 1998. *J. Controlled Release*, 50, 31.
66. Hármia, H., Speiser, P., and Kreuter, J. 1986. *J. Microencapsul.*, 3, 3.
67. Alléman, E. Gumy, E., and Doelkar, E. 1992. *Int. J. Pharm.*, 87, 247.
68. Fessi, H., Puisieux, F., Devissaguet, J. -P., Ammoury, N., and Benita, S. 1989. *Int. J. Pharm.*, 55, R1.
69. Chacon, M., Berges, L., Molpaceres, J., Aberturass, M. R., and Guzman, M. 1996. *Int. J. Pharm.*, 141, 81.
70. Chacon, M., Berges, L., Molpaceres, J., Aberturass, M. R., and Guzman, M. 1996. *Int. J. Pharm.*, 141, 81.
71. Siggs, L. J. and Mikos, A. G. 1996. Synthetic biodegradable polymers for medical applications. In *Physical Properties of Polymers Handbook*, J. E. Mark, ed., Chap. 44, Am. Inst. of Physics, Woodbury, New York.
72. Thomson, R. C., Wake, M. C., Yaszemski, M. J., and Mikos, A. G. 1995. *Adv. Polym. Sci.*, 122, 245.
73. Autian, J. 1970. *J. Fire Flammability*, 1, 239.
74. American Dental Association: Recommended standard procedures for biological evaluation of dental materials, *J. Am. Dent. Assoc.*, 84, 382, 1972.

2

Plastics Properties and Testing

2.1 Introduction

There are two stages in the process of becoming familiar with plastics. The first is rather general and involves an introduction to the unique molecular structures of polymers, their physical states, and transitions which have marked influence on their behavior. These have been dealt with in Chapter 1. The second stage, which will be treated in this chapter, is more specific in that it involves a study of the specific properties of plastics which dictate their applications. Besides the relative ease of molding and fabrication, many plastics offer a range of important advantages in terms of high strength/weight ratio, toughness, corrosion and abrasion resistance, low friction, and excellent electrical resistance. These qualities have made plastics acceptable as materials for a wide variety of engineering applications. It is important therefore that an engineer be aware of the performance characteristics and significant properties of plastics.

In this chapter plastics have been generally dealt with in respect to broad categories of properties, namely, mechanical, electrical, thermal, and optical. In this treatment the most characteristic features of plastic materials have been highlighted.

An important facet of materials development and proper materials selection is testing and standardization. The latter part of this chapter is therefore devoted to this aspect. It presents schematically (in simplified form) a number of standard test methods for plastics, highlighting the principles of the tests and the properties measured by them.

2.2 Mechanical Properties

Several unfamiliar aspects of material behavior of plastic need to be appreciated, the most important probably being that, in contrast to most metals at room temperature, the properties of plastics are time dependent [1-4]. Then superimposed on this aspect are the effects of the level of stress, the temperature of the material, and its structure (such as molecular weight, molecular orientation, and density). For example, with polypropylene an increase in temperature from 20 to 60°C may typically cause a 50% decrease in the allowable design stress. In addition, for each 0.001 g/cm^3 change in density of this material there is a corresponding 4% change in design stress. The material, moreover, will have enhanced strength in the direction of molecular alignment (that is, in the direction of flow in the mold) and less in the transverse direction.

Because of the influence of so many additional factors on the behavior of plastics, properties (such as modulus) quoted as a single value will be applicable only for the conditions at which they are measured. Properties measured as single values following standard test procedures are therefore useful only as a means of quality control. They would be useless as far as design is concerned, because to design a plastic component it is necessary to have complete information, at the relevant service temperature, on the time-dependent behavior (viscoelastic behavior) of the material over the whole range of stresses to be experienced by the component.

2.2.1 Stress and Strain

Any force or load acting on a body results in stress and strain in the body. Stress represents the intensity of the force at any point in the body and is measured as the force acting per unit area of a plane. The deformation or alteration in shape or dimensions of the body resulting from the stress is called *strain*. Strain is expressed in dimensionless units, such as *cm/cm, in./in., or* in *percentage.*

Corresponding to the three main types of stress—tensile, compressive, and shear—three types of strain can be distinguished. Thus, *tensile strain* is expressed as elongation per unit length (Figure 2.1a),

$$\varepsilon = \Delta\ell/\ell_0 = (\ell - \ell_0)/\ell_0 \tag{2.1}$$

and *compressive strain* as contraction per unit length (Figure 2.1b),

$$\varepsilon = \Delta\ell/\ell_0 = (\ell_0 - \ell)/\ell_0 \tag{2.2}$$

If the applied force or load, *F*, is tensile or compressive, the resulting tensile or compressive stress, σ, is defined by

$$\sigma = F/A \tag{2.3}$$

where *A* is the cross-sectional area perpendicular to the direction in which the force acts (Figure 2.1a).

The shearing stress is defined by a similar equation

$$\tau = F_s/A \tag{2.4}$$

where F_s is the shearing force acting on an area *A*, which is parallel to the direction of the applied force (Figure 2.1c).

Shear strain is measured by the magnitude of the angle representing the displacement of a certain plane relative to the other, due to the application of a pure shear stress, such as α in Figure 2.1c. The corresponding shear strain γ may be taken equal to the ratio *aa'*/*ab* (=tan α). A shear strain is produced in torsion, when, for example, a circular rod is twisted by tangential forces, as shown in

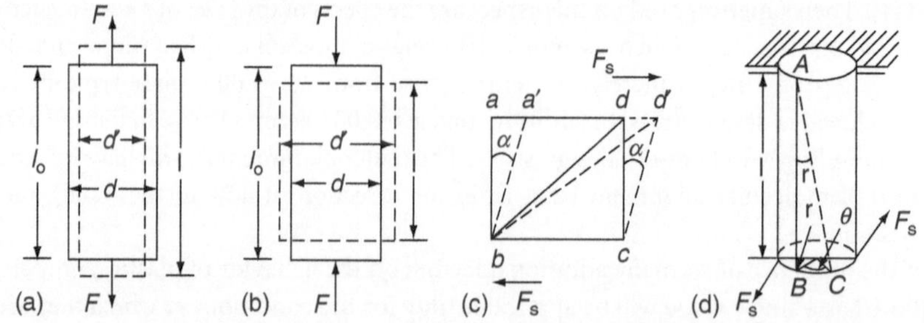

FIGURE 2.1 (a) Tensile or longitudinal strain, $\varepsilon = (\ell - \ell_0)/\ell_0$. (b) Compressive strain, $\varepsilon = (\ell_0 - \ell_0)/\ell_0$. (c) Shear strain, $\gamma = aa'/ab$. (d) Shear strain in torsion $\gamma = r\theta/\ell$.

Figure 2.1d. For small deformations the shear strain, γ, can be calculated from the triangle ABC

$$\gamma = BC/AB = r\theta/\ell \tag{2.5}$$

where r is the radius and θ is the angle of twist.

An ideal elastic material is one which exhibits no time effects. When a stress is applied the body deforms immediately, and it recovers its original dimensions completely and instantaneously when the stress is removed. When the ideal elastic body is subjected to tensile (or compressive) stress, the proportionality is expressed as

$$\sigma = E \cdot \varepsilon \tag{2.6}$$

where σ is the applied stress (tensile or compressive) in lbf/in.^{2}, kgf/cm^2 or other appropriate units of force per unit cross-sectional area (Equation 2.3), ε is the axial strain (Equation 2.1 and Equation 2.2), and E is the modulus of elasticity, commonly known as the Young's modulus. The proportionality law as defined above is known as Hooke's *law.*

Likewise, if the ideal solid is subjected to a shear stress (τ), then the shear strain (γ) developed as a function of stress applied is given by the expression

$$\tau = G \cdot \gamma \tag{2.7}$$

Here, the proportionality constant G is known as the shear modulus, also called the modulus of rigidity.

The elastic constants in tensile deformation and shear deformation are summarized and compared below:

	Tensile (Figure 2.1a)	Shear (Figure 2.1c)
Stress	$\sigma = F/A$	$\tau = F/A$
Strain	$\varepsilon = (\ell - \ell_0)/\ell$	$\gamma = \tan \alpha$
Modulus	$E = \sigma/\varepsilon$	$G = \tau/\gamma$
Compliance	$D = \varepsilon/\sigma$	$J = \gamma/\tau$

It may be noted that for an ideal elastic body compliance is the inverse of modulus.

The modulus of elasticity, E, and the modulus of rigidity, G, as defined above, apply under longitudinal and shear forces, respectively. When a hydrostatic force is applied, a third elastic modulus, the modulus of compressibility or bulk modulus, K, is used. It is the reciprocal of compressibility, β, and is defined as the ratio of the hydrostatic pressure, σ_h, to the volumetric strain, $\Delta V/V_0$:

$$K = \frac{\sigma_h}{\Delta V/V_0} = \frac{1}{\beta} \tag{2.8}$$

As indicated in Figure 2.1, an elongation (or compression) in one direction, due to an axial force, produces a contraction (or expansion) in the lateral direction, i.e., at right angles to the direction of the force. The ratio of the lateral strain to the longitudinal strain is called Poisson's ratio v. It is an important elastic constant, For instance, a tensile stress, σ_x, which produces a tensile strain, ε_x in the x-direction will also produce a contractive strain, ε_y, in the y-direction, the two being related by

$$\nu = -\varepsilon_y/\varepsilon_x \tag{2.9}$$

Combining Equation 2.9 with Equation 2.6 and rearranging yields

$$\varepsilon_y = -(\nu/E)\sigma_x \tag{2.10}$$

Equation 2.10 thus defines the contribution (ε_y) of the stress σ_x in the x-direction to the total strain in the y-direction.

Poisson's ratio, ν, as defined above, is a fourth elastic constant. For small deformations, the bulk modulus and modulus of rigidity can be calculated from the modulus of elasticity and Poisson's ratio by the following equations:

$$K = \frac{E}{3(1-2\nu)} \tag{2.11}$$

$$G = \frac{E}{2(1+\nu)} \tag{2.12}$$

The elastic modulus can also be calculated from the bulk modulus and the modulus of rigidity by the relation

$$\frac{1}{E} = \frac{1}{9K} + \frac{1}{3G} \tag{2.13}$$

For soft materials such as gels, pastes, putties and many colloidal systems, which do not compress to the extent to which they are deformed under stress, K is very large compared to G, and therefore from Equation 2.13, $E = 3G$. For other materials as metals, fibers, and some plastics, however, K must be considered.

2.2.2 Stress–Strain Behavior

The stress–strain behavior of plastics measured at a constant rate of loading provides a basis for quality control and comparative evaluation of various plastics. The diagram shown in Figure 2.2a is most typical of that obtained in tension for a constant rate of loading. For compression and shear the behavior is quite similar except that the magnitude and the extent to which the curve is followed are different.

In the diagram, load per unit cross section (stress) is plotted against deformation expressed as a fraction of the original dimension (strain). Even for different materials the nature of the curves will be similar, but they will differ in (1) the numerical values obtained and (2) how far the course of the typical curve is followed before failure occurs. Cellulose acetate and many other thermoplastics may follow the typical curve for almost its entire course. Thermosets like phenolics, on the other hand, have cross-linked molecules, and only a limited amount of intermolecular slippage can occur. As a result, they undergo fracture at low strains, and the stress–strain curve is followed no further than to some point below the knee, such as point 1.

Ultimate strength, elongation, and elastic modulus (Young's modulus) can be obtained from the stress–strain study (Figure 2.2a). For determining Young's modulus (E) the slope of the initial tangent, i.e., the steepest portion of the curve, is measured. Other moduli values are also used for plastics (see Figure 2.2b).

The appearance of a permanent set is said to mark a yield point, which indicates the upper limit of usefulness for any material. Unlike some metals, in particular, the ferrous alloys, the drop-of-beam effect and a sharp knee in the stress–strain diagram are not exhibited by plastics. An arbitrary yield point is usually assigned to them. Typical of these arbitrary values is the 0.2% or the 1% offset yield stress (Figure 2.3a).

Alternatively, a yield stress can be defined as that at which the ratio of total stress to total strain is some selected amount, say 50% or 70% of the elastic modulus (Figure 2.3b). In the first case the yield stress is conveniently located graphically by offsetting to the right the stated amount of 0.2% (or 1%) and drawing a line paralleling that drawn for the elastic modulus. The point at which this line intersects the observed stress–strain line defined the yield stress. In the second case also the point of intersection of the line drawn with a slope of 0.7E, for instance, with the observed stress–strain line determines the yield stress.

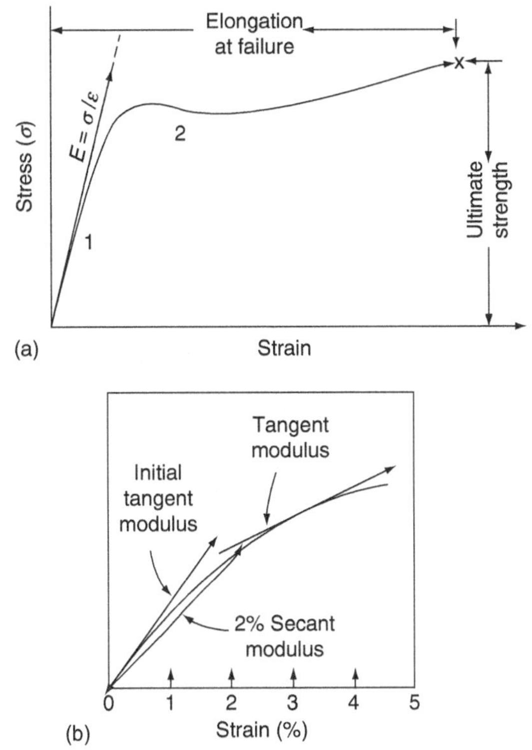

FIGURE 2.2 (a) Nominal stress–strain diagram. (b) Typical moduli values quoted for plastics.

Up to point 1 in Figure 2.2a, the material behaves as an elastic solid, and the deformation is recoverable. This deformation, which is small, is associated with the bending or stretching of the interatomic bonds between atoms of the polymer molecules (see Figure 2.4a). This type of deformation is nearly instantaneous and recoverable. There is no permanent displacement of the molecules relative to each other.

Between points 1 and 2 in Figure 2.2a, deformations have been associated with a straightening out of a kinked or coiled portion of the molecular chains (see Figure 2.4b), if loaded in tension. (For compression the reverse is true.) This can occur without intermolecular slippage. The deformation is recoverable ultimately but not instantaneously and hence is analogous to that of a nonlinear spring. Although the deformation occurs at stresses exceeding the stress at the proportional limit, there is no permanent change in intermolecular arrangement. This kind of deformation, characterized by recoverability and nonlinearity, is very pronounced in the rubber state.

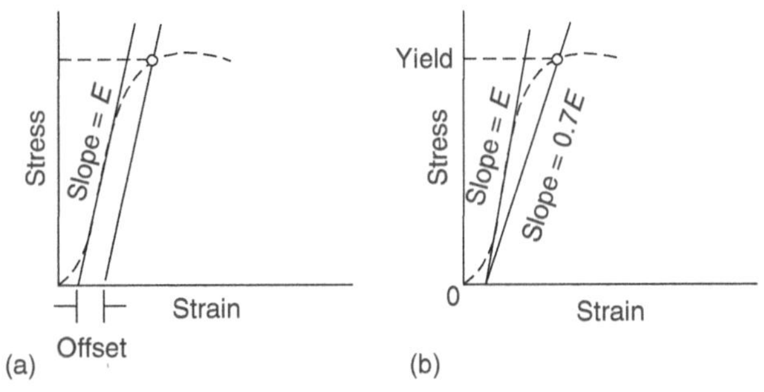

FIGURE 2.3 Location of a yield value.

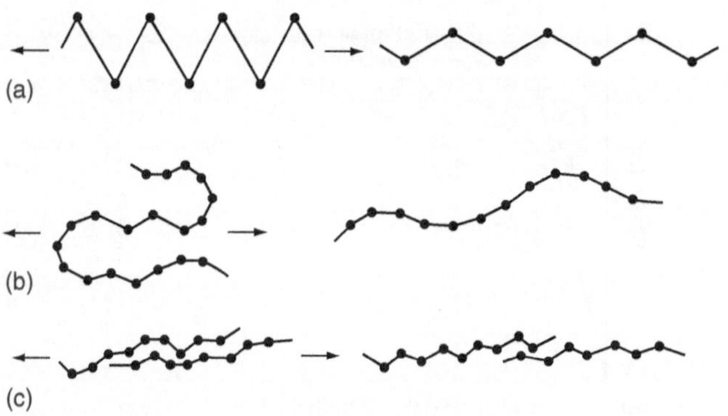

FIGURE 2.4 Deformation in plastics. (a) Stretching of polymer molecule. (b) Staightening out of a coiled molecular chain. (c) Intermolecular slippage.

The greatest extension that is recoverable marks the elastic limit for the material. Beyond this point extensions occur by displacement of molecules with respect to each other (Figure 2.4c), as in Newtonian flow of a liquid. The displaced molecules have no tendency to slip back to their original positions, therefore these deformations are permanent and not recoverable.

Poisson's ratio is a measure of the reduction in the cross section accompanying stretching and is the ratio of the transverse strain (a contraction for tensile stress) to longitudinal strain (elongation). Poisson's ratio for many of the more brittle plastics such as polystyrene, the acrylics, and the thermoset materials is about 0.3; for the more flexible plasticized materials, such as cellulose acetate, the value is somewhat higher, about 0.45. Poisson's ratio for rubber is 0.5 (characteristic of a liquid); it decrease to 0.4 for vulcanized rubber and to about 0.3 for ebonite. Poisson's ratio varies not only with the nature of the material but also with the magnitude of the strain for a given material. All values cited here are for zero strain.

Strain energy per unit volume is represented as the area under the stress–strain curve. It is another property that measures the ability of a material to withstand rough treatment and is related to toughness of the material. The stress–strain diagram thus serves as a basis for classification of plastics. Strong materials have higher ultimate strength than weak materials. Hard or unyielding materials have a higher modulus of elasticity (steeper initial slope) than soft materials. Tough materials have high elongations with large strain energy per unit volume. Stress–strain curves for type cases are shown in Figure 2.5.

It must be emphasized that the type behavior shown in Figure 2.5 depends not only on the material but also very definitely on conditions under which the test is made. For example, the bouncing putty silicone is putty-like under slow rates of loading (type curve a) but behaves as an elastic solid under high rates of impact (type curve b or d).

Figure 2.6a shows that at high extension rates (>1 mm/sec) unplasticized PVC is almost brittle with a relatively high modulus and strength. At low extension rates (<0.05 mm/sec) the same material exhibits a lower modulus and a high ductility, because at low extension rates the polymer molecular chains

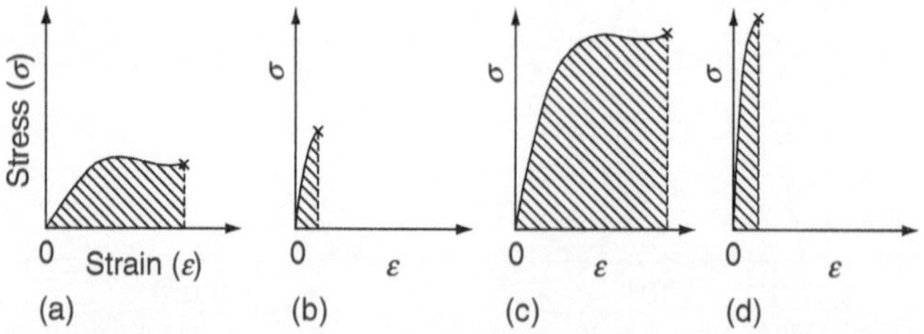

FIGURE 2.5 Classification of plastics on the basis of stress–strain diagram. (a) Soft and weak. (b) Weak and brittle. (c) Strong and tough. (d) Hard and strong.

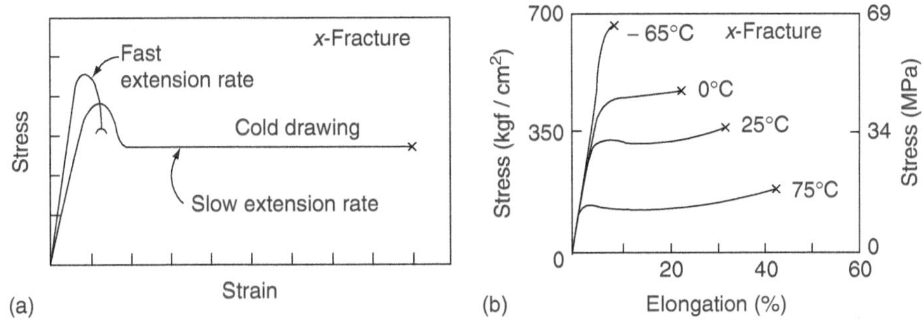

FIGURE 2.6 (a) Typical tensile behavior of unplasticized PVC. (b) Stress–strain curves of cellulose acetate at different temperatures.

have time to align themselves under the influence of the applied load. Thus the material is able to flow at the same rate as it is being strained. (This interesting phenomenon observed in some plastics is known as cold drawing.)

Further examples of the effect of conditions on the behavior of plastics are illustrated by the stress-strain curves for plasticized cellulose acetate when determined at different temperatures (Figure 2.6b). Thus the material is hard and strong at low temperatures, relatively tough at ordinary temperatures, and soft and weak at higher temperatures. This behavior may be attributed to variable molecular slippage effects associated with plasticizer action.

2.2.3 Viscoelastic Behavior of Plastics

In a perfectly elastic (Hookean) material the stress, σ, is directly proportional to the strain, ε. For uniaxial stress and strain the relationship may be written as

$$\sigma = \text{constant} \times \varepsilon \tag{2.14}$$

where the constant is referred to as the modulus of elasticity.

In a perfectly viscous (Newtonian) liquid the shear stress, τ, is directly proportional to the rate of strain, $\dot{\gamma}$, and the relationship may be written as

$$\tau = \text{constant} \times \dot{\gamma} \tag{2.15}$$

where the constant is referred to as the viscosity.

Polymeric materials exhibit stress–strain behavior which falls somewhere between these two ideal cases; hence, they are termed viscoelastic. In a viscoelastic material the stress is a function of both strain and time and so may be described by an equation of the form

$$\sigma = f(\varepsilon, t) \tag{2.16}$$

This equation represents nonlinear viscoelastic behavior. For simplicity of analysis it is often reduced to the form

$$\sigma = \varepsilon f(t) \tag{2.17}$$

which represents linear viscoelasticity. It means that in a tensile test on linear viscoelastic material, for a fixed value of elapsed time, the stress will be directly proportional to strain.

The most characteristics features of viscoelastic materials are that they exhibit time-dependent deformation or strain when subjected to a constant stress (creep) and a time-dependent stress when subjected to a constant strain (relaxation). Viscoelastic materials also have the ability to recover when the applied stress is removed. To a first approximation, this recovery can be considered as a reversal of creep.

2.2.3.1 Modulus and Compliance

Consider the tensile experiment of Figure 2.1a as a stress relaxation study in which the deformation is imposed suddenly and held fixed while the resulting stress, $\sigma(t)$, is followed with time. The tensile relaxation modulus, $E(t)$, is then obtained as

$$E(t) = \sigma(t)/\varepsilon_0 \tag{2.18}$$

with ε_0 being the constant strain and the parenthesis showing the functional dependence. Similarly a shear relaxation experiment measures the *shear* relaxation modulus $G(t)$:

$$G(t) = \tau(t)/\gamma_0 \tag{2.19}$$

where γ_0 is the constant strain.

Consider now the tensile experiment of Figure 2.1a as a creep study in which a steady stress σ_0 is suddenly applied to the polymer specimen and held constant. In general, the resulting strain $\varepsilon(t)$ will be a function of time starting from the imposition of load. The results of creep experiments are often expressed in terms of compliances rather than moduli. The tensile creep compliance $D(t)$ is

$$D(t) = \varepsilon(t)/\sigma_0 \tag{2.20}$$

The shear creep compliance $J(t)$ is similarly defined as

$$J(t) = \gamma(t)/\tau_0 \tag{2.21}$$

where τ_0 is the constant shear stress and $\gamma(t)$ is the resulting time-dependent strain.

A compliance is the inverse of a modulus for an ideal elastic body, but this is not true for viscoelastic materials. Consider, for example, two experiments carried out with identical samples of a viscoelastic material. In experiment (a) the sample is subjected to a tensile stress σ_1 for time t. The resulting tensile strain at t is ε_1, and the creep compliance measured at that time is $D_1(t) = \varepsilon_1/\sigma_1$. In experiment (b) a sample is stressed to a level σ_2 such that strain ε_1 is achieved immediately. The stress is then gradually decreased so that the strain remains at ε_1 for time t (i.e., the sample does not creep further). Let the stress on the material at time t be σ_3; the corresponding relaxation modulus will be $E_2(t) = \sigma_3/\varepsilon_1$. In measurements of this type, it can be expected that $\sigma_2 > \sigma_1 > \sigma_3$ and $E(t) \neq 1/D(t)$.

$E(t)$ and $G(t)$ are obtained directly only from stress relaxation measurements, while $D(t)$ and $J(t)$ require creep experiments for their direct observation.

2.2.4 Stress–Strain–Time Behavior

When a mass of polymer is stressed the deformation produced may be considered as a sum of the following three deformations (see Figure 2.4):

1. A deformation due to bond bending and stretching which is instantaneous and independent of temperature (ordinary elastic deformation, ε_{oe}).
2. A deformation due to chain uncoiling which is not instantaneous and whose rate depends on temperature (high elastic deformation, ε_{he}).
3. A deformation due to slippage of polymer molecules past one another (viscous deformation ε_v). It is often assumed that the rates of such viscous deformation do not change with time if the applied stress is constant.

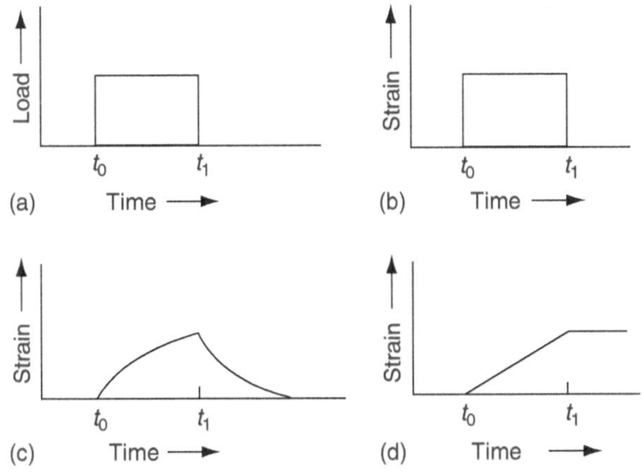

FIGURE 2.7 Types of deformational response as a result of (a) a fixed load being imposed between times t_0 and t_1: (b) ordinary elastic material; (c) highly elastic material; (d) viscous material.

Figure 2.7 shows schematically the above types of deformational response as a result of a fixed stress imposed on a body showing ordinary elastic deformation only (Figure 2.7b), a second body showing high elastic deformation only (Figure 2.7c), and a third body showing viscous deformation only (Figure 2.7d). In each case, the stress is imposed at time t_0 and held at a constant value until time t_1, when it is removed. Real polymers exhibit deformation-time curves which are a combination of the three basic responses, and a simple relationship for a combined or total strain ε

$$\varepsilon = \varepsilon_{oe} + \varepsilon_{he} + \varepsilon_{v}$$

can be used to analyze the deformation under a given stress.

The combined response, however, differs widely among polymers. Figure 2.8 shows typical deformation-time curves. It will be noted that, given sufficient time, ε_{he} will reach a constant value while ε_v continues to increase with time. On release of stress, ε_{he} will eventually disappear but ε_v will remain constant. An important conclusion resulting from this is that since both the high elastic and the viscous components of strain depend on both time and temperature the total strain also will depend on time and temperature. This has been shown to be an important factor affecting many polymer properties. It is therefore proposed to consider the background to this fact in greater detail in the following section.

2.2.4.1 The WLF Equations

For a polymeric segment to move from its position to its adjacent site there must be some holes in the mass of the material into which the segment can move—and simultaneously leave a vacant space into which another segment may move. The important point is that molecular motion cannot take place without the presence of holes. These holes, collectively, are called free volume.

One interpretation of the glass transition temperature (T_g) is that it is a temperature below which the free volume is really too small for much molecular movement. However, at or above T_g the molecules have sufficient energy for movement, jostling occurs and the free volume increases quite sharply with an increase in temperature.

It is usual to express the temperature coefficient of the free volume as being the difference between the thermal expansion coefficients above and below T_g. This may be expressed mathematically by the equation

$$f = f_g + (\alpha_a - \alpha_b)(T - T_g) = f_g + \Delta\alpha(T - T_g) \tag{2.22}$$

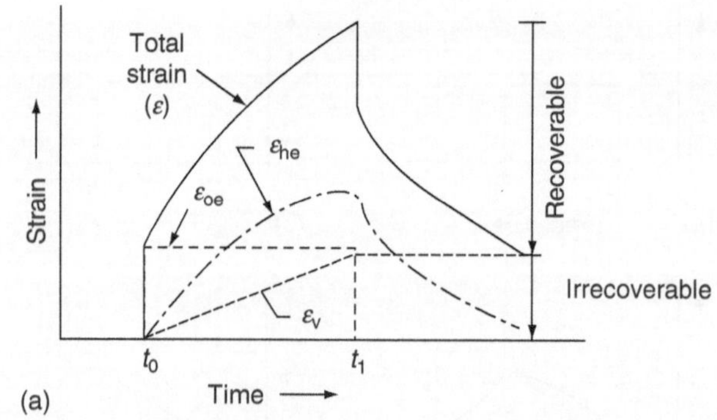

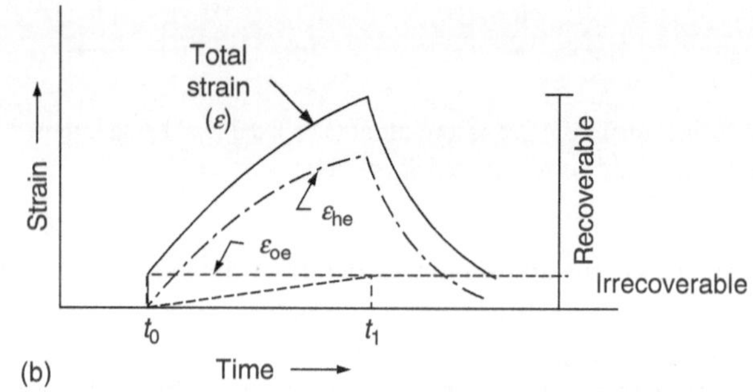

FIGURE 2.8 Strain-time curves: (a) material showing substantial ordinary elastic, high elastic, and viscous components of strain; (b) material in which high elastic deformation predominates.

where f is the fractional free volume at temperature T, f_g is the fractional free volume at T_g, and α_a and α_b are the coefficients of thermal expansion above and below the T_g, respectively. The value of $\Delta\alpha$ is simply $(\alpha_a - \alpha_b)$.

Now it has been shown that the viscosity is related to the fractional free volume by an expression of the form

$$\frac{1}{\eta_T} = Ke^{-A/f} \tag{2.23}$$

so that

$$\frac{1}{\eta_{T_g}} = Ke^{-A/f_g} \tag{2.24}$$

where K and A are constants. Combining these one may write

$$\log_e\left(\frac{\eta_T}{\eta_{T_g}}\right) = \frac{1}{f} - \frac{1}{f_g} \tag{2.25}$$

Substituting for f from Equation 2.22, this expression yields

$$\log_{10}\left(\frac{\eta_T}{\eta_{T_g}}\right) = \frac{-(T - T_g)}{2.303 f_g[(f_g/\Delta\alpha) + (T - T_g)]} \tag{2.26}$$

Experimental data on a large range of polymers have demonstrated the approximate general validity of the equation

$$\log_{10}\left(\frac{\eta_T}{\eta_{T_g}}\right) = \frac{-17.44(T - T_g)}{51.6 + (T - T_g)} \tag{2.27}$$

known as the Williams, Landel, and Ferry Equation (WLF equation). Solving Equation 2.26 and Equation 2.27 one obtains $f_g = 0.025$ and $\Delta\alpha = 4.8 \times 10^{-4}\ \text{deg}^{-1}$.

Equation 2.27 implies that if we know the viscosity at some temperature T we can estimate the viscosity at T_g, and from this estimate the viscosity at another temperature T_1. The WLF equation thus gives the effect of temperature on viscosity.

There are also other applications of the WLF equation. In essence, if the value of material property changes with temperature, and if this change arises from changes in the viscosity of the system, then it may well be possible to apply the WLF equation to the property change.

One example of this is in relation to stress relaxation. If a polymer is deformed to a fixed strain at constant temperature, the force required to maintain that strain will decay with time due to viscous slippage of molecules. One measure of this rate of decay or stress relaxation is the relaxation time λ, which may be defined as the time taken for the stress to decrease to $1/e$ of its initial value on application of strain (discussed later). In this case, it is found that

$$\log_{10}\left(\frac{\lambda_T}{\lambda_{T_g}}\right) = \frac{-17.44(T - T_g)}{51.6 + (T - T_g)} \tag{2.28}$$

which is of the same form as Equation 2.27.

For experiments performed in shear, there is a rather complicated relation between the time-dependent stress relaxation shear modulus $G(t)$ defined by Equation 2.19 and the time-dependent creep compliance $J(t)$ defined by Equation 2.21. But if the slope of $\log G(t)$ versus $\log t$ is $-m$, then, to a good approximation,

$$G(t) \cdot J(t) = \frac{\sin m\pi}{m\pi} \tag{2.29}$$

for $m < 0.8$. Not only are $G(t)$ and $J(t)$ related, but the former in turn is related to the tensile modulus which itself is related to the stress relaxation time λ. It is therefore possible in theory to predict creep-temperature relationships from WLF data, though in practice these are best determined by experiments.

2.2.5 Creep Behavior

Except for a few exceptions like lead, metals generally exhibit creep at higher temperatures. Plastics and rubbers, however, possess very temperature-sensitive creep behavior; they exhibit significant creep even at room temperature. In creep tests a constant load or stress is applied to the material, and the variation of deformation or strain with time is recorded. A typical creep curve plotted from such a creep test is shown in Figure 2.9a. The Figure shows that there is typically an almost instantaneous elastic strain AB followed by a time-dependent strain, which occurs in three stages: primary or transient creep BC (stage 1), secondary or steady-state creep CD (stage II), and tertiary or accelerated creep DE (stage III).

The primary creep has a rapidly decreasing strain rate. It is essentially similar in mechanism to retarded elasticity and, as such, is recoverable if the stress is removed. The secondary or steady-state creep is essentially viscous in character and is therefore nonrecoverable. The strain rate during this state is commonly referred to as the creep rate. It determines the useful life of the material. Tertiary

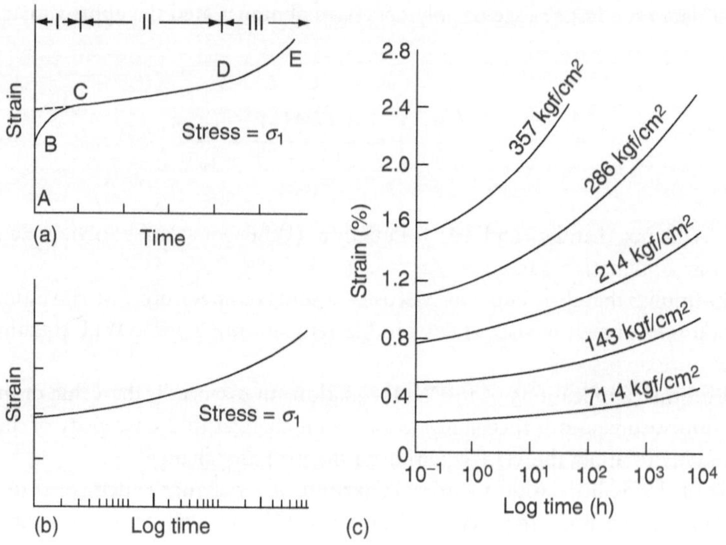

FIGURE 2.9 Typical creep curve (a) with linear time scale and (b) with logarithmic time scale. (c) Family of creep curves for poly(methyl methacrylate) at 20°C (1 kgf/cm^2=0.098 MPa). (Adapted from Crawford, R. J. 1981. *Plastics Engineering*, Pergamon, London.)

creep occurs at an accelerated rate because of an increase in the true stress due to necking of the specimen.

Normally a logarithmic time scale is used to plot the creep curve, as shown in Figure 2.9b, so that the time dependence of strain after long periods can be included. If a material is linearly viscoelastic (Equation 2.17), then at any selected time each line in a family of creep curves (with equally spaced stress levels) should be offset along the strain axis by the same amount. Although this type of behavior may be observed for plastics at low strains and short times, in most cases the behavior is nonlinear, as indicated in Figure 2.9c.

Plastics generally exhibit high rates of creep under relatively low stresses and temperatures, which limits their use for structural purposes. Creep behavior varies widely from one polymer to another; thermoset polymers, in general, are much more creep resistant than thermoplastic polymers. The steady-state creep in plastics and rubbers (often referred to as cold flow) is due to viscous flow, and increases continuously. Clearly, the material cannot continue to get larger indefinitely, and eventually fracture will occur. This behavior is referred to as creep rupture. Creep strength of these materials is defined as the maximum stress which may be applied for a specified time without causing fracture. The creep strength of plastics is considerably increased by adding fillers and other reinforcing materials, such as glass fibers and glass cloth, since they reduce the rate of flow.

The creep and recovery of plastics can be simulated by an appropriate combination of elementary mechanical models for ideal elastic and ideal viscous deformations. Although there are no discrete molecular structures which behave like individual elements of the models, they nevertheless aid in understanding the response of plastic materials.

2.2.6 Maxwell Model

The Maxwell model consists of a spring and dashpot connected in series (Figure 2.10a). When a load is applied, the elastic displacement of the spring occurs immediately and is followed by the viscous flow of liquid in the dashpot which requires time. After the load is removed, the elastic displacement is recovered immediately, but the viscous displacement is not recovered.

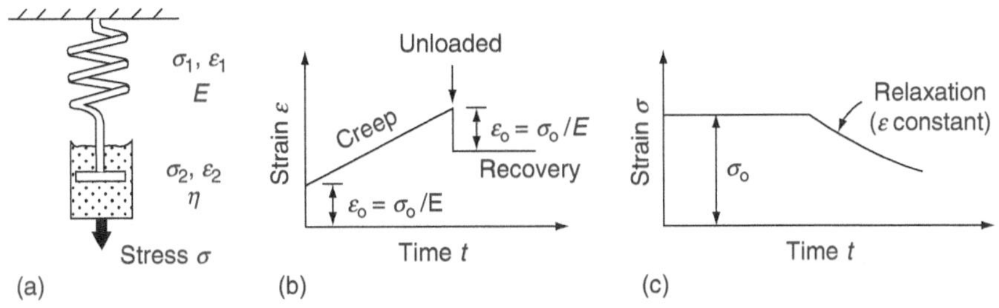

FIGURE 2.10 (a) The Maxwell model. (b), (c) Responses of the model under time-dependent modes of deformation.

2.2.6.1 Stress–Strain Relation

The spring is the elastic component of the response and obeys the relation

$$\sigma_1 = E\varepsilon_1 \tag{2.30}$$

σ_1 and ε_1 are the stress and strain, respectively, and E is a constant.

The dashpot (consisting of a piston loosely fitting in a cylindrical vessel containing a liquid) accounts for the viscous component of the response. In this case the stress σ_2, is proportional to the rate of strain $\dot{\varepsilon}_2$; i.e.,

$$\sigma_2 = \eta\dot{\varepsilon}_2 \tag{2.31}$$

where η is a material constant called the coefficient of viscous traction.

The total strain, ε, of the model under a given stress, σ, is distributed between the spring and the dashpot elements:

$$\varepsilon = \varepsilon_1 + \varepsilon_2 \tag{2.32}$$

From Equation 2.32 the rate of total displacement with time is

$$\dot{\varepsilon} = \dot{\varepsilon}_1 + \dot{\varepsilon}_2 \tag{2.33}$$

and from Equation 2.30 through Equation 2.32,

$$\dot{\varepsilon} = \frac{1}{E}\dot{\sigma}_1 + \frac{1}{\eta}\sigma_2 \tag{2.34}$$

But both elements are subjected to the entire stress, σ,

$$\sigma = \sigma_1 = \sigma_2 \tag{2.35}$$

Therefore Equation 2.34 can be written as

$$\dot{\varepsilon} = \frac{1}{E}\dot{\sigma} + \frac{1}{\eta}\sigma \tag{2.36}$$

which is the governing equation of the Maxwell model. It is interesting to consider the responses that this model predicts under three common time-dependent modes of deformation.

Equation 2.36 is commonly rearranged as follows:

$$\dot{\sigma} = E\dot{\varepsilon} - \frac{1}{\lambda}\sigma \tag{2.37}$$

where $\lambda = \eta/E$ is the ratio of the viscosity η of the dashpot and the tensile modulus E of the spring.

Note that λ has the units of time and that it characterizes the viscoelastic nature of the element very concisely, as the ratio of the viscous portion of the response to the elastic portion. This naturally occurring parameter is taken to be the response time or the relaxation time of the model.

Equation 2.37 is the governing equation of the Maxwell model. It is interesting to consider the responses that this model predicts under three common time-dependent modes of deformation.

1. *Creep.* If a constant stress σ_0 is applied, then Equation 2.37 becomes

$$\dot{\varepsilon} = \frac{\sigma_0}{E}\frac{1}{\lambda} = \frac{\varepsilon_0}{\lambda} \tag{2.38}$$

Integration yields

$$\varepsilon = \varepsilon_0(1 + t/\lambda) \tag{2.39}$$

which indicates a constant rate of increase of strain with time—i.e., steady-state creep (Figure 2.10b). Equation 2.39 describes the creep response of the Maxwell element.

2. *Relaxation.* If the strain is held constant, then Equation 2.37 becomes

$$\dot{\sigma} + \sigma/\lambda = 0 \tag{2.40}$$

Solving this differential equation with the initial condition $\sigma = \sigma_0$ at $t = t_0$ gives

$$\sigma = \sigma_0 \exp(-t/\lambda) \tag{2.41}$$

This indicates an exponential decay of stress with time (Figure 2.10c). The stress will relax and approach zero monotonically. The relaxation time λ is thus the time required for the stress to decay to $1/e$, or 0.37, of its initial value.

Since the strain remains constant in a relaxation experiment, Equation 2.41 can also be written as

$$E(t) = E_0 \exp(-t/\lambda) \tag{2.42}$$

where $E(t)$ is the tensile modulus of the Maxwell element at time t, and E_0 the modulus at the initial time of deformation.

The corresponding equation for the Maxwell element in shear is

$$G(t) = G_0 \exp(-t/\lambda) \tag{2.43}$$

where $G(t)$ is the shear modulus of the Maxwell element at time t, and G_0 is the modulus at $t=0$; the relaxation time λ is now the ratio of the viscosity η of the viscous component and the shear modulus G of the elastic component of the model.

3. *Recovery.* When the initial stress, σ_0, is removed, there is an instantaneous recovery of the elastic strain, ε_0, and then, as shown by Equation 2.36, the strain rate is zero, so there is no further recovery (Figure 2.10b).

2.2.6.2 Generalized Maxwell Model

The behavior of a polymer system is so complicated that we cannot represent it with the response time of a single Maxwell element. In other words, the simple model described above cannot approach the behavior of a real system. In 1893, Weichert showed that stress-relaxation experiments could be represented as a generalization of Maxwell's equation. The mechanical model according to Weichert's formulation is shown in Figure 2.11; it consists of a large number of Maxwell elements coupled in parallel.

Since the strain in each element is common, we sum the forces acting on the individual elements to obtain the total stress as a function of time, i.e.,

$$\sigma(t) = \sum \sigma_i \tag{2.44}$$

For relaxation with constant strain ε_0 we combine Equation 2.41 and Equation 2.44 to obtain

$$\sigma(t) = \varepsilon_0 \sum E_i \exp\left(-\frac{t}{\lambda_i}\right) \tag{2.45}$$

where $\lambda_i = \eta_i / E_i$ is the relaxation time of the *i*th element.

The overall modulus as a function of time, $E(t)$, is thus

$$E(t) = \frac{\sigma(t)}{\varepsilon_0} = \sum E_i \exp\left(-\frac{t}{\lambda_i}\right) \tag{2.46}$$

The synthesis of $E(t)$ from known values of E_i and λ_i is simplified by the use of semilog paper.

Example 2.1

Derive the overall time-dependent modulus $E(t)$ for $0 < t < 200$ sec when [4]

i	E_0 [(dynes/cm^2)$\times 10^{-8}$]	λ_i (sec)
1	1.000	100
2	0.667	50
3	0.333	25

Answer. For each element, $E_i(t)$ is given by a straight line on semilog paper (Equation 2.42) with intercept $(E_i)_0$ and a negative slope of λ_i (Figure 2.12). Adding the curves arithmetically gives $E(t)$ directly:

$$E_i(t) = (E_i)_0 \exp\left(-\frac{t}{\lambda_i}\right)$$

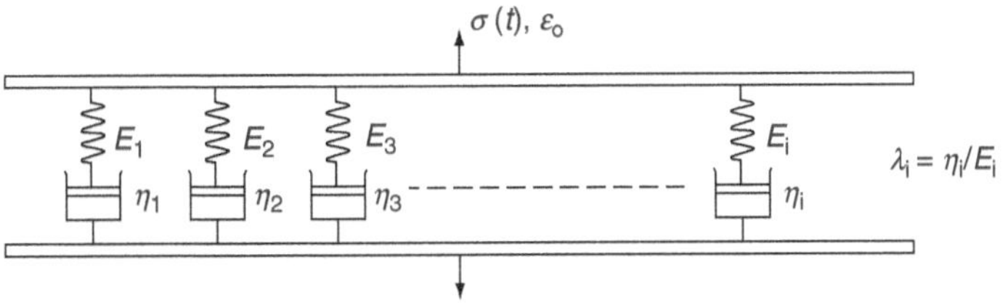

FIGURE 2.11 Generalized Maxwell model (Weichert's formulation).

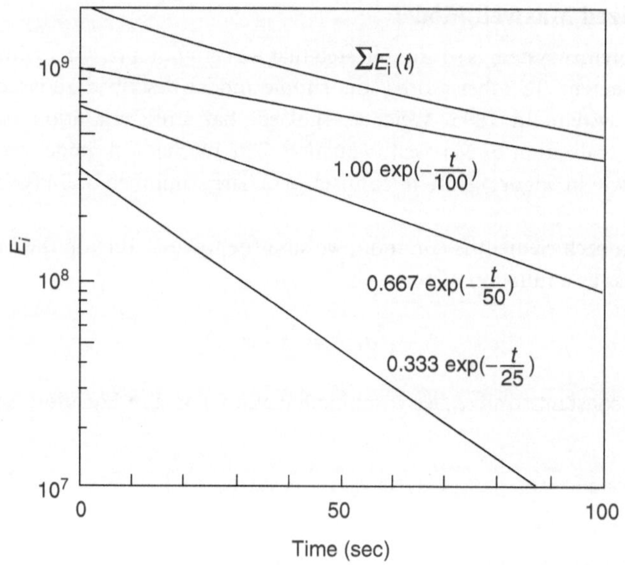

FIGURE 2.12 Time-dependent modulus for individual Maxwell elements and for the sum of three elements in parallel $\sum E_i(t)$.

$$E(t) = \sum_{i=1}^{3} E_i(t)$$

Note that $E(t)$ is not a straight line in Figure 2.12.

As can be seen, the Maxwell–Weichert model possesses many relaxation times. For real materials we postulate the existence of a continuous spectrum of relaxation times (λ_i). A spectrum-skewed toward lower times would be characteristic of a viscoelastic fluid, whereas a spectrum skewed toward longer times would be characteristic of a viscoelastic solid. For a real system containing crosslinks the spectrum would be skewed heavily toward very long or infinite relaxation times. In generalizing, λ may thus be allowed to range from zero to infinity. The concept that a continuous distribution of relaxation times should be required to represent the behavior of real systems would seem to follow naturally from the fact that real polymeric systems also exhibit distributions in conformational size, molecular weight, and distance between crosslinks.

If the number of units is allowed to become infinite, the summation over the differential units of the model (Equation 2.44) can be explained by an integration over all relaxation times. Thus as $i \rightarrow \infty$ in Figure 2.11, the range of allowable relaxation times becomes zero to infinity. From the notion that the stresses in the individual elements, σ_i, are functions of time and relaxation times, $\sigma_i = \sigma_i(t, \lambda_i)$, we define a continuous function $\sigma(t, \lambda)$ such that the total stress, $\sigma(t)$, is given by the following [compare Equation 2.44]:

$$\sigma(t) = \int_0^\infty \sigma(t, \lambda)\mathrm{d}\lambda \tag{2.47}$$

and for relaxation with constant strain, ε_0, it is given by the following (compare Equation 2.45):

$$\sigma(t) = \varepsilon_0 \int_0^\infty E(\lambda)\mathrm{e}^{-t/\lambda}\,\mathrm{d}\lambda \tag{2.48}$$

Since $E(t) = \sigma(t)/\varepsilon_0$, we find that we have developed an expression suitable for representing the time dependence of the relaxation modulus, i.e.,

$$E(t) = \int_0^{\infty} E(\lambda) e^{-t/\lambda} \, d\lambda \tag{2.49}$$

The function $E(\lambda)$ is referred to as the distribution of relaxation times or the relaxation spectrum. In principle, once $E(\lambda)$ is known, the result of any other type of mechanical experiment can be predicted. In practice $E(\lambda)$ is determined from experimental data on $E(t)$. Since the distribution of relaxation times is so broad, it is more convenient to consider ln λ. Hence we introduce the function $H(\ln \lambda)$, where the parenthesis denotes functional dependence, to replace $E(\lambda)$ as

$$E(\lambda) = \frac{H(\ln \lambda)}{\lambda} \tag{2.50}$$

Then Equation 2.49 becomes (note the change of limits):

$$E(t) = \int_{-\infty}^{+\infty} H(\ln \lambda) e^{-t/\lambda} \, d(\ln \lambda) \tag{2.51}$$

and all relaxation times are considered as ln λ.

What we desire now is a means to determine $H(\ln \lambda)$ from data obtained as $E(t)$ versus ln t. This is virtually impossible to do directly, and a number of approximate methods have been devised. These approximation are discussed in the advanced reference works of Ferry [5] and Tobolsky [6].

2.2.7 Kelvin or Voigt Model

In the Kelvin or Voigt model the spring and dashpot elements are connected in parallel, as shown in Figure 2.13a. This model roughly approximates the behavior of rubber. When the load is applied at zero time, the elastic deformation cannot occur immediately because the rate of flow is limited by the dashpot. Displacements continue until the strain equals the elastic deformation of the spring and it resists further movement. On removal of the load the spring recovers the displacement by reversing the

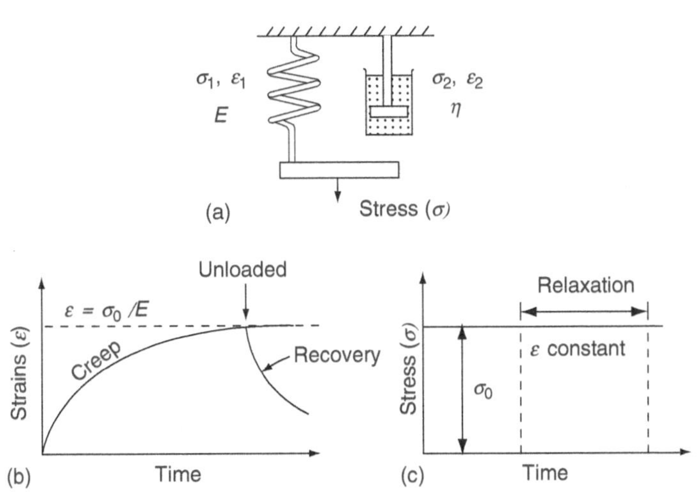

FIGURE 2.13 (a) The Kelvin or Voigt model. (b), (c) Responses of the model under time-dependent modes of deformation.

flow through the dashpot, and ultimately there is no strain. The mathematical relations are derived next.

2.2.7.1 Stress–Strain Relation

Since the two elements are connected in parallel, the total stress will be distributed between them, but any deformation will take place equally and simultaneously in both of them; that is,

$$\sigma = \sigma_1 + \sigma_2 \tag{2.52}$$

$$\varepsilon = \varepsilon_1 + \varepsilon_2 \tag{2.53}$$

From Equation 2.30, Equation 2.31, and Equation 2.52,

$$\sigma = E\varepsilon_1 + \eta\dot{\varepsilon}_2$$

or, using Equation 2.53,

$$\sigma = E\varepsilon + \eta\dot{\varepsilon} \tag{2.54}$$

which is the governing equation for the Kelvin (or Voigt) model. Its predictions for the common time-dependent deformations are derived next.

1. *Creep.* If a constant stress, σ_0, is applied, Equation 2.54 becomes

$$\sigma_0 = E\varepsilon + \eta\dot{\varepsilon} \tag{2.55}$$

The differential equation may be solved for the total strain, ε, to give

$$\varepsilon = \frac{\sigma_0}{E}\left[1 - \exp\left(-\frac{E}{\eta}\right)t\right] \tag{2.56}$$

This equation indicates that the deformation does not appear instantaneously on application of stress, but it increases gradually, attaining asymptotically its maximum value $\varepsilon = \sigma_0/E$ at infinite time (Figure 2.13b). The Voigt model is thus said to exhibit retarded elastic deformation in creep experiments. The quantity $\eta/E = \lambda$ is called a retardation time. It is the time ($t = \lambda$) at which the deformation is retarded by $1/e$ of its maximum value. (The physical meaning of η/E for Maxwell and Voigt models should not be confused.)

By comparison with Equation 2.56 the creep equation under a constant shear stress τ_0 may be written as

$$\gamma = \frac{\tau_0}{G}[1 - \exp(-t/\lambda)] \tag{2.57}$$

where γ is the time-dependent shear strain of the Voigt element and G is the shear modulus of its elastic component; λ is the retardation time ($=\eta/G$).

Equation 2.57 is conveniently written as

$$J = J_0[1 - \exp(-t/\lambda)] \tag{2.58}$$

where J is the shear creep compliance ($=\gamma/\tau_0$) at time t, and J_0 is the shear creep compliance at the time of stress application.

2. *Relaxation.* If the strain is held constant; then Equation 2.54 becomes:

$$\sigma = E\varepsilon$$

3. *Recovery.* If the stress is removed, Equation 2.54 becomes

$$0 = E\varepsilon + \eta\dot{\varepsilon}$$

This differential equation may be solved with the initial condition $\varepsilon = \varepsilon_0$ at the time of stress removal to give

$$\varepsilon = \varepsilon_0 \exp\left(-\frac{E}{\eta}\right)t \tag{2.59}$$

This equation represents an exponential recovery of strain which, as a comparison with Equation 2.56 shows, is a reversal of the predicted creep.

The Kelvin (or Voigt) model therefore gives an acceptable first approximation to creep and recovery behavior but does not predict relaxation. By comparison, the previous model (Maxwell model) could account for relaxation but was poor in relation to creep and recovery. It is evident therefore that a better simulation of viscoelastic materials may be achieved by combining the two models.

2.2.8 Four-Element Model

A combination of the Maxwell and Kelvin models comprising four elements is shown in Figure 2.14a. The total strain is

$$\varepsilon = \varepsilon_1 + \varepsilon_2 + \varepsilon_k \tag{2.60}$$

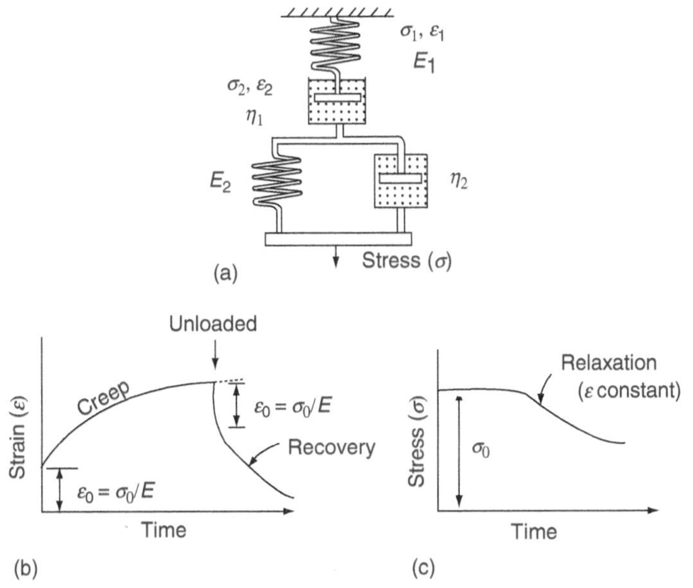

FIGURE 2.14 (a) Four-element model. (b), (c) Responses of the model under time-dependent modes of deformation.

where ε_k is the strain response of the Kelvin model. From Equation 2.30, Equation 2.31, and Equation 2.56,

$$\varepsilon = \frac{\sigma_0}{E_1} + \frac{\sigma_0 t}{\eta_1} + \frac{\sigma_0}{E_2}\left[1 - \exp\left(-\frac{E_2}{\eta_2}\right)t\right] \tag{2.61}$$

Thus the strain rate is

$$\dot{\varepsilon} = \frac{\sigma_0}{\eta_1} + \frac{\sigma_0}{\eta_2}\exp\left(-\frac{E_2}{\eta_2}\right)t \tag{2.62}$$

The response of this model to creep, relaxation, and recovery situations is thus the sum of the effects described previously for the Maxwell and Kelvin models and is illustrated in Figure 2.14b.

Though the model is not a true representation of the complex viscoelastic response of polymeric materials, it is nonetheless an acceptable approximation to the actual behavior. The simulation becomes better as more and more elements are added to the model, but the mathematics also becomes more complex.

2.2.9 Zener Model

Another model, attributed to Zener, consists of three elements connected in series and parallel, as illustrated in Figure 2.15, and known as the standard linear solid. Following the procedure already given, we derive the governing equation of this model:

$$\eta_3\dot{\sigma} + E_1\sigma = \eta_3(E_1 + E_2)\dot{\varepsilon} + E_1E_2\varepsilon \tag{2.63}$$

This equation may be written in the form

$$a_1\dot{\sigma} + a_0\sigma = b_1\dot{\varepsilon} + b_0\varepsilon \tag{2.64}$$

where a_1, a_0, b_1, and b_0 are all material constants. A more general form of Equation 2.64 is

$$a_n\frac{\partial^n\sigma}{\partial t^n} + a_{n-1}\frac{\partial^{n-1}\sigma}{\partial t^{n-1}} + \cdots + a_0\sigma = b_m\frac{\partial^m\varepsilon}{\partial t^m} + \cdots + b_0\varepsilon \tag{2.65}$$

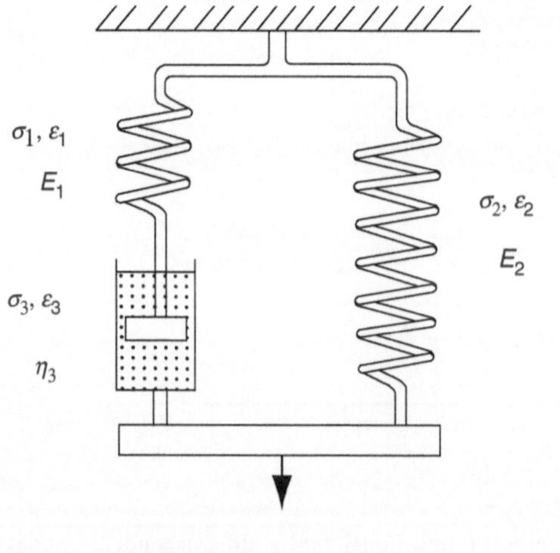

FIGURE 2.15 The standard linear solid.

The modern theory of viscoelasticity favors this type of equation. The models described earlier are special cases of this equation.

Hookean body. All constants a and b except a_0 and b_0 are zero. Equation 2.65 becomes

$$a_0\sigma = b_0\varepsilon \tag{2.65a}$$

Maxwell element. All constant a and b except a_0, a_1, and b_1 are zero. Equation 2.65 becomes

$$a_0\sigma + a_1\frac{\partial\sigma}{\partial t} = b_1\frac{\partial\varepsilon}{\partial t} \tag{2.65b}$$

This is the spring and dashpot in series and applies to stress relaxation at constant strain.

Voigt element. All constants a and b except a_0, b_0, and b_1 are zero. Equation 2.65 becomes

$$a_0\sigma = b_0\varepsilon + b_1\frac{\partial\varepsilon}{\partial t} \tag{2.65c}$$

This is the spring and dashpot in parallel and applies to strain retardation at constant stress.

2.2.10 Superposition Principle

Each of the creep curves in Figure 2.9c depicts the strain response of a material under a constant stress. However, in service, materials are often subjected to a complex sequence of stresses or stress histories, and obviously it is not practical to obtain experimental creep data for all combinations of loading. In such cases a theoretical model can be very useful for describing the response of a material to a given loading pattern.

The most commonly used model is the Boltzmann superposition principle, which proposes that for a linear viscoelastic material the entire loading history contributes to the strain response, and the latter is simply given by the algebraic sum of the strains due to each step in the load. The principle may be expressed as follows. If an equation for the strain is obtained as a function of time under a constant stress, then the modulus as a function of time may be expressed as

$$E(t) = \frac{\sigma}{\varepsilon(t)} \tag{2.66}$$

Thus if the applied stress is σ_0 at zero time, the creep strain at any time, t, will be given by

$$\varepsilon(t) = \frac{1}{E(t)}\sigma_0 \tag{2.67}$$

On the other hand, if the stress, σ_0, was applied at zero time and an additional stress, σ_1, at time u, the Boltzmann superposition principle says that the total strain at time t is the algebraic sum of two independent responses; that is,

$$\varepsilon(t) = \frac{1}{E(t)}\sigma_0 + \frac{1}{E(t-u)}\sigma_1 \tag{2.68}$$

For any series of stress increments this equation can be generalized to

$$\varepsilon(t) = \sum_{u=-\infty}^{u=t} \sigma_i \frac{1}{E(t-u)} \tag{2.69}$$

The lower limit of the summation is taken as $-\infty$ since the entire stress history contributes to the response.

As an illustration, for a series of step changes in stress as in Figure 2.16a, the strain response predicted by the model is shown schematically in Figure 2.16b. The time-dependent strain response (creep curve) due to the stress σ_0 applied at zero time is predicted by Equation 2.66 with $\sigma = \sigma_0$. When a second stress, σ_1, is added to σ_0, the new curve will be obtained, as illustrated in Figure 2.16b, by adding the creep due to σ_1 to the anticipated creep due to σ_0. Removal of all stress at a subsequent time u_2 is then equivalent to removing the creep stain due to σ_0 and σ_1, independently, as shown in Figure 2.16b. The procedure is repeated in a similar way for other stress changes.

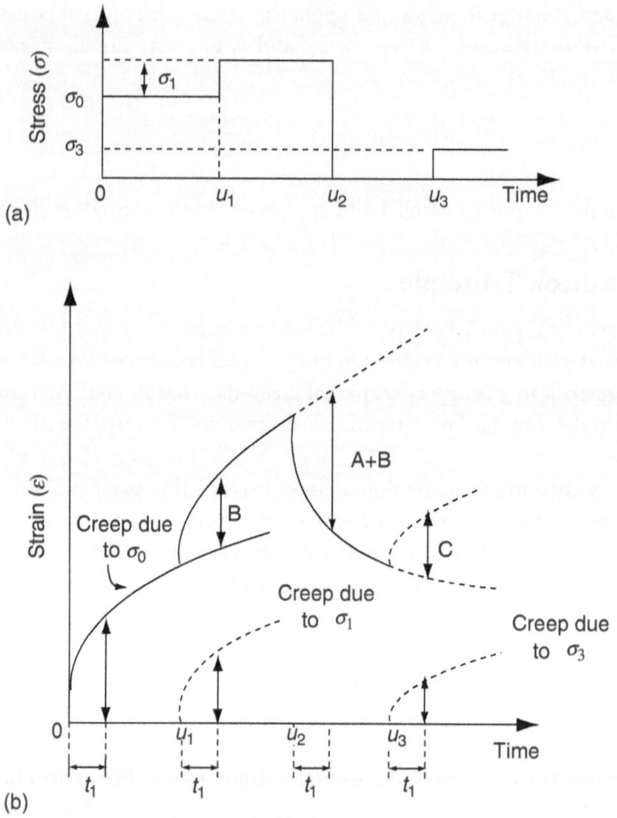

FIGURE 2.16 (a) Stress history. (b) Predicted Strain response using Boltzmann's superposition principle. (Adapted from Crawford, R. J. 1981. *Plastics Engineering*, Pergamon, London.)

To take into account a continuous loading cycle, we can further generalize Equation 2.69 to

$$\varepsilon(t) = \int_{-\infty}^{t} \frac{1}{E(t-u)} \frac{\mathrm{d}\sigma(u)}{\mathrm{d}u} \mathrm{d}u \tag{2.70}$$

In the same way the stress response to a complex strain history may be derived as

$$\sigma(t) = \int_{-\infty}^{t} E(t-u) \frac{\mathrm{d}\varepsilon(u)}{\mathrm{d}u} \mathrm{d}u \tag{2.71}$$

When the stress history has been defined mathematically, substitution in Equation 2.70 and integration within limits gives the strain at the given time. The stress at a given time is similarly obtained from Equation 2.71.

2.2.11 Isometric and Isochronous Curves

Isometric curves are obtained by plotting stress vs. time for a constant strain; *isochronous curves* are obtained by plotting stress vs. strain for a constant time of loading. These curves may be obtained from the creep curves by taking a constant-strain section and a constant-time section, respectively, through the creep curves and replotting the data, as shown in Figure 2.17.

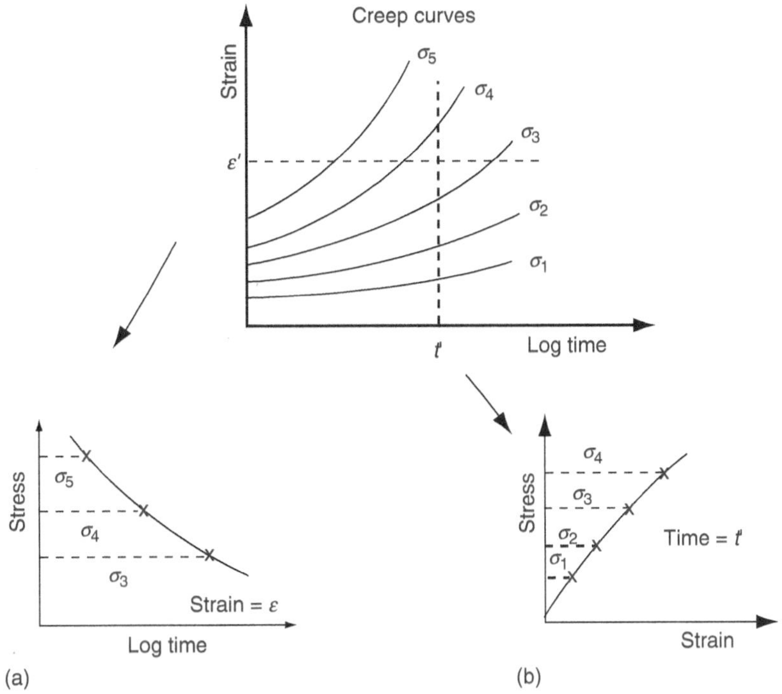

FIGURE 2.17 (a) Isometric and (b) isochronous curves from creep curves. (Adapted from Crawford, R. J. 1981. *Plastics Engineering*, Pergamon, London.)

An isometric curve provides an indication of the relaxation of stress in the material when the strain is kept constant. Since stress relaxation is a less common experimental procedure than creep testing, an isometric curve, derived like the preceding curves from creep curves, is often used as a good approximation of this property.

Isochronous curves, on the other hand, are more advantageously obtained by direct experiments because they are less time consuming and require less specimen preparation than creep testing. The experiments actually involve a series of mini creep and recovery tests on the material. Thus a stress is applied to a specimen of the material, and the strain is recorded after a time t (typically 100 sec). The stress is then removed and the material is allowed to recover. This procedure is repeated unit there are sufficient points to plot the isochronous curve.

Note that the isochronous test method is quite similar to that of a conventional incremental loading tensile test and differs only in that the presence of creep is recognized and the "memory" of the material for its stress history is overcome by the recovery periods. Isochronous data are often presented on log–log scales because this provides a more precise indication of the nonlinearity of the data by yielding a straight-line plot of slope less than unity.

2.2.12 Pseudoelastic Design Method

Due to the viscoelastic nature of plastics, deformations depend on such factors as the time under load and the temperature. Therefore the classical equations available for the design of structural components, such as springs, beams, plates, and cylinders, and derived under the assumptions that (1) the modulus is constant and (2) the strains are small and independent of loading rate or history and are immediately reversible, cannot be used indiscriminately. For example, classical equations are derived using the relation

$$\text{Stress} = \text{constant} \times \text{strain}$$

where the constant is the modulus. From the nature of the creep curves shown in Figure 2.17a, it is clear that the modulus of a plastic is not constant. Several approaches have been developed to allow for this fact, and some of them also give very accurate results; but mathematically they are quite complex, and this has limited their use. However, one method that has been widely accepted is the *pseudoelastic design method.* In this method appropriate values are chosen for the time-dependent properties, such as modulus, and substituted into the classical equations.

The method has been found to give sufficiently accurate results, provided that the value of the modulus is chosen judiciously, taking into account the service life of the component and the limiting strain of the plastic. Unfortunately, however, there is no straightforward method for finding the limiting strain of a plastic. The value may differ for various plastics and even for the same plastic in different applications. The value is often arbitrarily chosen, although several methods have been suggested for arriving at an appropriate value.

One method is to draw a secant modulus which is 0.85 of the initial tangent modulus and to note the strain at which this intersects the stress-strain curve (see Figure 2.2b). But this method may be too restrictive for many plastics, particularly those which are highly crystalline. In most situations the maximum allowable strain is therefore decided in consultations between designer and product manufacturer.

Once an appropriate value for the maximum strain is chosen, design methods based on creep curves and the classical equations are quite straightforward, as shown in the following examples.

Example 2.2

A plastic beam, 200 mm long and simply supported at each end, is subjected to a point load of 10 kg at its mid-span. If the width of the beam is 14 mm, calculate a suitable depth so that the central deflection does not exceed 5 mm in a service life of 20,000 h. The creep curves for the material at the service temperature of 20°C are shown in Figure 2.18a [4]. The maximum permissible strain in this material is assumed to be 1%.

Answer. The linear elastic equation for the central deflection, δ, of the beam is

$$\delta = \frac{PL^3}{48EI}$$

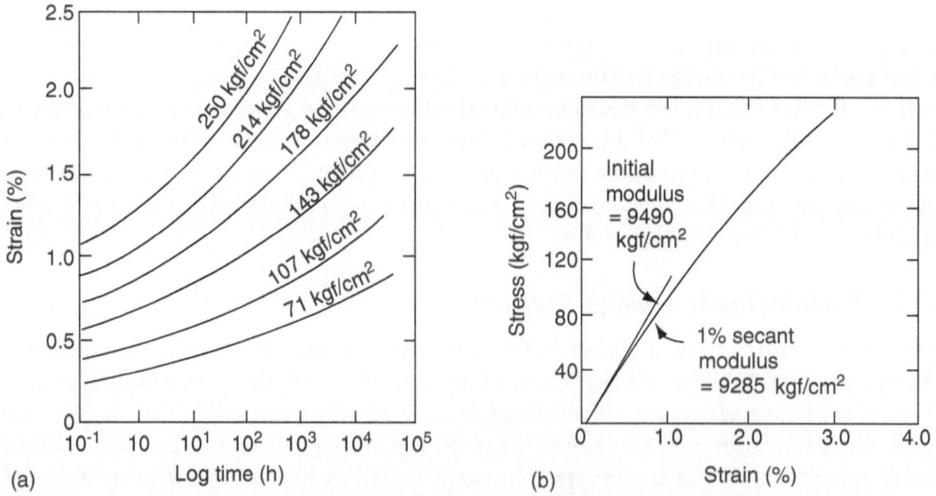

FIGURE 2.18 (a) Creep curves for material used in illustrative examples. (b) Isochronus curve at 20,000 h service life (1 kgf/cm^2 = 0.098 MPa). (Adapted from Crawford, R. J. 1981. *Plastics Engineering*, Pergamon, London.)

where *P*, load at mid-span; *L*, length of beam; *E*, modulus of beam material; *I*, second moment of area of beam cross section

The second moment of area is

$$I = \frac{bd^3}{12} = \frac{14d^3}{12}\,\text{mm}^4$$

So from the expression for δ,

$$d^3 = \frac{PL^3}{56E\delta}$$

The only unknown on the right side is *E*. For plastic this is time dependent, but a suitable value corresponding to the maximum permissible strain may be obtained by referring to the creep curves in Figure 2.18a. A constant-time section across these curves at 20,000 h gives the isochronous curve shown in Figure 2.18b. Since the maximum strain is recommended as 1%, a secant modulus may be taken at this value. It is 9285 kgf/cm^2 (=92.85 kgf/mm^2). Using this value in the above equation gives

$$d^3 = \frac{10(200)^3}{56 \times 92.85 \times 5}$$

$$d = 14.5\text{ mm}$$

Example 2.3

A thin-wall plastic pipe of diameter 150 mm is subjected to an internal pressure of 8 kgf/cm^2 at 20°C. It is suggested that the service life of the pipe should be 20,000 h with a maximum strain of 2%. The creep curves for the plastic material are shown in Figure 2.18a [4]. Calculate a suitable wall thickness for the pipe.

Answer. The hoop stress, σ, in a thin-wall pipe of diameter *d* and thickness *h*, subjected to an internal pressure, *P*, is given by

$$\sigma = \frac{Pd}{2h} \quad \text{so} \quad h = \frac{Pd}{2\sigma}$$

A suitable design stress may be obtained from the creep curves in Figure 2.18a. By referring to the 20,000-h isochronous curve (Figure 2.18b) derived from these curves, the design stress at 2% strain is obtained as 167.7 kg/cm^2. (Note that a similar result could have been obtained by plotting a 2% isometric curve from the creep curves and reading the design stress at a service life of 20,000 h.) Substituting the design stress into the equation for *h* gives

$$h = \frac{8 \times 150}{2 \times 167.7} = 3.58\text{ mm}$$

It may be seen from the creep curves (Figure 2.18a) that when the pipe is first pressurized, the strain is less than 1%. Then as the material creeps, the strain increases steadily to reach its limit of 2% at 20,000 h.

In both examples it has been assumed that the service temperature is 20°C. If this is not the case, then creep curves at the appropriate temperature should be used. However, if none are available, a linear extrapolation between available temperatures may be sufficient for most purposes.

Again, for some materials like nylon the moisture content of the material has a significant effect on its creep behavior. In such a case creep curves are normally available for the material in both wet and dry states, and appropriate data should be used, depending on the service conditions.

2.2.13 Effect of Temperature

Many attempts have been made to obtain mathematical expressions which describe the time and temperature dependence of the strength of plastics. Since for many plastics at constant temperature a plot of stress, σ, against the logarithm of time to failure (creep rupture), t, is approximately linear, one of the expressions most commonly used is

$$t = Ae^{-B\sigma} \tag{2.72}$$

where A and B are constants. In reality, however, they depend on factors such as material structure and on temperature.

The most successful attempts to include the effects of temperature in a relatively simple expression have been made by Zhurkov and Bueche, who used an equation of the form [7]

$$t = t_0 \exp\left(\frac{U_0^{-\gamma\sigma}}{RT}\right) \tag{2.73}$$

where t_0 is a constant which has approximately the same value for most plastics, U_0 is the activation energy of the fracture process, γ is a coefficient which depends on the structure of the material, R is the molar gas constant, and T is the absolute temperature.

A series of creep rupture tests on a given material at a fixed temperature would permit the values for U_0 and γ for the material to be determined from this expression. The times to failure at other stresses and temperatures could then be predicted.

The relative effects of temperature rises on different plastic materials depend on the structure of each material and, particularly, whether it is crystalline or amorphous. If a plastic is largely amorphous (e.g., polymethyl methacrylate, polystyrene), then it is the glass transition temperature (T_g) which will determine the maximum service temperature, since above T_g the material passes into the rubbery region (see Figure 1.19 in *Plastics Fundamentals, Properties, and Testing*).

On the other hand, in plastics which have a high degree of crystallinity (e.g., polyethylene, polypropylene), the amorphous regions are small, so T_g is only of secondary importance. For them it is the melting temperature which will limit the maximum service temperature. The lowest service temperatures which can be used are normally limited by the brittleness introduced into the material. The behavior of plastics materials at room temperature is related to their respective T_g values. This aspect has been dealt with in Chapter 1 of *Plastics Fundamentals, Properties, and Testing.*

2.2.14 Time–Temperature Superposition

In engineering practice, it is often necessary to design for the use of a material over a long period of time—many years, for example. A common parameter to use in design work is the elastic modulus. We know, however, that for polymers the modulus decreases with increasing time under load.

Accumulation of long-term data for design with plastics can be very inconvenient and expensive. A method is thus needed to extrapolate data from shorter time studies at higher temperature to longer times over several decades of time scale at the desired temperature so that a lower limit of the modulus can be determined for use in design. On the other hand, it is sometimes difficult to obtain data over a very short time scale. One must then extrapolate data obtained under practicable experimental conditions to these short time scales. An empirical method for such extrapolations is available for amorphous polymer systems and, in general, for polymer systems where structure does not change during the period of testing.

The aforesaid extrapolations make use of a time-temperature superposition principle which is based on the fact that time and temperature have essentially equivalent effects on the modulus values of amorphous polymers. Figure 2.19 shows modulus data taken at several temperatures for poly(methyl methacrylate) [8]. Because of the equivalent effect of time and temperature, data at different

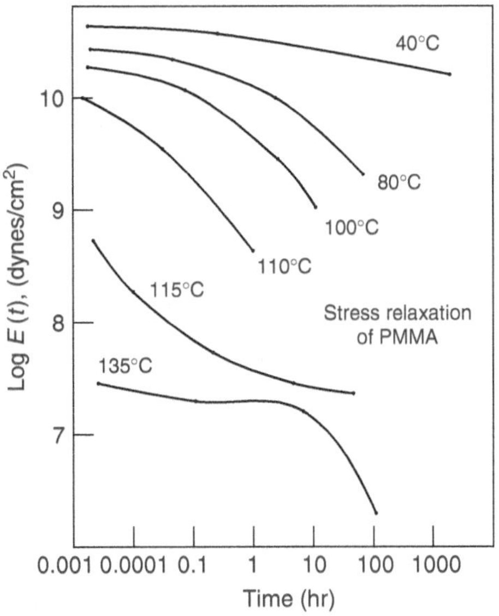

FIGURE 2.19 Logarithm of tensile relaxation modulus versus logarithm of time for unfractionated poly(methyl methacrylate) of $\bar{M}_v = 3.6 \times 10^6$. (After McLoughlin, J. R. and Tobolsky, A. V. 1952. *J. Colloid Sci.*, 7, 555.)

temperatures can be superposed on data taken at a specified reference temperature merely by shifting individual curves one at a time and consecutively along the log *t* axis about the reference temperature.

This time–temperature superposition procedure has the effect of producing a single continuous curve of modulus values extending over many decades of log *t* at the reference temperature. A curve constructed in this way, as shown in Figure 2.20 (with a reference temperature 115°C), is known as the master curve.

The time-temperature superposition can be expressed mathematically as

$$E(T_1,t) = E(T_2,t/a_T) \tag{2.74}$$

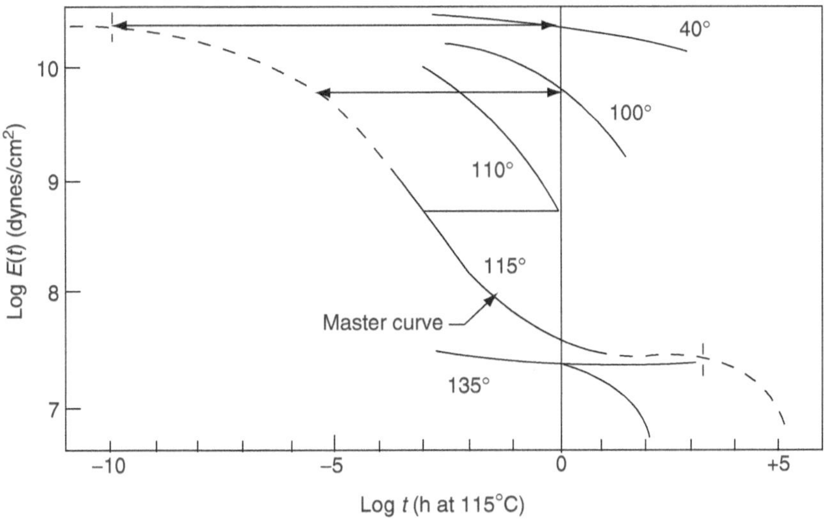

FIGURE 2.20 Modulus-time master curve based on time–temperature superposition of data in Figure 2.19. Times referred to temperature of 115°C.

for a tensile stress relaxation experiment ($T_2 > T_1$). The procedure asserts that the effect of changing the test temperature on viscoelastic properties is the same as that of multiplying or dividing the time scale by a constant quantity (a_T) at each temperature. The quantity a_T is called the *shift factor*, and it must be obtained directly from the experimental curve by measuring the amount of shift along the log t scale necessary to match the curve. The parameter a_T is chosen as unity at the reference temperature and is a function of the temperature alone, decreasing with increasing temperature.

It is common practice now to use the glass transition temperature (T_g) as the reference temperature for master curve construction. For most amorphous polymers, the shift factor at any other temperature T is then given fairly well by

$$\log_{10} a_T = \log \frac{t(T)}{t(T_g)} = \frac{-17.44(T - T_g)}{51.6 + (T - T_g)} \tag{2.75}$$

Equation 2.75 is known as the WLF equation (see Equation 2.27) after the initials of the researchers who proposed it [9]. The expression given holds between T_g and $T_g + 100°C$. However, if a different reference temperature is chosen an equation with the same form as Equation 2.75 can be used, but the constants on the right hand side must be re-evaluated.

The significance of the WLF generalization (Equation 2.75) cannot be over-emphasized. Again and again, one finds in the literature methods of superposing time and temperature for mechanical and other properties in amorphous and partially amorphous materials. Whatever modifications are introduced usually reduce the behavior back in the direction of Equation 2.75.

2.2.15 Dynamic Mechanical Properties

A complete description of the viscoelastic properties of a material requires information over very long times. To supplement creep and stress relaxation measurements which are limited by experimental limitations, experiments are therefore performed in which an oscillating stress or strain is applied to the specimen. These constitute an important class of experiments for studying the viscoelastic behavior of polymeric solids. In addition to elastic modulus, it is possible to measure by these methods the viscous behavior of the material in terms of characteristic damping parameters.

Damping is an engineering material property and the observed response is much more sensitive to the polymer constitution than in step-function experiments. Oscillatory experiments (also referred to as dynamic mechanical experiments) thus offer a powerful technique to study molecular structure and morphology. A significant feature is the breadth of the time-scale spectrum available with these methods, e.g., 10^{-5}–10^{8} cycles/sec.

In a dynamic experiment, the stress will be directly proportional to the strain if the magnitude of the strain is small enough. Then, if the stress is applied sinusoidally the resulting strain will also vary sinusoidally. (The same holds true if the strain is the input and the stress the output.) At sufficiently low frequencies, the strain will follow the stress in phase. However, in the general case the strain will be out of phase.

In the last instance, the strain can be factored into two components—one of which is in phase with the stress and the other which lags behind the stress by $\pi/2$ rad. Alternatively, the stress can be decomposed into a component in phase with the strain and one which leads the strain by $\pi/2$ rad. This is accomplished by use of a rotating vector scheme, as shown in Figure 2.21.

The magnitude of the stress at any time is represented by the projection OC of the vector **OA** on the vertical axis. Vector **OA** rotates with a frequency ω equal to that of the sinusoidally varying stress. The length of **OA** is the stress amplitude (maximum stress) involved in the experiment. The strain is represented by the projection OD of vector **OB** on the vertical axis. The strain vector **OB** rotates in the same direction as **OA** with frequency ω but it lags **OA** by an angle δ. The loss tangent (discussed later) is defined as tan δ.

The strain vector **OB** can be resolved into vector **OE** along the direction of **OA** and **OF** perpendicular to **OA**. Then the projection OH of **OE** on the vertical axis is the magnitude of the strain which is in phase

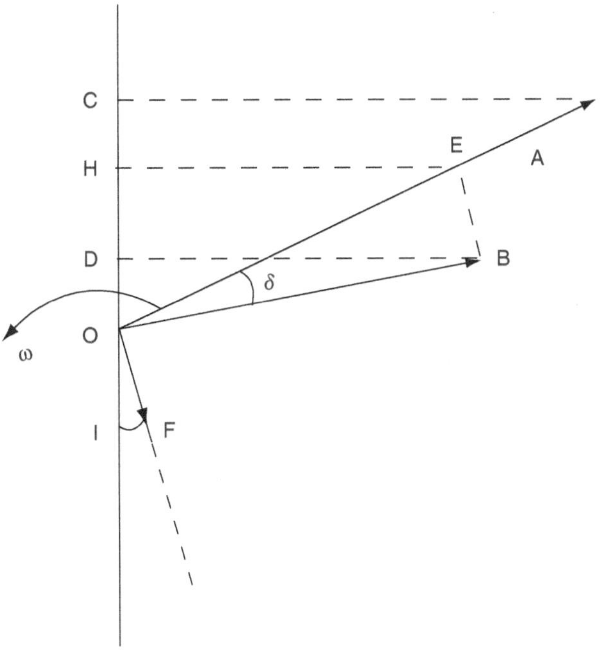

FIGURE 2.21 Decomposition of strain vector into two components in a dynamic experiment.

with the stress at any time. Similarly, projection OI of vector **OF** is the magnitude of the strain which is $\pi/2$ rad (one quarter cycle) out of phase with the stress.

The stress can be similarly resolved into two components with one along the direction of **OB** and one leading the strain vector by $\pi/2$ rad. The ratio of the in-phase stress to the strain amplitude (maximum strain) is called the storage modulus. In a shear deformation experiment this quantity is labeled $G'(\omega)$. The ratio of the out-of-phase stress to the strain is the loss modulus $G''(\omega)$.

If, on the other hand, the strain vector is resolved into its components, the ratio of the in-phase strain to stress amplitude (maximum stress) is the storage compliance $J'(\omega)$ and the ratio of the out-of-phase strain to the stress amplitude is the loss compliance $J''(\omega)$.

It is evident from the above description that $G'(\omega)$ and $J'(\omega)$ are associated with the periodic storage and complete release of energy in the sinusoidal deformation process. The loss parameters $G''(\omega)$ and $J''(\omega)$, on the other hand, reflect the nonrecoverable use of applied mechanical energy to cause viscous flow in the material. At a specified frequency and temperature, the dynamic response of a polymer in shear deformation can be summarized by any one of the following pairs of parameters: $G'(\omega)$ and $G''(\omega)$, $J'(\omega)$ and $J''(\omega)$, or absolute modulus $|G|$ and tan δ.

2.2.15.1 Maxwell Element

As an example, we shall consider the behavior of a Maxwell element in harmonic oscillation. If a sinusoidal force acts on a Maxwell element, the resulting strain will be sinusoidal at the same frequency, but out of phase. The same holds true if the strain is the input and the stress the output. For example, let the strain be a sinusoidal function of time t with a frequency ω rad/sec:

$$\varepsilon = \varepsilon_0 \sin \omega t \tag{2.76}$$

where ε_0 is the maximum or peak strain.

The motion of the Maxwell element (with modulus E and relaxation time $\lambda = \eta/E$) is given by (see Equation 2.36)

$$\frac{d\varepsilon}{dt} = \frac{1}{E}\frac{d\sigma}{dt} + \frac{\sigma}{\lambda E} \tag{2.77}$$

The resulting sinusoidal stress (see Appendix A7) is given by

$$\sigma = \varepsilon_0 E \frac{\omega\lambda}{(1+\omega^2\lambda^2)^{1/2}} \sin(\omega t + \delta) \tag{2.78}$$

where

$$\delta = \cot^{-1}\omega\lambda \tag{2.79}$$

Thus while the stress will be directly proportional to E, the magnitude will be affected by the products $\omega\lambda$ and will lead the strain by an angle δ. This situation is shown in Figure 2.22.

Instead of working with Equation 2.76 and Equation 2.78, it is easier to work with complex functions. Thus the strain may be written as a complex oscillating function of time with maximum amplitude ε_0 and frequency ω:

$$\varepsilon^* = \varepsilon_0 \exp(i\omega t) \tag{2.80}$$

The real strain is the real part of the complex strain ε^*.

The resulting complex stress σ^* is given (see Appendix A7) by

$$\sigma^* = \varepsilon^* \left[\frac{E\omega^2\lambda^2}{1+\omega^2\lambda^2} + \frac{i(E\omega\lambda)}{1+\omega^2\lambda^2} \right] \tag{2.81}$$

The complex elastic (Young's) modulus E^* is expressed as

$$E^* = E' + iE'' = \frac{\sigma^*}{\varepsilon^*} \tag{2.82}$$

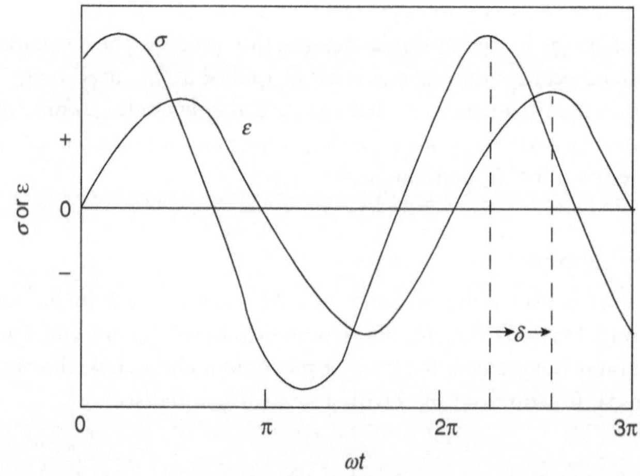

FIGURE 2.22 Strain (ε) and stress (σ) as a function of time (t) on application of a sinusoidal strain to a viscoelastic specimen at a frequency ω.

Identifying Equation 2.81 and Equation 2.82 together, we conclude that

$$E' = \frac{E\omega^2\lambda^2}{1+\omega^2\lambda^2} \text{ and } E'' = \frac{E\omega\lambda}{1+\omega^2\lambda^2} \tag{2.83}$$

By the conventions of complex notation we find a loss angle δ, where

$$E^* = [(E')^2 + (E'')^2]^{1/2}\exp(i\delta) \quad \text{and} \quad \tan\delta = \frac{E''}{E'} \tag{2.84}$$

From Equation 2.83 and Equation 2.84,

$$\tan\delta = 1/\omega\lambda \tag{2.85}$$

$$[(E')^2 + (E'')^2]^{1/2} = \frac{E\omega\lambda}{(1+\omega^2\lambda^2)^{1/2}} \tag{2.86}$$

Another term sometimes used is the dynamic tensile viscosity, $\eta' = E''/\omega$. Using Equation 2.80, Equation 2.84, and Equation 2.86, one further obtains

$$\begin{aligned}\sigma^* = E^*\varepsilon^* &= \frac{E\omega\lambda\varepsilon^*}{(1+\omega^2\lambda^2)^{1/2}}\exp(i\delta)\\ &= \frac{E\omega\lambda\varepsilon_0}{(1+\omega^2\lambda^2)^{1/2}}\exp(i\omega t + i\delta)\end{aligned} \tag{2.87}$$

The actual stress σ is the real part of σ^*. Equation 2.87 shows that σ^* depends on E (Hookean model constant) and ε^* but leads by an angle δ. The maximum value of stress σ_0 occurs at $\omega t + \delta = 0$, that is,

$$\sigma_0 = \frac{E\omega\lambda\varepsilon_0}{(1+\omega^2\lambda^2)^{1/2}} \tag{2.88}$$

$$\sigma^* = \sigma_0 \exp(i\omega t + i\delta) \tag{2.89}$$

By substituting Equation 2.88 in Equation 2.83 we can also write

$$E' = \frac{\sigma_0\omega\lambda}{\varepsilon_0(1+\omega^2\lambda^2)^{1/2}} \tag{2.90}$$

Thus we can evaluate E, λ, E', and E'' from experimental measurement of ε_0, σ_0, and δ at a known value of ω. In experiments involving shear stress (τ) and shear strain (γ), one evaluates G, λ, G', and G'' from experimental measurements of γ_0, τ_0, and δ at a known value of ω.

The dynamic modulus E' (or G'), which is the real component of E^* (or G^*), is associated with energy storage and release in the periodic deformation and is therefore called the storage modulus. The imaginary part of the modulus, E'' (or G''), is associated with viscous energy dissipation and is a measure of the energy lost per cycle per unit volume (u_e), since

$$u_e = \pi E''\varepsilon_0^2 \tag{2.91}$$

where, ε_0 is the maximum amplitude of strain. Hence E'' and G'' are called the *loss moduli.* These are the damping terms that give a measure of the dissipation of energy into heat as the material is deformed. The useful damping term called the dissipation factor or loss tangent ($\tan\delta$) is given by (G''/G') or (E''/E') (see Equation 2.84). The value of $\tan\delta$ thus gives a measure of the relative importance of the viscous and elastic behavior of the material.

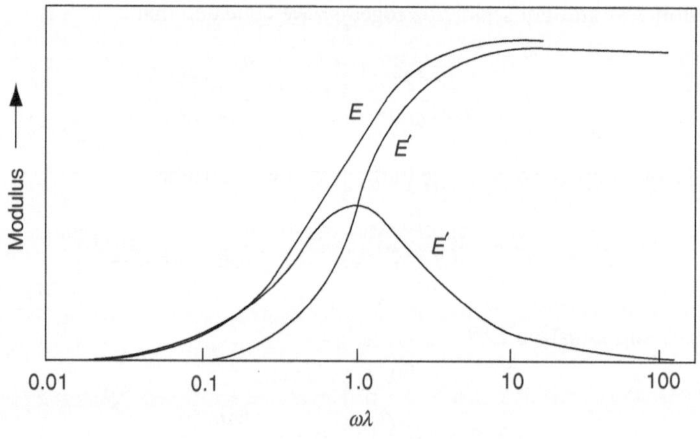

FIGURE 2.23 Dynamic behavior of a Maxwell unit as a function of frequency ω.

The dynamic modulus and damping of a Maxwell unit as a function of frequency are schematically shown in Figure 2.23. The nature of the curves in Figure 2.23 can be explained qualitatively as follows.

At low frequencies, the dynamic modulus is very low as most of the deformation or motion comes from the dashpot (viscous component). At very high frequencies, enough time is not available to allow appreciable flow in the dashpot in a given cycle, and hence the deformation or motion is mostly due to the stretching of the spring such that the dynamic modulus nearly equals the modulus of the spring (elastic component).

At intermediate frequencies, the deformations or motions of both the spring and dashpot take place measurably and in this frequency range, the dynamic modulus (E') increases rapidly with frequency and so also the loss modulus (E''). The latter, however, approaches zero at low as well as high frequencies, and passes through a maximum when the frequency $\omega = 1/\lambda$, where λ is the relaxation time. The absolute value of the modulus E is equal to E'' at low and to E' at high frequencies.

The above discussion also applies to dynamic experiments in shear for which the corresponding moduli are represented by G, G' and G''.

The analysis presented above for the Maxwell element to explain the significance of dynamic testing can be extended to the Voigt element and corresponding expressions for moduli can be derived. However, models comprised of single elements are useful only as pedagogical tools. They can be combined in series or parallel to yield useful models.

2.2.15.2 Terminology of Dynamic Mechanical Experiments

Dynamic testing is the most commonly used one in the study of polymer properties. Because of its widespread usage, dynamic data have been analyzed in terms of six different functions, as opposed to a single one in stress relaxation and creep. The experiment involves (1) imposition (on a specimen of the material) of either a shear stress or a shear strain which varies sinusoidally with time, and (2) study of the corresponding response.

Using complex functions, the oscillating imposed shear stress and the corresponding shear strain may be written as

$$\tau^*(t) = \tau_0 \exp(i\omega t) \tag{2.92}$$

$$\gamma^*(t) = \gamma_0(\omega) \exp\{i[\omega t - \delta(\omega)]\} \tag{2.93}$$

where $\tau^*(t)$ and $\gamma^*(t)$ are complex functions and $i = \sqrt{-1}$; ω is the frequency of oscillation and τ_0 is its amplitude; $\gamma_0(\omega)$ is the amplitude of the sinusoidal response and $\delta(\omega)$ is the phase difference between stress and strain; and the parenthesis indicates functional dependence.

The dynamic modulus, $G^*(\omega)$, is defined as the ratio of the complex stress to the complex strain

$$G^*(\omega) \equiv \frac{\tau^*(t)}{\gamma^*(t)} = \frac{\tau_0}{\gamma_0(\omega)} e^{i\delta(\omega)} \equiv G_0(\omega) e^{i\delta(\omega)} \tag{2.94}$$

$G_0(\omega)$ and $\delta(\omega)$ represent the first two material properties that can be obtained from dynamic experiments.

The next two material properties are obtained by writing $G^*(\omega)$ in terms of real and imaginary parts as

$$G^*(\omega) = G'(\omega) + iG''(\omega) \tag{2.95}$$

$G'(\omega)$ is referred to as the storage modulus and $G''(\omega)$ as the loss modulus since, as explained before, they are related, respectively, to the energy stored and energy lost in the material. Relationships between $G'(\omega)$, $G''(\omega)$, $G_0(\omega)$ and $\delta(\omega)$ can be written from the fundamentals of complex variables as follows:

$$\begin{aligned} G'(\omega) &= G_0(\omega) \cos[\delta(\omega)] \\ G''(\omega) &= G_0(\omega) \sin[\delta(\omega)] \\ \tan[\delta(\omega)] &= G''(\omega)/G'(\omega) \\ G_0(\omega) &= [\{G'(\omega)\}^2 + \{G''(\omega)\}^2]^{1/2} \end{aligned} \tag{2.96}$$

Alternatively, $\gamma^*(t)$ and $\tau^*(t)$ may be related in terms of complex dynamic compliance, $J^*(\omega)$, as

$$J^*(\omega) \equiv \frac{\gamma^*(t)}{\tau^*(t)} = \frac{\gamma_0(\omega)}{\tau_0} \mathrm{e}^{-i\delta(\omega)} \equiv J_0(\omega) \mathrm{e}^{-i\delta(w)} \tag{2.97}$$

and the complex dynamic compliance may be written in terms of real and imaginary parts, $J'(\omega)$ and $J''(\omega)$, which are termed storage compliance and loss compliance, respectively, as

$$J^*(\omega) = J'(\omega) - iJ''(\omega) \tag{2.98}$$

Again a series of simple algebraic manipulations leads to the following interrelationships:

$$\begin{aligned} G^*(\omega) &= \frac{1}{J^*(\omega)} \\ J'(\omega) &= \frac{G'(\omega)}{\{G'(\omega)\}^2 + \{G''(\omega)\}^2} = \frac{\cos[\delta(\omega)]}{G_0(\omega)} \\ J''(\omega) &= \frac{G''(\omega)}{\{G''(\omega)\}^2 + \{G''(\omega)\}^2} = \frac{\sin[\delta(\omega)]}{G_0(\omega)} \\ \tan[\delta(\omega)] &= J''(\omega)/J'(\omega) \\ G_o^2(\omega) = \{G'(\omega)\}^2 + \{G''(\omega)\}^2 &= \frac{1}{\{J_o(\omega)\}^2} = \frac{2}{\{J'(\omega)\}^2 + \{J''(\omega)\}^2} \end{aligned} \tag{2.99}$$

In the above discussion, six functions $G_0(\omega)$, $\delta(\omega)$, $G'(\omega)$, $G''(\omega)$, $J'(\omega)$, and $J''(\omega)$ have been defined in terms of an idealized dynamic testing, while earlier we defined shear stress relaxation modulus $G(t)$ (see Equation 2.19) and shear creep compliance $J(t)$ (see Equation 2.21) in terms of an idealized stress relaxation experiment and an idealized creep test, respectively. Mathematical relationships relating any one of these eight functions to any other can be derived. Such relationships for interconversion of viscoelastic function are described by Ferry [5], and interested readers are referred to this treatise for the same.

Two reasons can be given as to why so many functions have been defined, especially since the entire information could be presented in terms of only a single one and all the remaining seven functions are derived from it. First, no single instrument can give the entire dependence of any function over a large enough period of time or frequency extending over several decades. Indeed experimental data are taken with different instruments yielding different material properties and the results are converted in terms of any one desired viscoelastic function.

The second and more important reason for defining so many different functions is that each of these emphasizes different aspects of molecular motions associated with the macroscope distortion of the material. Thus, in order to gain sufficient insight into these molecular mechanisms, several or all of these functions should be studied.

It should be noted that the treatment in this section is restricted to the study of small deformations; that is, only when the material is subjected either to low shear strains (below about 1%) or to low shear stresses. Under such conditions, the eight material properties are functions of time or frequency only and are independent of the magnitude of the applied shear stress or strain. The material is then said to be in the linear viscoelastic region.

Though most practical applications of polymers would seem to involve much higher stresses, the importance of the study of linear viscoelasticity stems from the fact that it leads to a very sound understanding of the molecular mechanisms of flow and sufficient insight is gained by its study so as to be of immense advantage in the study of large deformations. The study of large deformations, however, is almost empirical, and the reader is referred to several excellent works [10,11].

Rheovibron (dynamic) viscometer is widely used for measurements of dynamic mechanical properties such as loss modulus, storage modulus, and dissipation factor, each as a function of temperature. In this instrument, the test specimen is clamped between strain gauges and subjected to low order of sinusoidal strain at a specified frequency. The value of tan δ is directly read and the storage and loss moduli are calculated using sample dimensions and instrument readings.

In a rheometric dynamic spectrometer such as RDS7700 manufactured by Rheometrics, Inc., the torque and normal force generated in response to an imposed motion are measured by a transducer. A microcomputer determines stresses from these values with measured sample motion to calculate strains and viscoelastic functions such as G', G'', and tan δ [12].

2.2.15.3 Dynamic Mechanical Behavior at Thermal Transitions

Modulus-temperature relations for amorphous polymers in static test reveal a sharp drop of modulus in the glass-to-rubber transition region (see Figure 1.19). Since the storage modulus $G'(\omega)$ behaves like a modulus measured in a static test, it decreases in the glass transition region. However, the loss modulus $G''(\omega)$ and tan δ go through a maximum under the same conditions.

Figure 2.24 shows some typical experimental data. It is seen that T_g is easily identifiable as a peak in the tan δ or the loss modulus trace. These maxima do not coincide exactly. The maximum in tan δ is at a higher temperature than that in $G''(\omega)$, because tan δ is the ratio of $G''(\omega)$ and $G'(\omega)$ (see Equation 2.96) and both these moduli are changing in the transition region. At low frequencies (about 1 Hz) the peak in tan δ is about 5°C higher than T_g from static measurements or the maximum in the loss modulus-temperature curve.

The development of a maximum in tan δ or the loss modulus at the glass-to-rubber transition can be explained as follows. At temperatures below T_g the polymer behaves elastically, and there occurs little or no flow to convert the applied energy into internal work in the material. To explain more explicitly, let h be the energy dissipated as heat per unit volume of material per unit time because of flow in shear deformation. Then

$$h = \sigma(d\gamma/dt) = \eta(d\gamma/dt)^2 \tag{2.100}$$

Thus the work dissipated is proportional to the viscosity (η) of the material at fixed straining rate $d\gamma/dt$.

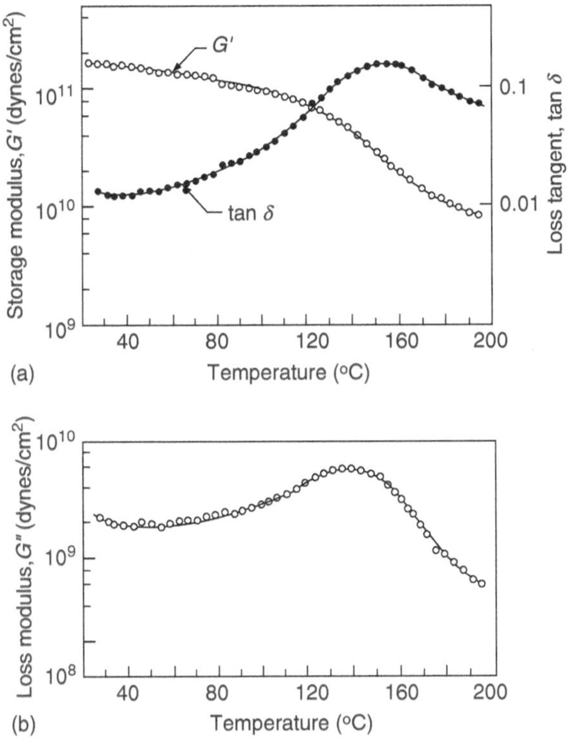

FIGURE 2.24 (a) Storage modulus (G') and loss tangent (tan δ) for an oriented poly(ethylene terephthalate) fiber; 11 Hz frequency. (b) Loss modulus (G'') for the same fiber.

At low temperatures, η is very high but γ and $d\gamma/dt$ are vanishingly small and h is negligible. As the structure is loosened in the transition region, η decreases but $d\gamma/dt$ becomes much more significant so that h (and hence the loss modulus and tan δ) increases.

The effective straining rate of polymer segments continues to increase somewhat with temperature above T_g, but η decreases at the same time. The net result is a decrease in damping and a fall-off of the magnitudes of the storage modulus and tan δ.

2.2.16 Rheological Behavior

Rheology is the science of deformation and flow of matter. Essentially, all thermoplastic resins (and many thermosetting resins) are required to undergo flow in the molten state during the course of product manufacture. Important fabrication processes such as injection, extrusion, and calendering all involve the flow of molten polymers. In plastics fabrication, it is important to understand the effect, on melt viscosity, of such factors as temperature, pressure, rate of shear, molecular weight, and structure. It is also equally important to have reliable means of measuring viscous properties of materials.

The flow behavior of polymeric melts cannot be considered to be purely viscous in character. The response of such materials is more complex, involving characteristics that are both viscous and elastic. This is only to be expected when one is trying to deform variously entangled long-chain molecules with a distribution of molecular weights.

During flow, polymer molecules not only slide past each other, but also tend to uncoil—or at least they are deformed from their equilibrium, random coiled-up configuration. On release of the deforming stresses, these molecules tend to revert to random coiled-up forms. Since molecular entanglements cause the molecules to act in a cooperative manner, some recovery of shape corresponding to the recoiling occurs. In phenomenological terms, we say that the melt shows elasticity in addition to viscous flow. The elastic—that is to say, time-dependent—effects play a most important part in die swell, extrusion defects, and melt fracture, which will be considered later in this section.

2.2.16.1 Classification of Fluid Behavior

Although one can measure deformation in a solid, one cannot normally do this in a liquid since it undergoes a continuously increasing amount of deformation when a shear stress is applied. But one can determine the deformation rate (the shear rate) caused by an applied shear stress or vice versa, and fluid behavior can be classified on this basis.

We begin by making a reference to Figure 2.25, which schematically illustrates two parallel plates of very large area A separated by a distance r with the space in between filled with a liquid. The lower plate is fixed and a shear force F_s is applied to the top plate of area A producing a shear stress ($\tau = F_s/A$) that causes the plate to move at a uniform velocity v in a direction parallel to the direction of the force.

It may be assumed that the liquid wets the plates and that the molecular layer of liquid adjacent to the stationary plate is stationary while the layer adjacent to the top plate moves at the same velocity as the plate. Intermediate layers of liquid move at intermediate velocities, and at steady state in laminar flow a velocity distribution is established as indicated by the arrows in the diagram. The velocity gradient between the two plates is dv/dr. It is defined as the *shear rate* and is commonly given the symbol $\dot{\gamma}$ i.e.,

$$\dot{\gamma} = dv/dr \tag{2.101}$$

If the liquid is ideal and it is maintained at a constant temperature, the shear stress is linearly and directly proportional to the shear rate such that one may write

$$\tau = \eta(dv/dr) = \eta\dot{\gamma} \quad \text{or} \tag{2.102}$$

$$\eta = \tau/\dot{\gamma} \tag{2.103}$$

where η is the coefficient of viscosity or simply the viscosity or internal friction of the liquid. The linear relationship between τ and $\dot{\gamma}$ given by Equation 2.102 or Equation 2.103 is known as Newton's law and liquids which behave in this manner are called Newtonian fluids or ideal fluids. Other fluids which deviate from Newton's law are described as non-Newtonian. For such fluids, the viscosity defined by Equation 2.103 is also known as the apparent viscosity.

In practice, the Newtonian behavior is confined to low molecular weight liquids. Polymer melts obey Newton's law only at shear rates close to zero and polymer solutions only at concentrations close to zero. The most general rheological equation is

$$\eta = f(\dot{\gamma}, T, t, P, c, \ldots\ldots) \tag{2.104}$$

where the variables are $\dot{\gamma}$ = shear rate (itself a function of the shear stress), T = temperature, t = time, P = pressure (itself a function of volume), c = concentration, and the multiple dots which follow include, for example, molecular parameters (such as molecular weight and molecular weight distribution), compositional variables (crystallinity and the presence of additives), and factors that relate to processing history. Such an equation is clearly unrealistic, so we shall consider here some of the principal variables, one at a time, assuming that the others remain constant.

Several common types of rheological behavior are shown in Figure 2.26 based upon τ vs. $\dot{\gamma}$ curves. These flow phenotypes are named Newtonian,

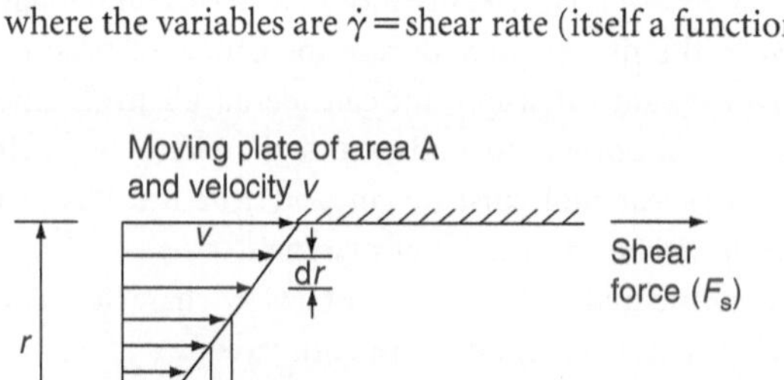

FIGURE 2.25 Velocity distribution of a liquid between two parallel plates, one stationary and the other moving.

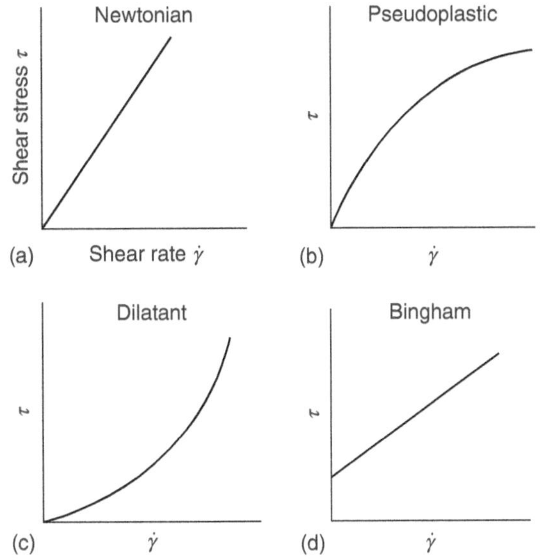

FIGURE 2.26 Flow curves (τ versus $\dot{\gamma}$) for different types of fluid material.

pseudoplastic, dilatant, and Bingham. In Newtonian liquids, the viscosity is constant and independent of shear rate.

In pseudoplastic and dilatant liquids the viscosity is no longer constant. In the former it decreases and in the latter it increases with increasing shear rate; that is to say, the shear stress increases with increasing shear rate less than proportionately in a pseudoplastic and more than proportionately in a dilatant. Pseudoplastics are thus described as shear-thinning and dilatants as shear-thickening fluid systems. These two flow phenotypes can be described by an Equation the 'power law':

$$\tau = \eta_N \dot{\gamma}^n \tag{2.105}$$

where η_N is the zero shear (Newtonian) viscosity. The exponent n is greater than unity for a dilatant and less than unity for a pseudoplastic. A Newtonian is then seen to be a special case, with $n=1$.

Equation 2.105 gives a linear relationship between log τ and log $\dot{\gamma}$ and the slope of the experimental plot (Figure 2.27) gives the value of n. For the analysis of flow behavior of many systems, the power law relationship has been useful as from a plot of data (Figure 2.27), and measuring the slope, one can readily get an idea of just how non-Newtonian the fluid is.

A Bingham body would be described by the equation

$$\tau - \tau_y = \eta \dot{\gamma} \tag{2.106}$$

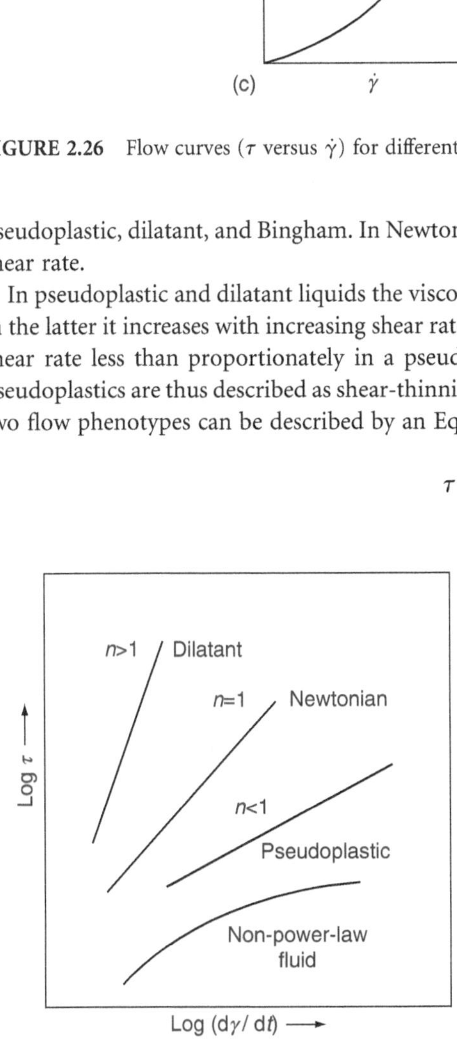

FIGURE 2.27 Power-law plot showing log τ versus log $(d\gamma/dt)$ for different types of fluid material (schematic).

where τ_y is the yield stress or yield value. Below τ_y the material will not flow at all (hence $\dot{\gamma} = 0$ and $\eta = \infty$) so that the material, to all intents and

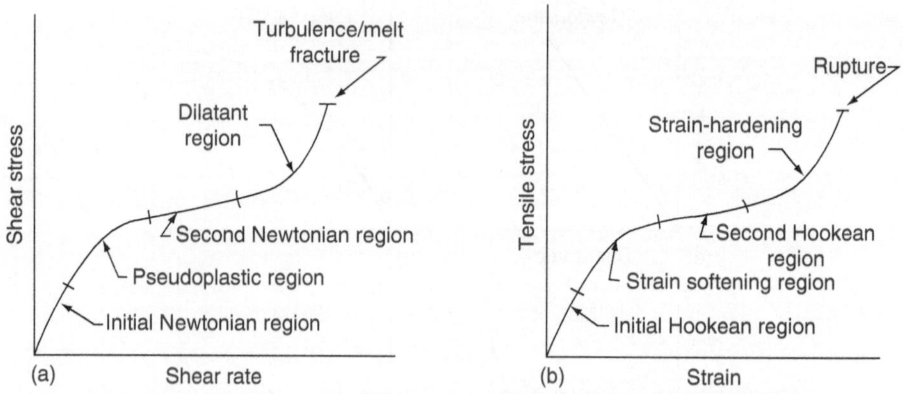

FIGURE 2.28 (a) Generalized flow curve. (b) Typical fully developed stress–strain curve as found in tough plastics under appropriate conditions. The conventional stress has been converted to the true stress.

purposes, is a solid. However, as soon as τ exceeds τ_y the material suddenly behaves like a liquid with a viscosity that remains constant with increasing shear rate. Materials which exhibit this type of behavior include drilling muds, sewage sludge, toothpaste, greases and fats, as well as the clay slurries originally observed by Bingham.

Lenk [13] has shown that the flow phenotypes form part of a general response pattern which may be summarized in a general flow curve. The generalized flow curve is shown in Figure 2.28 alongside a fully developed stress–strain curve for a typical tough solid in tension after conversion of the conventional stress (i.e., force per unit of original cross-sectional area) to true stress (i.e., force per unit of actual cross-sectional area after deformation). This conversion can be effected if a continuous record of the changes in cross-sectional area of the specimen under test is kept. It may be seen that the shapes of the two curves are absolutely identical and the regions into which the two curves divide have also analogous physical significance.

It is well known that at extremely low shear rates the slope of the $\tau/\dot{\gamma}$ curve (Figure 2.26) is constant and that there exists some very low but finite threshold shear rate beyond which deviation from linearity commences. The slope of the initial linear portion of the curve is known as the 'limiting viscosity,' the zero shear viscosity, or the Newtonian viscosity. Beyond this low shear rate region (initial Newtonian regime) the material is shear-softened (i.e., becomes pseudoplastic), a phenomenon which has its counterpart in the solid state where it is known as strain-softening.

Continuing in the pseudoplastic region it is often found that an upper threshold can be reached beyond which no further reduction in viscosity occurs. The curve then enters a second linear region of proportionality the slope of which is the second Newtonian viscosity.

Polymer melts are almost invariably of the pseudoplastic type, and the existence of first and second Newtonian regions has long been recognized. The pseudoplastic behavior appears to arise from the elastic nature of the melt and from the fact that under shear, polymers tend to be oriented.

At low shear rates Brownian motion of the segments occurs so polymers can coil up (re-entangle) at a faster rate than they are oriented. At higher shear rates such re-entangling rates are slower than the orientation rates and the polymer is hence apparently less viscous. However at very high shear rates (beyond the range of usual interest in the polymer processing industries) the degree of orientation reaches a maximum and so a further decrease in effective viscosity cannot occur; the polymer in this range again becomes Newtonian.

Generally speaking, the larger the polymer molecule the longer the recoiling (re-entangling, relaxation) time so that high-molecular-weight materials tend to be more non-Newtonian at lower shear rates than lower-molecular-weight polymers.

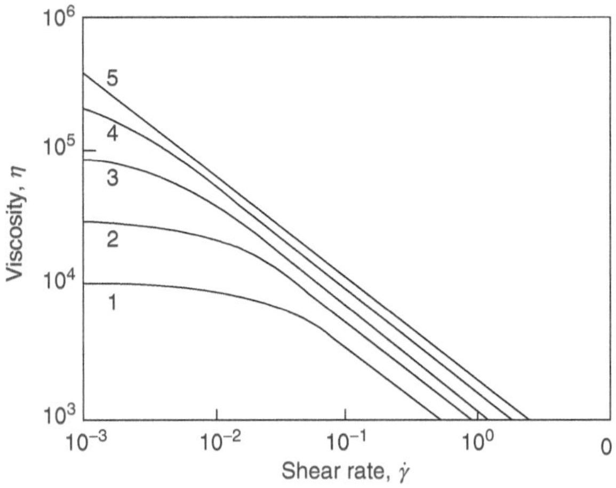

FIGURE 2.29 Typical log viscosity-log shear rate curves at five different temperatures. Curve 1 is for the highest temperature and curve 5 is for the lowest temperature. For a typical polymer, the temperature difference between each curve is approximately 10°C.

The dilatant type is less common among plastics under ordinary conditions, but it can be found in heavily filled systems and in some PVC pastes. The dilatant type (Figure 2.26c) represents the entire generalized flow curve in which the initial Newtonian and pseudoplastic regions have degenerated to a vanishingly small portion of the curve as a whole. In some cases of dilatancy a linear region may be distinguished before the curvature appears. In other cases no distinctly linear portion can be seen at the low-shear-rate end. The Bingham type is not common among plastics.

2.2.16.2 Effect of Shear Rate on Viscosity

Most polymer melts exhibit non-Newtonian behavior with the apparent viscosity decreasing with increasing shear rate (*shear thinning*). Viscosity of polymer at high shear rates may be several orders of magnitude smaller than the viscosity at low shear rates. Typical shear rate vs. viscosity curves are shown in Figure 2.29. The polymer melt has a Newtonian viscosity which is high at very low shear rates. Viscosity decreases nearly linearly with shear rate when plotted on a log–log scale (Figure 2.29). In this linear range, the power law equation (Equation 2.105) with $n<1$ is applicable.

The reduction of viscosity with increasing rate of shear [14,15] is taken advantage of in achieving desirable and optimum viscosity for polymers in processing machines and equipment without raising the temperature to detrimental levels, simply by raising the shear rate to as high a level as economically and otherwise possible. A reduction in viscosity with increasing shear rate is also taken advantage of in brushing and spraying of paints which are polymer solutions/suspensions containing pigments.

2.2.16.3 Effect of Molecular Weight on Viscosity

The molecular weight of a polymer is the most important factor affecting rheology. For most polymers the zero-shear viscosity is approximately proportional to the weight-average molecular weight (M_w) below a critical value (M_c) and depends on M_w to a power equal to 3.5 at molecular weights above M_c:

$$\eta = K_1 M_w \quad \text{for} \quad M_w < M_c \tag{2.107}$$

and

$$\eta = K_2 M_w^{3.5} \quad \text{for} \quad M_w > M_c \tag{2.108}$$

The relationship between viscosity and molecular weight shown by a graphical logarithmic plot in Figure 2.30 is characterized by a sharp change at the critical molecular weight (M_c). The value of M_c varies from one polymer to another. For most polymers M_c is between 5,000 and 15,000.

The critical molecular weight corresponding to the transition in the viscosity behavior at M_c points to additional hindrance to flow from this point onwards due to chain entanglements. Below this point the molecules usually move independently as in low-molecular-weight liquids, but above this point mutual entanglements of chain molecules is so prominent that movement of one involves dragging of others along with it; as a consequence, increasing molecular weight is associated with a very high rate of viscosity increase.

Considering the fact that the viscosity is increasing logarithmically with molecular weight in Figure 2.30, it should be clear why molecular weight control is important in polymer processing. One needs a molecular weight high enough to attain good mechanical properties but not so high that the molten polymer is too viscous to be processed economically.

The distribution of molecular weights in a polymer also influences its rheology. In general, the broader the range, the lower the shear rate at which shear thinning (decrease in viscosity) develops. Thus polymers with broad molecular-weight distribution are easier to extrude than those with narrow distribution.

Chain branching is another factor that influences flow. The more highly branched a given polymer, the lower will be its hydrodynamic volume and the lower its degree of entanglement at a given molecular weight. One can make the general observation, therefore, that viscosity is higher with linear than with branched polymers at a given shear rate and molecular weight. This does not mean that chain branching is necessarily desirable. In fact, branching results in weaker secondary bonding forces and possibly poorer mechanical properties.

2.2.16.4 Effect of Temperature on Polymer Viscosity

The viscosity of most polymers changes with temperature. An Arrhenius equation of the form

$$\eta = Ae^{E/RT} \tag{2.109}$$

where A is a constant and E is the activation energy, has often been used to relate viscosity and temperature. Constants in the Arrhenius equation can be evaluated by plotting the logarithm of viscosity against the reciprocal of absolute temperature, using shear stress or shear rate as a parameter. The data for most materials give straight line over reasonably large range of temperature.

Whilst the Arrhenius equation can be made to fit experimental data quite well it does nothing to explain the difference between polymers. In this regard, the WLF equation (see Equation 2.27):

$$\log\left(\frac{\eta_T}{\eta_{T_g}}\right) = \frac{-17.44(T - T_g)}{51.6 + (T - T_g)}$$

is more useful. Melt viscosity, according to this equation, is a function of $(T - T_g)$. Thus, for example, a major cause of the difference between the viscosity of poly(methacrylate) at its processing temperature (where $T - T_g = 100°C$) and the

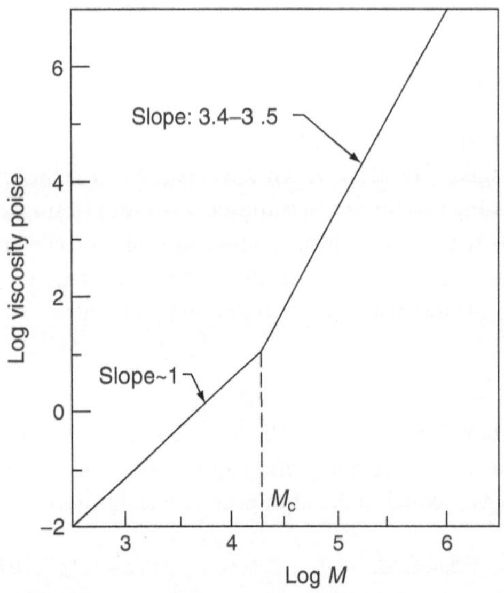

FIGURE 2.30 Dependence of polymer (melt) viscosity on molecular weight (M): a typical plot of log viscosity against log M.

viscosity of polyethylene at its processing temperature (where $T - T_g = 200°C$) is explicable by this relationship. The WLF equation also explains why viscosity is more temperature sensitive with materials processed closer to their T_g, for example, poly(methyl methacrylate), compared with nylon 6.

2.2.16.5 Effect of Pressure on Viscosity

While temperature rises at constant pressure cause a decrease in viscosity, pressure rises at constant temperature cause an increase in viscosity since this causes a decrease in free volume. It is commonly found that

$$\left(\frac{\partial\eta}{\partial T}\right)_v \simeq 0 \tag{2.110}$$

In other words, if the volume and hence free volume are made constant by increasing pressure as temperature is increased then the viscosity also remains constant. It is in fact found that within the normal processing temperature range for a polymer it is possible to consider an increase in pressure as equivalent, in its effect on viscosity, to a decrease in temperature.

For most polymers an increase in pressure of 100 atm is equivalent to a drop of temperature in the range 30–50°C. It is also found that those polymers most sensitive to temperature changes in their normal processing range are the most sensitive to pressure.

2.2.16.6 Weissenberg Effects

An elastic aftereffect is generally found in high-molecular-weight fluid materials after extrusion under high shear stress through an orifice or die, and this is seen to happen within a fraction of a second after extrusion. Herzog and Weissenberg [16] observed the existence of a "normal" force in polymers subjected to shear stress. In the polymeric melt systems, the entangled polymer chains get deformed elastically during flow and a different kind of force is generated within the flowing melt in addition to the force applied. The additional force generated is manifested as tensile force perpendicular to the shear plane.

This is visualized when a cone with vertical channels and with its axis normal to a plate is rotated with a viscous liquid placed between cone and plate. Liquid will climb into these channels (Figure 2.31a). The arrows in Figure 2.31a indicate that a force exists which acts on the liquid, normal to the shear plane. This force can be measured, without drilling channels into the cone, by placing pressure transducers in contact with the liquid at the cone face and by measuring the pressure exerted on the cone by the liquid.

The effect, known as the Weissenberg effect, is more easily demonstrated by subjecting a viscous liquid mass to shear in a coaxial cylinder system by rotating one while keeping the other fixed (Figure 2.31b). It

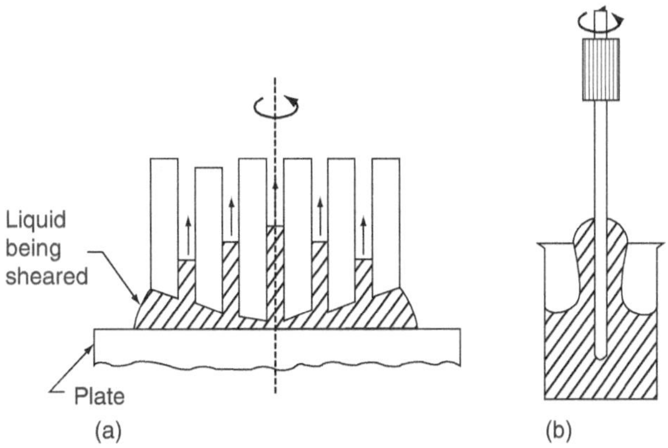

FIGURE 2.31 Experiments demonstrating the normal force (Weissenberg) effect. Liquid climb on rotation (a) in channels drilled into a cone and (b) in a coaxial cylinder.

is most easily demonstrated by rotating the inner cylinder. On rotation, the liquid climbs up the rotating inner rod or cylinder to a significant height.

The Weissenberg effect is clearly manifested in the increase in diameter of extruded profiles of a variety of molten polymers. The extrusion swelling (Figure 2.32), more commonly known as die swell, arises probably due to a combination of normal stress effects and a possible elastic recovery consequent to prior compression before the melt or liquid enters the die.

A practical aspect of the Weissenberg effect or die swell is that, on extrusion or calendaring, the melt coming out is larger than the die or nip size. The control of size is done by partly using a smaller aperture and by partly having an increased draw down. The extrusion, the swelling may be largely minimized by having a conical die design with the narrow end towards the interior and by lowering of melt temperatures, extrusion rates, or molecular weight of the polymer.

2.2.16.7 Irregular Flow or Melt Fracture

Above a certain shear stress considered as the critical shear stress, flow instabilities of many polymer melts exhibited by an abrupt change in the shape of the molten extrudate are found to occur. The irregularities in flow may be due to Reynolds turbulence or structural turbulence and thermo-mechanical breakdown of the polymer [17].

Above the critical shear stress, the material near the wall relaxes very much faster than the core material leading to the flow irregularities, which, according to Tordella [18], is caused by a fracture of the melt before its entry into the die. Polymers that show flow irregularities at low output rates are those that have comparatively long relaxation times. The origin of flow irregularities and the site of melt fracture is near the entrance to the die since this is the zone of greatest shear stress. The stress enhancement is so great compared to the relaxation time that the polymer melt fractures much like a solid.

Melt fracture depends on die geometry, molecular weight, molecular-weight distribution, and chain branching. A linear polymer such as high-density polyethylene is characterized by a higher critical shear stress than the corresponding branched polymer (low-density) polyethylene) of comparable molecular weight.

2.2.17 Measurement of Viscosity

Perhaps the most important factor to a process engineer in predicting extrusion or molding behavior is melt viscosity. Several methods are used to obtain the viscosity of polymer solutions and melts experimentally as a function of shear rate [19]. Instruments for making such measurements must necessarily accomplish two things: (1) the fluid must be sheared at measurable rates, and (2) the stress developed must be known. Two kinds of instruments having simple geometry and wide use a rotational viscometer and capillary or extrusion rheometer.

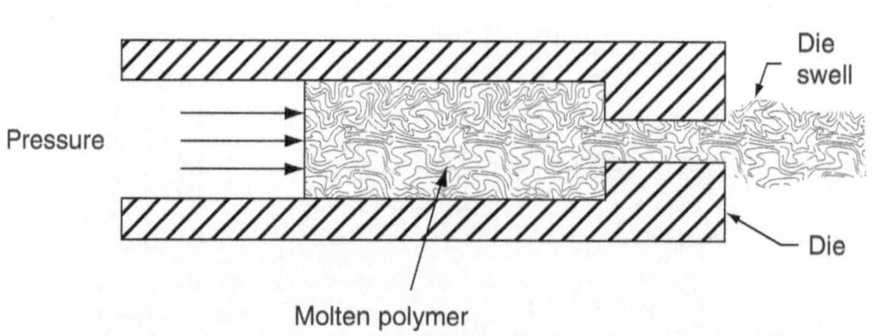

FIGURE 2.32 Schematic representation of polymer flow through a die orifice.

2.2.17.1 Rotational Viscometers

In a rotational viscometer (using cylinders, cones, spheres, and discs) the fluid is sheared at a given temperature in the annular or enclosed space due to rotation of the inner cylinder or the like device while the outer cylinder or device is kept stationary or vice versa. In either case the torque required for the rotation is a measure of the shearing stress and the speed of rotation gives a measure of the rate of shear. Rotational viscometers using coaxial cylinders (Figure 2.33a) measure relatively low viscosity liquids. Typical is the Haake-Rotovisco. In this device, the cup is stationary and the bob is driven through a torsion spring. In a cone-and-plate rotational viscometer (Figure 2.33b), the molten polymer is contained between the bottom plate and the cone, which is rotated at a constant velocity (Ω). Shear stress (τ) is defined as

$$\tau = \frac{3\Phi}{2\pi R_c^3} \tag{2.111}$$

where Φ is the torque in dynes per centimeter (CGS) or in Newtons per meter (SI), and R_c is the cone radius in centimeters or meters. Shear rate ($\dot{\gamma}$) is given by

$$\dot{\gamma} = \frac{\Omega}{\alpha} \tag{2.112}$$

where Ω is the angular velocity in degrees per second (CGS) or in radians per second (SI) and α is the cone angle in degrees or radians. Viscosity is then

$$\eta = \tau/\dot{\gamma} = \frac{3\alpha\Phi}{2\pi R_c^3 \Omega} = \frac{k\Phi}{\Omega} \tag{2.113}$$

where ($K = 3\alpha/2\pi R_c^3$) is a constant defined by viscometer design.

An analogous result is obtained if the plate rotates and the cone and plate viscometer is the Weissenberg Rheogoniometer. It consists of a plate that can be rotated at different speeds by means of a constant speed motor-cum-gear assembly (Figure 2.34). The speed or rotation is measured accurately by means of a transducer.

A cone is placed concentrically above the plate, the cone angle being around 1–5°. The cone is supported vertically by a frictionless air bearing and is attached to a firm support through a calibrated torsional spring. Any torque experienced by the cone leads to an equilibrium deflection of this spring

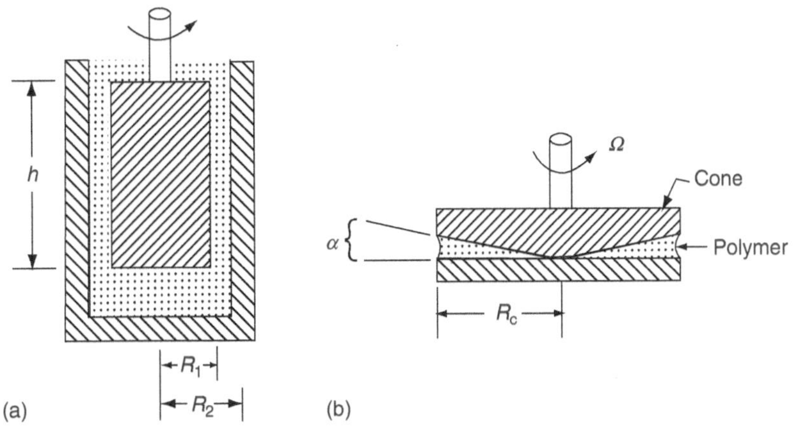

FIGURE 2.33 (a) Coaxial cylinder viscometer. (b) Cone and plate rheometer.

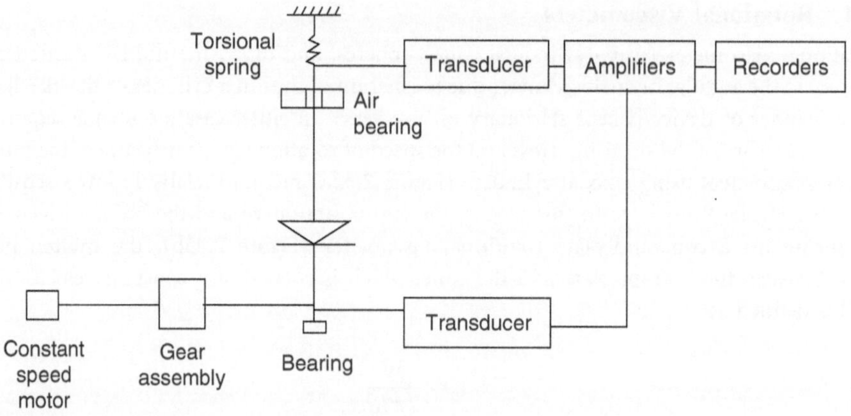

FIGURE 2.34 Scheme of a Weissenberg Rheogoniometer.

which is measured by means of a transducer. The polymer sample is placed in the space between the cone and plate and the torque experienced by the stationary cone is measured for different rotational speeds of the plate.

Relating the shear stress at the cone surface to the measured torque and the shear rate to the angular velocity of the plate, the expression for the viscosity (η) is obtained as

$$\eta = \frac{3K\theta \sin \alpha}{2\pi R_{\mathrm{p}}^3 \omega} \qquad (2.114)$$

where K is the torsional constant and θ is the deflection of the spring; R_{p} is the radius and ω is the angular velocity of the plate; and α is the angle of the cone. While θ and ω are experimentally determined quantities, K and α are obtained by calibration on other materials.

The cone and plate viscometer gives reliable experimental data over an extensive range of shear rates (10^{-4}–10^{4} $\sec^{-1}$). Not only can it be used to measure viscosities in simple shear, but it can also be used to determine the dynamic properties of viscoelastic materials. The unit is also set up to measure the normal stresses exhibited by viscoelastics, i.e., those perpendicular to the plane of shear.

2.2.17.2 Capillary Rheometers

These rheometers are widely used to study the rheological behavior of molten polymers. As shown in Figure 2.35 the fluid is forced from a reservoir into and through a fine-bore tube, or capillary, by either mechanical or pneumatic means. The fluid is maintained at isothermal conditions by electrical temperature control methods. Either the extrusion pressure or volumetric flow rate can be controlled as the independent variable with the other being the measured dependent variable.

Under steady flow and isothermal conditions for an incompressible fluid (assuming only axial flow and no slip at the wall), the viscous force resisting the motion of a column of fluid in the capillary is equal to the applied force tending to move the column in the direction of flow. Thus,

$$\tau = \frac{R\Delta P}{2L} \qquad (2.115)$$

where R and L are the radius and length of the column and ΔP is the pressure drop across the capillary. The shear stress τ is therefore zero at the center of the capillary and increases to a maximum value at the capillary wall. This maximum value is the one generally used for the shear stress in capillary flow.

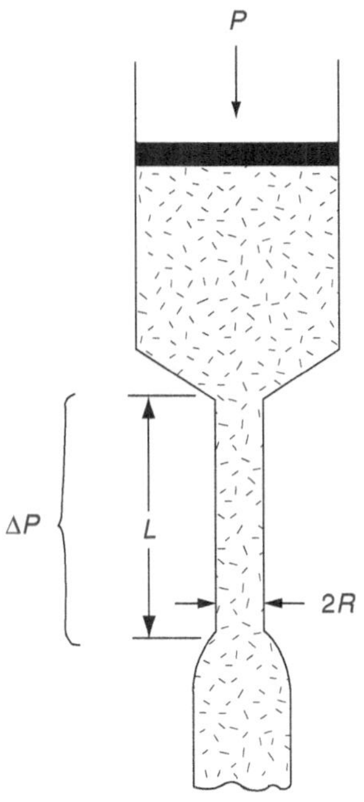

FIGURE 2.35 Capillary rheometer (schematic).

In normal capillary rheometry for polymer melts, the flowing stream exits into the atmosphere, and the driving static pressure in the reservoir is taken to be ΔP. In such cases, end effects involving viscous and elastic deformations at the entrance and exit of the capillary should be taken into account when calculating the true shear stress at the capillary wall, particularly if the ratio of capillary length to radius (L/R) is small.

For a fluid showing Newtonian behavior the shear rate ($\dot{\gamma}$) at the wall is given by

$$\dot{\gamma} = 4Q/\pi R^3 \tag{2.116}$$

where Q is the volumetric flow rate through the capillary under a pressure drop ΔP. Melt viscosity is expressed as

$$\eta = \tau/\dot{\gamma} = \frac{\pi R^4 \Delta P}{8LQ} \tag{2.117}$$

The measured values of polymer flow taken by capillary rheometers are often presented as plots of shear stress versus shear rate at certain temperatures. These values are called apparent shear stress and apparent shear rate at the tube wall. Corrections must be applied to these values in order to obtain true values. The corrected value of shear stress is determined by the Bagley correction [20]

$$\tau_c = \frac{R\Delta P}{2(L+e)} \tag{2.118}$$

where τ_c is the corrected value and e is the length correction expressed as a function of radius.

Correction to the shear rate is necessitated by the fact that unlike in isothermal Newtonian flow where the velocity distribution from wall to wall in a tube is parabolic, nonparabolic velocity profile develops in non-Newtonian flow. The Rabinowitsch correction [21] is applied to shear rate to eliminate this error as follows:

$$\dot{\gamma}_c = \frac{3n+1}{4n}\dot{\gamma}_a \tag{2.119}$$

where subscript c stands for corrected value and subscript a stands for apparent value of shear rate at tube wall; the correction term n is given by $d \log \tau_a / d\dot{\gamma}_a$ and is 1 for Newtonian flow.

There are three main reasons why the capillary rheometer is widely used in the plastics industry: (1) shear rate and flow geometry in capillary rheometer are very similar to conditions actually encountered in extrusion and injection modeling; (2) a capillary rheometer typically covers the widest shear rate ranges (10^{-6} sec^{-1} to 10^{6} sec^{-1}); and (3) a capillary rheometer provides good practical data and information on the die swell, melt instability, and extrudate defects.

2.2.18 Plastics Fractures

The principal causes of fracture of a plastic part are the prolonged action of a steady stress (creep rupture), the application of a stress in a very short period of time (impact), and the continuous application of a cyclically varying stress (fatigue). In all cases the process of failure will be accelerated if the plastic is in a aggressive environment.

Two basic types of fracture under mechanical stresses are recognized; brittle fracture and ductile fracture. These terms refer to the type of deformation that precedes fracture.

Brittle fractures are potentially more dangerous because there occurs no observable deformation of the material. In a ductile failure, on the other hand, large nonrecoverable deformations occur before rupture actually takes place and serve as a valuable warning. A material thus absorbs more energy when it fractures in a ductile fashion than in a brittle fashion.

In polymeric materials fracture may be ductile or brittle, depending on several variables, the most important of which are the straining rate, the stress system, and the temperature. Both types of failures may thus be observed in the one material, depending on the service conditions.

2.2.19 Impact Behavior of Plastics

Tests of brittleness make use of impact tests. The main causes of brittle failure in materials have been found to be (1) triaxiality of stress, (2) high strain rates, and (3) low temperatures. Test methods developed for determining the impact behavior of materials thus involve striking a notched bar with a pendulum. This is the most convenient way of subjecting the material to triaxiality of stress (at the notch tip) and a high strain rate so as to promote brittle failures.

The standard test methods are the *Charpy* and *Izod* tests, which employ the pendulum principle (Figure 2.36a). The test procedures are illustrated in Figure 2.36b and c. The specimen has a standard notch on the tension side.

In the Charpy test the specimen is supported as a simple beam and is loaded at the midpoint (Figure 2.36b). In the Izod test it is supported as a cantilever and is loaded at its end (Figure 2.36c). The standard energy absorbed in breaking the specimen is recorded.

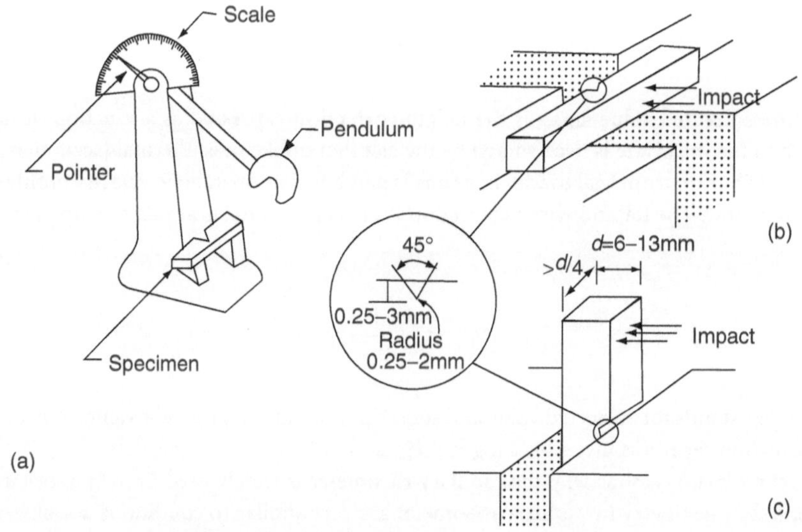

FIGURE 2.36 Impact test. (a) Schematic diagram of Charpy impact testing machine. (b) Arrangement of Charpy impact specimen. (c) Mounting of Izod impact specimen.

The results of impact tests are often scattered, even with the most careful test procedure. A normal practice in such cases is to quote the median strength rather than the average strength, because the median is more representative of the bulk of the results if there is a wide scatter with a few very high or very low results. Impact strengths are normally expressed as

$$\text{Impact} = \frac{\text{Energy abosorbed to break}}{\text{Area at notch section}}$$

(ft-lbf/in^2, cm-kgf/cm^2, or J/m^2)

Occasionally, the less satisfactory term of energy to break per unit width may be quoted in units of ft-lbf/in, cm-kgf/cm or J/m.

The choice of notch depth and tip radius will affect the impact strength observed. A sharp notch is usually taken as a 0.25-mm radius, a blunt notch as a 2-mm radius. The typical variation of impact strength with notch-tip variation for several thermoplastics is presented in Figure 2.37. It is evident that the use of a sharp notch may even rank plastic materials in an order different from that obtained by using a blunt notch. This fact may be explained by considering the total energy absorbed to break the specimen as consisting of energy necessary for crack initiation and for crack propagation.

When the sharp notch (0.25-mm radius) is used, it may be assumed that the energy necessary to initiate the crack is small, and the main contribution to the impact strength is the propagation energy. On this basis Figure 2.37 would suggest that high-density polyethylene and ABS have relatively high crack-propagation energies, whereas materials such as PVC, nylon, polystyrene, and acrylics have low values. The large improvement in impact strength observed for PVC and nylon when a blunt notch is used would imply that their crack-initiation energies are high. On the other hand, the smaller improvement in the impact strength of ABS with a blunt notch would suggest that the crack-initiation energy is low. Thus

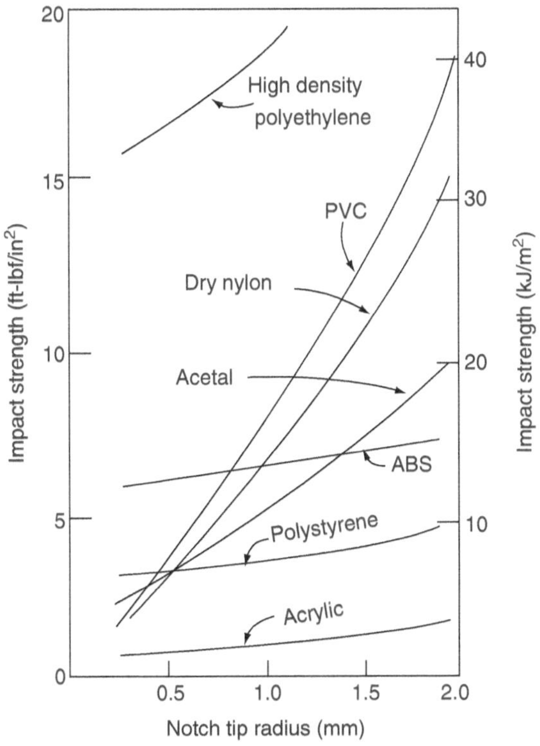

FIGURE 2.37 Variation of impact strength with notch radius for several thermoplastics.

the benefit derived from using rounded corners would be much less for ABS than for materials such as nylon or PVC.

Temperature has a pronounced effect on the impact strength of plastics. In common with metals, many plastic materials exhibit a transition from ductile behavior to brittle as the temperature is reduced. The variation of impact strength with temperature for several common thermoplastics is shown in Figure 2.38.

The ranking of the materials with regard to impact strength is seen to be influenced by the test temperature. Thus, at room temperature (approximately 20°C) polypropylene is superior to acetal; at subzero temperatures (e.g., −20°C) polypropylene does not perform as well as acetal. This comparison pertains to impact behavior measured with a sharp (0.25-mm) notch. Note that notch sharpness can influence the impact strength variation with temperature quite significantly. Figure 2.39 shows that when a blunt (2-mm) notch is used, there is indeed very little difference between acetal and polypropylene at 20°C, whereas at −20°C acetal is much superior to polypropylene.

It may be seen from Figure 2.38 and Figure 2.39 that some plastics undergo a change from ductile or tough (high impact strength) to brittle (low impact strength) behavior over a relatively narrow temperature change. This allows a temperature for ductile-brittle transition to be cited. In other plastic materials this transition is much more gradual, so it is not possible to cite a single value for transition temperature. It is common to quote in such cases a brittleness temperature, $T_B(1/4)$.

This temperature is defined as the value at which the impact strength of the material with a sharp notch (1/4-mm tip radius) is 10 kJ/m^2 (4.7 ft-lbf/in^2). When quoted, it provides an indication of the temperature above which there should be no problem of brittle failure. However, it does not mean that a material should never be used below its $T_B(1/4)$, because this temperature, by definition, refers only to the impact behavior with a sharp notch. When the material is unnotched or has a blunt notch, it may still have satisfactory impact behavior well below $T_B(1/4)$.

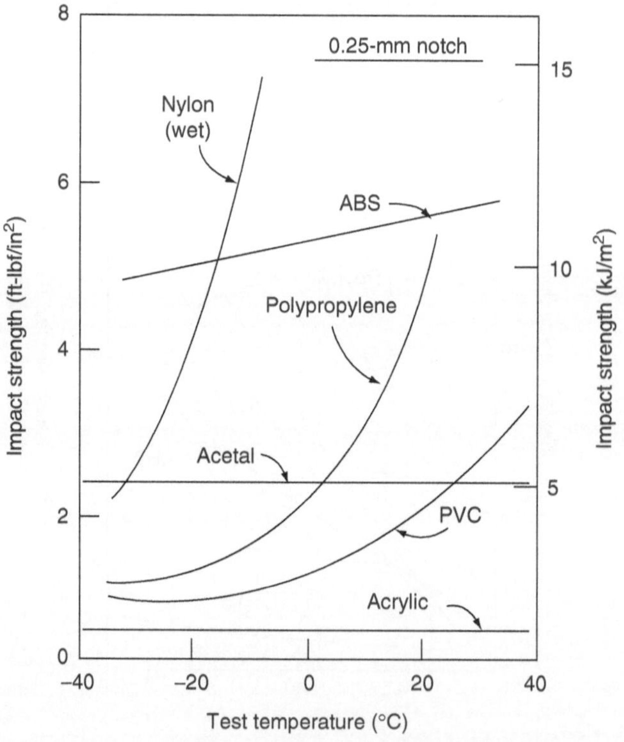

FIGURE 2.38 Variation of impact strength with temperature for several thermoplastics with sharp notch.

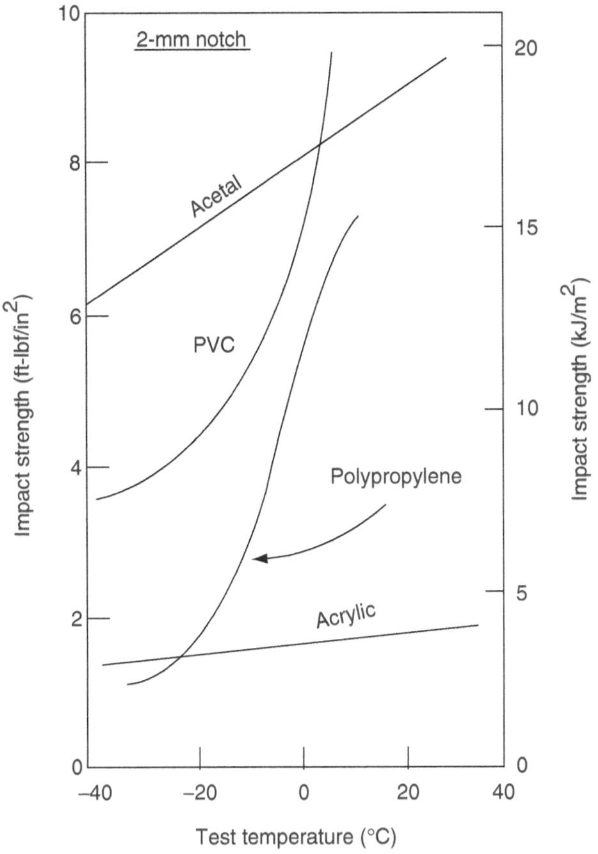

FIGURE 2.39 Variation of impact strength with temperature for several thermoplastics with blunt notch.

Other environmental factors besides temperature may also affect impact behavior. For example, if the material is in the vicinity of a fluid which attacks it, then the crack-initiation energies may be reduced, resulting in lower impact strength. Some materials, particularly nylon, are significantly affected by water, as illustrated in Figure 2.40. The absorption of water produces a spectacular improvement in the impact behavior of nylon.

Note that the method of making the plastic sample and the test specimen can have significant effect on the measured values of the properties of the material. Test specimens may be molded directly or machined from samples which have been compression molded, injection molded, or extruded. Each processing method involves a range of variables, such as melt temperature, mold or die temperature, and shear rate, which influence the properties of the material.

Fabrication defects can affect impact behavior for example, internal voids, inclusion, and additives, such as pigments, which can produce stress concentrations within the material. The surface finish of the specimen may also affect impact behavior. All these account for the large variation usually observed in the results of testing one material processed and/or fabricated in different ways. It also emphasizes the point that if design data are needed for a particular application, then the test specimen must match as closely as possible the component to be designed.

In some applications impact properties of plastics may not be critical, and only a general knowledge of their impact behavior is needed. In these circumstances the information provided in Table 2.1 would be adequate. The table lists the impact behavior of a number of commonly used thermoplastics over a range of temperatures in three broad categories [22].

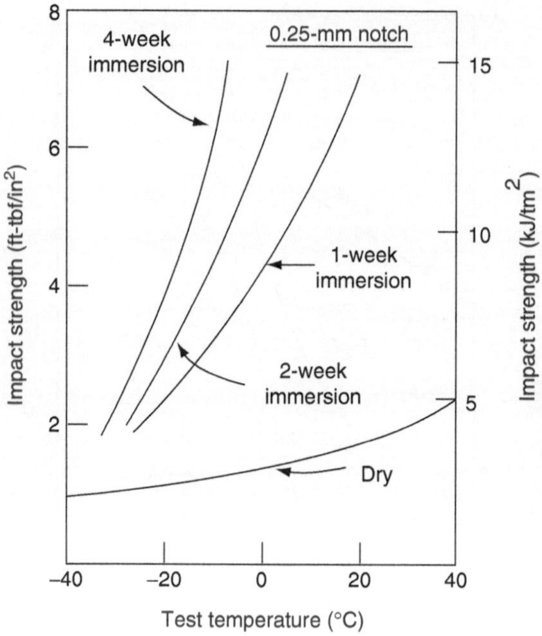

FIGURE 2.40 Effect of water content on impact strength of nylon.

2.2.20 Fatigue of Plastics

A material subject to alternating stresses over long periods may fracture at stresses much below its maximum strength under static loading (tensile strength) due to the phenomenon called fatigue. Fatigue has been recognized as one of the major causes of fracture in metals. Although plastics are susceptible to a wider range of failure mechanisms, it is likely that fatigue still plays an important part in plastics failure.

TABLE 2.1 Impact Behavior of Common Thermoplastics over a Range of Temperatures

Plastic Material	Temperature (°C)							
	−20	−10	0	10	20	30	40	50
Polyethylene (low density)	A	A	A	A	A	A	A	A
Polyethylene (high density)	B	B	B	B	B	B	B	B
Polypropylene	C	C	C	C	B	B	B	B
Polystyrene	C	C	C	C	C	C	C	C
Poly(methyl methacrylate)	C	C	C	C	C	C	C	C
ABS	B	B	B	B	B	B	A	A
Acetal	B	B	B	B	B	B	B	B
Teflon	B	A	A	A	A	A	A	A
PVC (rigid)	B	B	B	B	B	B	A	A
Polycarbonate	B	B	B	B	A	A	A	A
Poly(phenylene oxide)	B	B	B	B	B	B	A	A
Poly(ethylene terephthalate)	B	B	B	B	B	B	B	B
Nylon (dry)	B	B	B	B	B	B	B	B
Nylon (wet)	B	B	B	A	A	A	A	A
Glass-filled nylon (dry)	C	C	C	C	C	C	C	B
Polysulfone	B	B	B	B	B	B	B	B

A, tough (specimens do not break completely even when sharply notched); B, notch brittle; C, brittle even when unnotched.

For metals the fatigue process is generally well understood and is divided into three stages crack initiation, crack growth, and fracture. Fatigue theory of metals is well developed, but the fatigue theory of polymers is not. The completely different molecular structure of polymers means that there is unlikely to be a similar type of crack initiation process as in metals, though it is possible that once a crack is initiated the subsequent phase of propagation and failure may be similar.

Fatigue cracks may develop in plastics in several ways. If the plastic article has been machined, surface flaws capable of propagation may be introduced. However, if the article has been molded, it is more probable that fatigue cracks will develop from within the bulk of the material. In a crystalline polymer the initiation of cracks capable of propagation may occur through slip of molecules. In addition to acting as a path for crack propagation, the boundaries of spherulites (see Chapter 1), being areas of weakness, may thus develop cracks during straining. In amorphous polymers cracks may develop at the voids formed during viscous flow.

A number of features are peculiar to plastics, which make their fatigue behavior a complex subject not simply analyzed. Included are viscoelastic behavior, inherent damping, and low thermal conductivity. Consider, for example, a sample of plastic subjected to a cyclic stress of fixed amplitude. Because of the high damping and low thermal conductivity of the material, some of the input energy will be dissipated in each cycle and appear as heat. The temperature of the material will therefore rise, and eventually a stage will be reached when the heat transfer to the surroundings equals the heat generation in the material. The temperature of the material will stabilize at this point until a conventional metal-type fatigue failure occurs.

If, in the next test, the stress amplitude is increased to a higher value, the material temperature will rise further and stabilize, followed again by a metal-type fatigue failure. In Figure 2.41, where the stress amplitude has been plotted against the logarithm of the number of cycles to failure, failures of this type have been labeled as fatigue failures. This pattern will be repeated at higher stress amplitudes until a point is reached when the temperature rise no longer stabilizes but continues to rise, resulting in a short-term thermal softening failure in the material. At stress amplitudes above this crossover point there will be thermal failures in an even shorter time. Failures of this type have been labeled as thermal failures in Figure 2.41. The fatigue curves in Figure 2.41 thus have two distinct regimes—one for the long-term conventional fatigue failures, and one for the relatively short-term thermal softening failures.

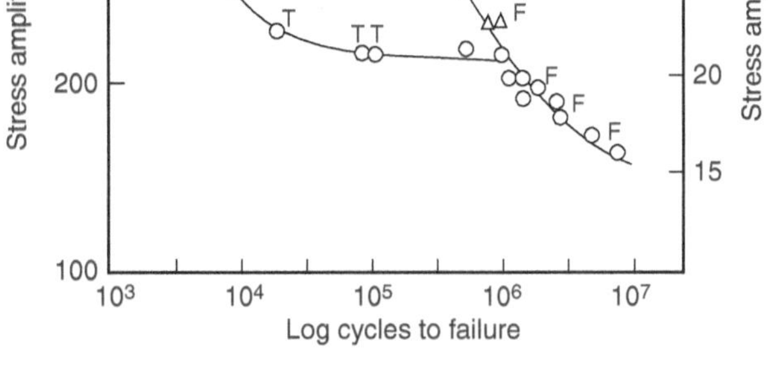

FIGURE 2.41 Typical fatigue behavior of a thermoplastic at several frequencies, *F*, fatigue failure; *T*, thermal failure, ○, 5.0 Hz; Δ, 1.67 Hz; □, 0.5 Hz. (Adapted from Crawford, R. J. 1981. *Plastics Engineering*, Pergamon, London.)

The frequency of the cyclic stress would be expected to have a pronounced effect on the fatigue behavior of plastics, a lower frequency promoting the conventional-type fatigue failure rather than thermal softening failure. Thus it is evident from Figure 2.41 that if the frequency of cycling is reduced, then stress amplitudes which would have produced thermal softening failures at a higher frequency may now result in temperature stabilization and eventually fatigue failure. Therefore, if fatigue failures are required at relatively high stresses, the frequency of cycling must be reduced.

Normally, fatigue failures at one frequency on the extrapolated curve fall from the fatigue failures at the previous frequency (Figure 2.41). As the cyclic stress amplitude is further reduced in some plastics, the frequency remaining constant, the fatigue failure curve becomes almost horizontal at large values of the number of stress cycles (N). The stress amplitude at which this leveling off occurs is clearly important for design purposes and is known as the fatigue limit. For plastics in which fatigue failure continues to occur even at relatively low stress amplitudes, it is necessary to define an endurance limit—that is, the stress amplitude which would not cause fatigue failure up to an acceptably large value of N.

2.2.21 Hardness

Hardness of a material may be determined in several ways: (1) resistance to indentation, (2) rebound efficiency, and (3) resistance to scratching. The first method is the most commonly used technique for plastics. Numerous test methods are available for measuring the resistance of a material to indentation, but they differ only in detail. Basically they all use the size of an indent produced by a hardened steel or diamond indentor in the material as an indication of its hardness—the smaller the indent produced, the harder the material, and so the greater the hardness number. Hardness tests are simple, quick, and nondestructive, which account for their wide use for quality control purposes.

2.2.22 Indentation Hardness

The test methods used for plastics are similar to those used for metals. The main difference is that because plastics are viscoelastic allowance must be made for the creep and the time-dependent recovery which occurs as a result of the applied indentation load.

2.2.22.1 Brinell Hardness Number

A hardened steel ball 10 mm in diameter is pressed into the flat surface of the test specimen under load of 500 kg for 30 sec. The load is then removed, and the diameter of the indent produced is measured (Figure 2.42). The Brinell hardness number (BHN) is then defined as

$$\text{BHN} = \frac{\text{Load applied to indentor (kgf)}}{\text{Contact area of indentation (mm}^2\text{)}} = \frac{2P}{\pi D(D - \sqrt{D^2 - d^2})} \tag{2.120}$$

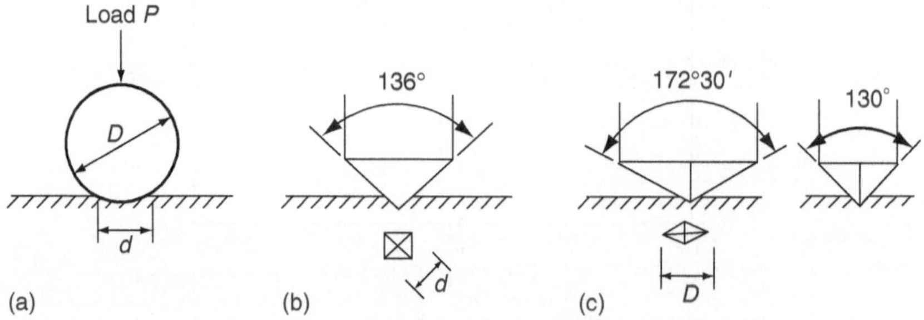

FIGURE 2.42 Indentation hardness tests. (a) Brinell test. (b) Vickers test. (c) Knoop test.

where D is the diameter of the ball and d is the diameter of the indent. Tables are available to convert the diameter of the indent into BHN.

Although the units of Brinell hardness are kgf/mm^2, it is quoted only as a number. A disadvantage of the Brinell hardness test when used for plastics is that the edge of the indent is usually not well defined. This problem is overcome in the following test.

2.2.22.2 Vickers Hardness Number

The Vickers hardness test differs from the Brinell test in that the indentor is a diamond (square-based) pyramid (Figure 2.42) having an apex angle of 136°. If the average diagonal of the indent is d, the hardness number is calculated from

$$\text{Vickers hardness number} = \frac{\text{Load applied to indentor (kgf)}}{\text{Contact area of indentation (mm}^2\text{)}} = 1.854\left(\frac{P}{d^2}\right) \qquad (2.121)$$

As in the Brinell test, tables are available to convert the average diagonal into Vickers number.

2.2.22.3 Knoop Hardness Number

The indentor used in the Knoop hardness is a diamond pyramid, but the lengths of the two diagonals, as shown in Figure 2.42, are different. If the long diagonal of the indent is measured as D, the hardness number is obtained from

$$\text{Knoop hardness number} = 14.23\left(\frac{P}{D^2}\right) \qquad (2.122)$$

Time-dependent recovery of the indentation in plastics is a problem common to all three tests. To overcome this problem, allow a fixed time before making measurements on the indent.

2.2.22.4 Rockwell Hardness Number

The Rockwell test differs from the other three tests because the depth of the indent rather than its surface area is taken as a measure of hardness. A hardened steel ball is used as the indentor. A major advantage of the Rockwell test is that no visual measurement of the indentation is necessary, and the depth of the indent is read directly as a hardness value on the scale.

The test involves three steps, as shown in Figure 2.43. A minor load of 10 kg is applied on the steel ball, and the scale pointer is set to zero within 10 sec of applying the load. In addition to this minor load, a major load is applied for 15 sec. A further 15 sec after removal of the major load (with the minor load still on the ball), the hardness value is read off the scale. Since creep and recovery effects can influence readings, it is essential to follow a defined time cycle for the test.

Several Rockwell scales (Table 2.2) are used, depending on the hardness of the material under test (Table 2.3). The scale letter is quoted along with the hardness number e.g., Rockwell R60. Scales R and L are used for low-hardness number, e.g., Rockwell R60. Scales R and L are used for low-hardness materials, and scales M and E when the hardness value is high. When the hardness number exceeds 115 on any scale, the sensitivity is lost, so another scale should be used.

2.2.22.5 Barcol Hardness

The Barcol hardness tester is a hand-operated hardness measuring device. Its general construction is shown in Figure 2.44. With the back support leg placed on the surface, the instrument is gripped in the hand and its measuring head is pressed firmly and steadily onto the surface until the instrument rests on the stop ring. The depth of penetration of the spring-loaded indentor is transferred by a lever system to an indicating dial, which is calibrated from 0 to 100 to indicate increasing hardness. To allow for creep, one normally takes readings after 10 sec.

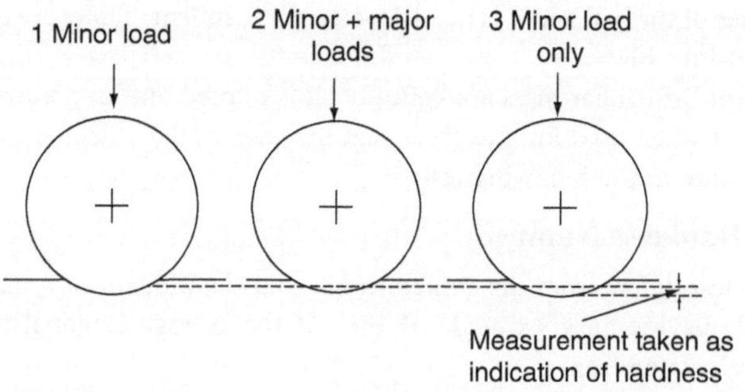

FIGURE 2.43 Stages in Rockwell hardness test: 1, minor load; 2, minor and major loads; 3, minor load only.

The indentor in the Barcol tester Model No. 934-1 is a truncated steel cone having an included angle of 26° with a flat tip of 0.157 mm (0.0062 in.) in diameter. The values obtained using this instrument are found to correlate well to Rockwell values on the M scale. This instrument is used for metals and plastics. Two other models, No. 935 and No. 936, are used for plastics and very soft materials, respectively.

2.2.22.6 Durometer Hardness

A durometer is an instrument for measuring hardness by pressing a needle-like indentor into the specimen. Operationally, a durometer resembles the Barcol tester in that the instrument is pressed onto the sample surface until it reaches a stop ring. This forces the indentor into the material, and a system of levers transforms the depth of penetration into a pointer movement on an indicating dial, which is calibrated from 0–100.

The two most common types of durometers used for plastics are the Shore Type A and Shore Type D. They differ in the spring force and the geometry of the indentor, as shown in Figure 2.45. Due to creep, readings should be taken after a fixed time interval, often chosen as 10 sec. Typical hardness values of some of the common plastics measured by different test methods are shown in Table 2.4.

2.2.23 Rebound Hardness

The energy absorbed when an object strikes a surface is related to the hardness of the surface: the harder the surface, the less the energy absorbed, and the greater the rebound height of the object after impact. Several methods have been developed to measure hardness in this way. The most common method uses a *Shore scleroscope*, in which the hardness is determined from the rebound height after the impact of a diamond cone dropped onto the surface of the test piece. Typical values of Scleroscope hardness together with the Rockwell M values (in parentheses) for some common plastics are as follows: PMMA 99 (M 102), LDPE 45 (M 25), polystyrene 70 (M 83), and PVC 75 (M 60).

TABLE 2.2 Rockwell Hardness Scales

Scale	Major Load (kg)	Dia. of Indentor (in.)
R	60	1/2
L	60	1/4
M	100	1/4
E	100	1/8

TABLE 2.3 Choice of Hardness Test Methods Based on Modulus Range of Plastics

	Material	Test Method
Low modulus	Rubber	Shore A or BS 903
↑	Plasticized PVC	Shore A or BS 2782
	Low-density polyethylene	Shore D
	Medium-density polyethylene	Shore D
	High-density polyethylene	Shore D
	Polypropylene	Rockwell R
	Toughened polystyrene	Rockwell R
	ABS	Rockwell R
	Polystyrene	Rockwell M
↓	Poly(methyl methacrylate)	Rockwell M
High modulus		

2.2.24 Scratch Hardness

Basically, scratch hardness is a measure of the resistance the test sample has to being scratched by other materials. The most common way of qualifying this property is by means of the Mohs scale. On this scale various materials are classified from 1 to 10. The materials used, as shown in Figure 2.46, range from talc (1) to diamond (10). Each material on the scale can scratch the materials that have a lower Mohs number; however, the Mohs scale is not of much value for classifying plastic materials, because most common plastics fall in the 2–3 Mohs range. However, the basic technique of scratch hardness may be used to establish the relative merits of different plastic materials from their ability to scratch one another.

Scratch hardness is particularly important in plastics used for their optical properties and is usually determined by some of mar-resistance test. In one type of test a specimen is subjected to an abrasive treatment by allowing exposure to a controlled stream of abrasive, and its gloss (specular reflection) is measured before and after the treatment. In some tests the light transmission property of the plastic is measured before and after marring.

2.2.25 Stress Corrosion Cracking of Polymers

Stress corrosion cracking of polymers occurs in a corrosive environment and also under stress [23,24]. This kind of crack starts at the surface and proceeds at right angles to the direction of stress. The amount

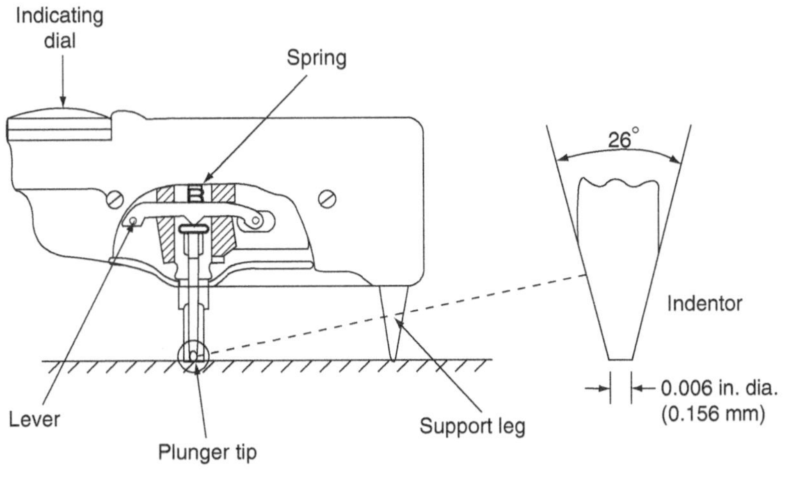

FIGURE 2.44 General construction of Barcol hardness tester.

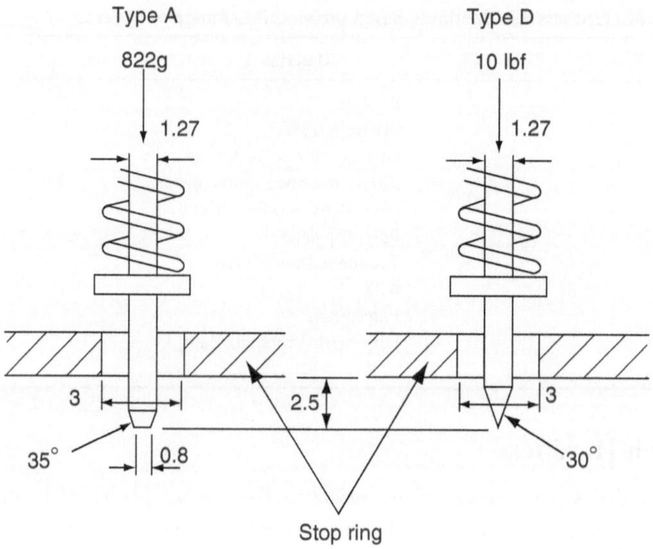

FIGURE 2.45 Two types of Shore durometer.

of stress necessary to cause stress corrosion cracking is much lower than the normal fracture stress, although there is a minimum stress below which no stress corrosion cracking occurs.

The stress corrosion resistance of polymers depends on the magnitude of the stress, the nature of the environment, the temperature, and the molecular weight of the specimen. Ozone cracking is a typical example of stress corrosion cracking of polymers. The critical energy for crack propagation (τ_c) in ozone cracking varies very little from one polymer to another and is about 100 erg/cm^2 (0.1 J/m^2). This value is much lower that the τ_c values for mechanical fracture, which are about 10^7 erg/cm^2 (10^4 J/m^2).

In ozone cracking very little energy is dissipated in plastic or viscoelastic deformations at the propagating crack, and that is why τ_c is about the same as the true surface energy. The only energy supplied to the crack is that necessary to provide for the fresh surfaces due to propagation of the crack, because in ozone cracking chemical bonds at the crack tip are broken by chemical reaction, so no high stress is necessary at the tip.

The critical energy τ_c is about 4,000 erg/cm^2 for PMMA in methylated spirits at room temperature, but the value is lower in benzene and higher in petroleum ether. Thus τ_c in this case is much higher than the true surface energy but still much lower than that for mechanical crack propagation.

TABLE 2.4 Some Typical Hardness Values for Plastics

Material	Brinell	Vickers	Knoop	Rockwell	Barcol	Shore D
High-density Polyethylene	4	2		R40		70
Polypropylene	7	6		R100		74
Polystyrene	25	7	17	M83	76	74
Poly(methyl methacrylate)	20	5	16	M102	80	90
Poly(vinyl chloride)	11	9		M60		80
Poly(vinyl chloride-*co*-vinyl acetate)	20	5	14	M75		
Polycarbonate		7		M70	70	60
Nylon		5	15	M75		80
Cellulose acetate	12	4	12	M64		70

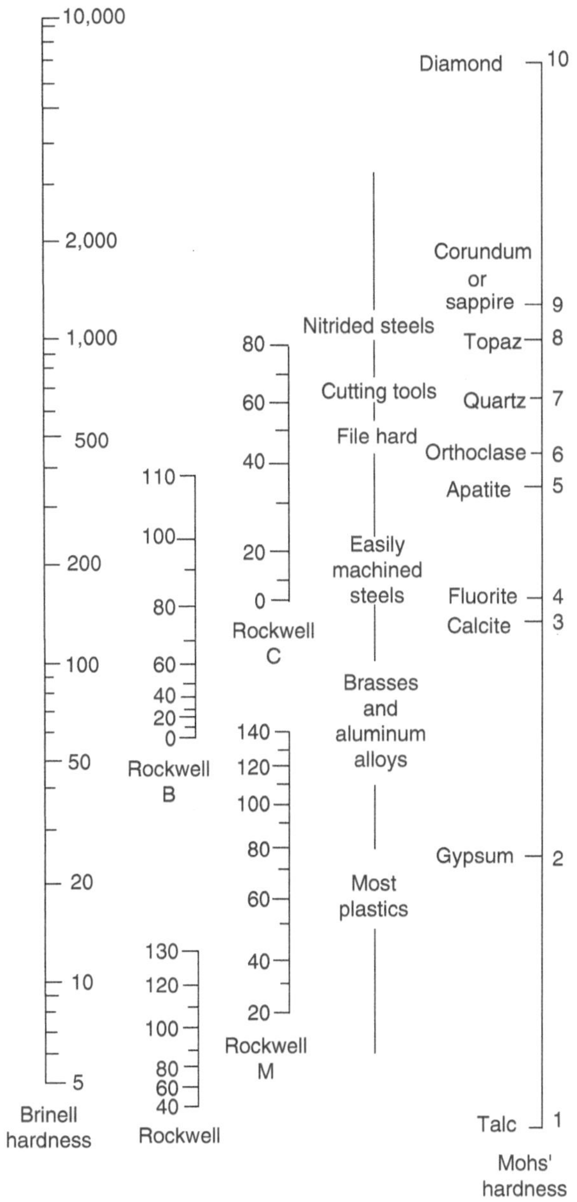

FIGURE 2.46 Comparison of hardness scales (approximate).

2.3 Reinforced Plastics

The modulus and strength of plastics can be increased significantly by means of reinforcement [25–27]. A reinforced plastic consists of two main components—a matrix, which may be either a thermoplastic or thermosetting resin, and a reinforcing filler, which is mostly used in the form of fibers (but particles, for example glass spheres, are also used).

The greater tensile strength and stiffness of fibers as compared with the polymer matrix is utilized in producing such composites. In general, the fibers are the load-carrying members, and the main role of the matrix is to transmit the load to the fibers, to protect their surface, and to raise the energy for crack

TABLE 2.5 Typical Properties of Reinforcing Fibers

		Tensile Strength		Tensile Modulus	
Fiber	Density (g/cm^3)	10^4 kgf/cm^2	GPa	10^5 kgf/cm^2	GPa
E Glass	2.54–2.56	3.5–3.7	3.4–3.6	7.1–7.7	70–76
Carbon	1.75–2.0	2.1–2.8	2.1–2.8	24.5–40.8	240–400
Asbestos	2.5–3.3	2.1–3.6	2.1–3.5	14.3–19.4	140–190
Boron	2.6	3.0–3.6	3.0–3.5	40.8–45.9	400–450
Silicon carbide	3.2–3.4	3.0–3.7	3.0–3.6	46.9–50.0	460–490
Kevlar-49	1.45	3.0–3.6	3.0–3.6	13.2	130

Source: Crawford, R. J. 1981. *Plastics Engineering*, Pergamon, London.

propagation, thus preventing a brittle-type fracture. The strength of the fiber-reinforced plastics is determined by the strength of the fiber and by the nature and strength of the bond between the fibers and the matrix.

2.3.1 Types of Reinforcement

The reinforcing filler usually takes the form of fibers, since it is in this form that the maximum strengthening of the composite is attained. A wide range of amorphous and crystalline materials can be used as reinforcing fibers, including glass, carbon, asbestos, boron, silicon carbide, and more recently, synthetic polymers (e.g., Kevlar fibers from aromatic polyamides). Some typical properties of these reinforcing fibers are given in Table 2.5.

Glass is relatively inexpensive, and in fiber form it is the principal form of reinforcement used in plastics. The earliest successful glass reinforcement had a low-alkali calcium–alumina borosilicate composition (E glass) developed specifically for electrical insulation systems. Although glasses of other compositions were developed subsequently for other applications, no commercial glass better than E glass has been found for plastics reinforcement. However, certain special glasses having extra high-strength properties or modulus have been produced in small quantities for specific applications (e.g., aerospace technology).

Glass fibers are usually treated with finishes. The function of a finish is to secure good wetting and to provide a bond between the inorganic glass and the organic resin. The most important finishes are based on silane compounds—e.g., vinyltrichlorosilane or vinyltriethoxysilane.

2.3.2 Types of Matrix

The matrix in reinforced plastics may be either a thermosetting or thermoplastic resin. The major thermosetting resins used in conjunction with glass-fiber reinforcement are unsaturated polyester resins and, to a lesser extent, epoxy resins. These resins have the advantage that they can be cured (cross-linked) at room temperature, and no volatiles are liberated during curing.

Among thermoplastic resins used as the matrix in reinforced plastics, the largest tonnage group is the polyolefins, followed by nylon, polystyrene, thermoplastic polyesters, acetal, polycarbonate, and polysulfone. The choice of any thermoplastic is dictated by the type of application, the service environment, and the cost.

2.3.3 Analysis of Reinforced Plastics

Fibers exert their effect by restraining the deformation of the matrix while the latter transfers the external loading to the fibers by shear at the interface. The resultant stress distributions in the fiber and the matrix tend to be complex. Theoretical analysis becomes further complicated because fiber length, diameter, and orientation are all factors. A simplified analysis follows for two types of fiber reinforcement commonly used, namely, (1) continuous fibers and (2) discontinuous fibers.

2.3.3.1 Continuous Fibers

We will examine what happens when a load is applied to an ideal fiber composite in which the matrix material is reinforced by fibers which are uniform, continuous, and arranged uniaxially, as shown in Figure 2.47a.

Let us assume that the fibers are gripped firmly by the matrix so that there is no slippage at the fiber-matrix interface and both phases act as a unit. Under these conditions the strains in the matrix and in the fiber under a load are the same (Figure 2.47b), and the total load is shared by the fiber and the matrix:

$$P_c = P_m + P_f \qquad (2.123)$$

where P is the load and the subscripts c, m, and f refer, respectively, to composite, matrix and fiber.

Since the load $P = \sigma A$, Equation 2.123, expressed in terms of stresses (σ) and cross-sectional areas (A), becomes

$$\sigma_c A_c = \sigma_m A_m + \sigma_f A_f \qquad (2.124)$$

Rearranging gives

$$\sigma_c = \sigma_m \left(\frac{A_m}{A_c}\right) + \sigma_f \left(\frac{A_f}{A_c}\right) \qquad (2.125)$$

Since the fibers run throughout the length of the specimen, the ratio A_m/A_c can be replaced by the volume fraction $\Phi_m = V_m/V_c$, and similarly A_f/A_c by Φ_f. Equation 2.125 thus becomes

$$\sigma_c = \sigma_m \Phi_m + \sigma_f \Phi_f \qquad (2.126)$$

Equation 2.126 represents the rule of mixture for stresses. It is valid only for the linear elastic region of the stress–strain curve (see Figure 2.2). Since $\Phi_m + \Phi_f = 1$, we can write

$$\sigma_c = \sigma_m (1 - \Phi_f) + \sigma_f \Phi_f \qquad (2.127)$$

Since the strains on the components are equal,

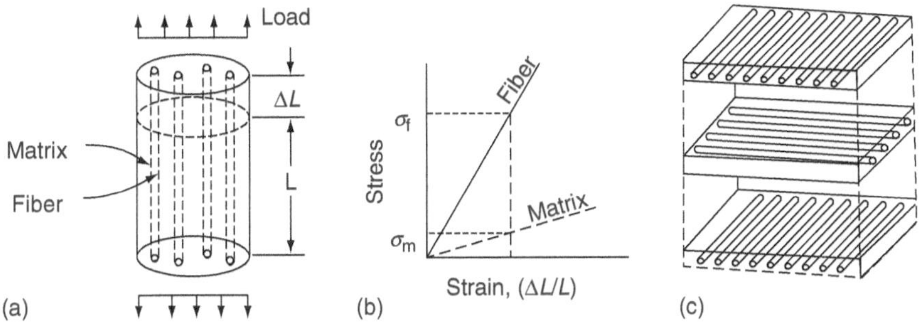

FIGURE 2.47 (a) Continuous-fiber reinforced composite under tensile load. (b) Iso-strain assumption in a composite. (c) Arrangement of fibers in a cross-plied laminate.

$$\varepsilon_c = \varepsilon_m = \varepsilon_f \tag{2.128}$$

Equation 2.126 can now be rewritten to give the rule of mixture for moduli

$$E_c\varepsilon_c = E_m\varepsilon_m\Phi_m + E_f\varepsilon_f\Phi_f$$

i.e.,

$$E_c = E_m\Phi_m + E_f\Phi_f \tag{2.129}$$

Equation 2.128 also affords a comparison of loads carried by the fiber and the matrix. Thus for elastic deformation Equation 2.128 can be rewritten as

$$\frac{\sigma_c}{E_c} = \frac{\sigma_m}{E_m} = \frac{\sigma_f}{E_f}$$

or

$$\frac{P_f}{P_m} = \frac{E_f}{E_m}\left(\frac{A_f}{A_m}\right) = \frac{E_f}{E_m}\left(\frac{\Phi_f}{\Phi_m}\right) \tag{2.130}$$

Because the modulus of fibers is usually much higher than that of the matrix, the load on a composite will therefore be carried mostly by its fiber component (see Example 2.4). However, a critical volume fraction of fibers (Φ_{crit}) is required to realize matrix reinforcement. Thus for Equation 2.127 and Equation 2.130 to be valid, $\Phi_f > \Phi_{crit}$.

The efficiency of reinforcement is related to the fiber direction in the composite and to the direction of the applied stress. The maximum strength and modulus are realized in a composite along the direction of the fiber. However, if the load is applied at 90° to the filament direction, tensile failure occurs at very low stresses, and this transverse strength is not much different than the matrix strength. To counteract this situation, one uses cross-plied laminates having alternate layers of unidirectional fibers rotated at 90°, as shown in Figure 2.47c. (A more isotropic composite results if 45° plies are also inserted.) The stress–strain behavior for several types of fiber reinforcement is compared in Figure 2.48.

As already noted, if the load is applied perpendicularly to the longitudinal direction of the fibers, the fibers exert a relatively small effect. The strains in the fibers and the matrix are then different, because they act independently, and the total deformation is thus equal to the sum of the deformations of the two phases.

$$V_c\varepsilon_c = V_m\varepsilon_m + V_f\varepsilon_f \tag{2.131}$$

Dividing Equation 2.131 by V_c and applying Hooke's law, since the stress is constant, give

$$\frac{\sigma}{E_c} = \frac{\sigma\Phi_m}{E_m} + \frac{\sigma\Phi_f}{E_f} \tag{2.132}$$

Dividing by σ and rearranging, we get

$$E_c = \frac{E_m E_f}{E_m\Phi_f + E_f\Phi_m} \tag{2.133}$$

The fiber composite thus has a lower modulus in transverse loading than in longitudinal loading.

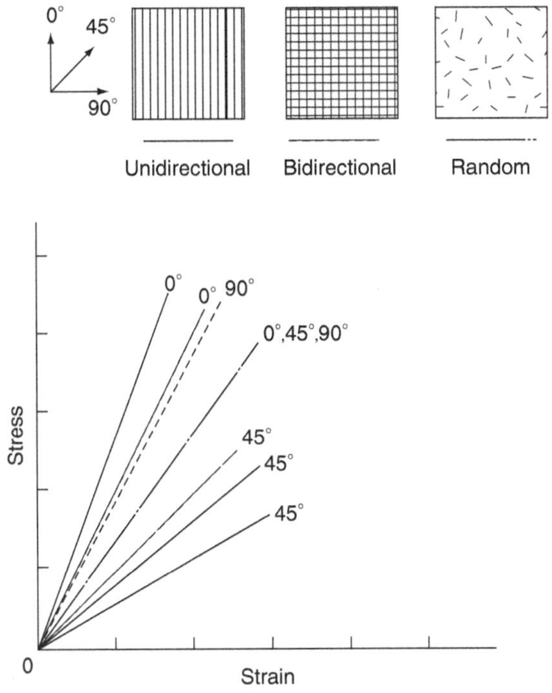

FIGURE 2.48 Stress–strain behavior for several types of fiber reinforcement.

Example 2.4

A unidirectional fiber composite is made by using 75% by weight of E glass continuous fibers (sp. gr. 2.4) having a Young's modulus of 7×10^5 kg/cm^2 (68.6 GPa), practical fracture strength of 2×10^4 kg/cm^2 (1.9 GPa), and an epoxy resin (sp. gr. 1.2) whose modulus and tensile strength, on curing, are found to be 10^5 kg/cm^2 (9.8 GPa) and 6×10^2 kg/cm^2 (58.8 MPa), respectively. Estimate the modulus of the composite, its tensile strength, and the fractional load carried by the fiber under tensile loading. What will be the value of the modulus of the composite under transverse loading?

Answer. Volume fraction of glass fibers (Φ_f)

$$= \frac{0.75/2.4}{0.75/2.4 + 0.25/1.2} = 0.60$$

$$\Phi_m = 1 - 0.6 = 0.4$$

From Equation 2.129,

$$E_c = 0.4(10^5) + 0.6(7 \times 10^5) = 4.6 \times 10^5 \text{ kg/cm}^2 (45 \text{ GPa})$$

From Equation 2.126,

$$\sigma_c = 0.4(6 \times 10^2) + 0.6(2 \times 10^4) = 1.22 \times 10^4 \text{ kg/cm}^2 (1.2 \text{ GPa})$$

Equation 2.130, on rearranging, gives

$$\frac{P_f}{P_c} = \frac{E_t}{E_c}\Phi_f = \frac{7\times 10^5}{4.6\times 10^5}\times 0.6 = 0.91$$

Thus, nearly 90% of the load is carried by the fiber, and the weakness of the plastic matrix is relatively unimportant.

For transverse loading, from Equation 2.133,

$$E_c = \frac{10^5(7\times 10^5)}{0.6\times 10^5 + 0.4(7\times 10^5)} = 2\times 10^5\ \text{kg/cm}^2\ (19.6\ \text{GPa})$$

Equation 2.127 and Equation 2.129 apply to ideal fiber composites having uni-axial arrangement of fibers. In practice, however, not all the fibers are aligned in the direction of the load. This practice reduces the efficiency of the reinforcement, so Equation 2.127 and Equation 2.129 are modified to the forms

$$\sigma_c = \sigma_m(1-\phi_f) + k_1\sigma_f\phi_f \tag{2.134}$$

$$E_c = E_m(1-\phi_f) + k_2 E_f\phi_f \tag{2.135}$$

If the fibers are bi-directional (see Figure 2.48), then the strength and modulus factors, k_1 and k_2, are about 0.3 and 0.65, respectively.

2.3.3.2 Discontinuous Fibers

If the fibers are discontinuous, the bond between the fiber and the matrix is broken at the fibers ends, which thus carry less stress than the middle part of the fiber. The stress in a discontinuous fiber therefore varies along its length. A useful approximation pictures the stress as being zero at the end of the filler and as reaching the maximum stress in the fiber at a distance from the end (Figure 2.49a).

The length over which the load is transferred to the fiber is called the transfer length. As the stress on the composite is increased, the maximum fiber stress as well as the transfer length increase, as shown in Figure 2.49a, until a limit is reached, because the transfer regions from the two ends meet at the middle of the fiber (and so no further transfer of stress can take place), or because the fiber fractures. For the latter objective to be reached, so as to attain the maximum strength of the composite, the fiber length must be greater than a minimum value called the critical fiber length, l_c.

Consider a fiber of length l embedded in a polymer matrix, as shown in Figure 2.49b. One can then write, equating the tensile load on the fiber with the shear load on the interface,

$$\frac{\sigma\pi d^2}{4} = \tau\pi d l \tag{2.136}$$

where σ is the applied stress, d is the fiber diameter, and τ is the shear stress at the interface.

The critical fiber length, l_c, can be derived from a similar force balance for an embedded length of $l_c/2$. Thus,

$$l_c = \frac{\sigma_{ff} d}{2\tau_i}$$

and

$$\frac{l_c}{d} = \frac{\sigma_{ff}}{2\tau_i} \tag{2.137}$$

where σ_{ff} is the fiber strength and τ_i is the shear strength of the interface or the matrix, whichever is smaller.

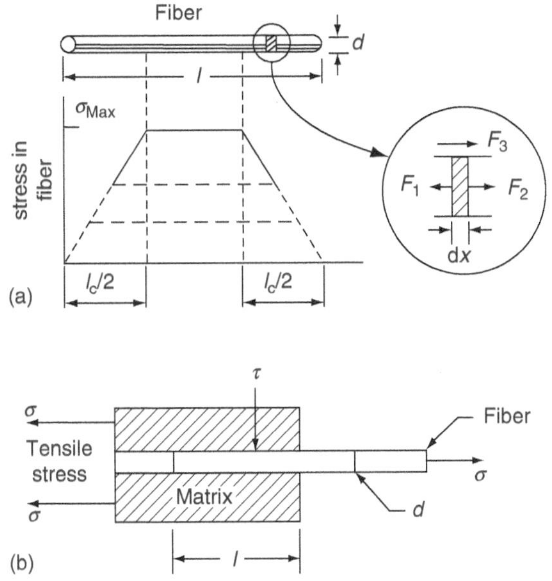

FIGURE 2.49 Composite reinforced with discontinuous fibers. (a) A total length l_c, at the two ends of a fiber carries less than the maximum stress. (b) Interfacial strength of the matrix fiber.

So if the composite is to fail through tensile fracture of the fiber rather than shear failure due to matrix flow at the interface between the fiber and the matrix, the ratio l_c/d, known as the critical aspect ratio, must be exceeded, or, in other words, for a given diameter of fiber, d, the critical fiber length, l_c, must be exceeded.

If the fiber length is less than l_c, the matrix will flow around the fiber, and maximum transfer of stress from matrix to fiber will not occur. Using Equation 2.137, we can estimate the value of l_c/d from the values of σ_{ff} and τ_i, and vice versa. Typical values of l_c/d for glass fiber and carbon fiber in an epoxy resin matrix are 30–100 and 70, respectively.

If the fibers are discontinuous, then, since the stress is zero at the end of the fiber, the average stress in the fibers will be less than the value σ_{fmax}, which it would have achieved if the fibers had been continuous over the whole length of the matrix. The value of the average stress will depend on the stress distribution in the end portions of the fibers and also on their lengths. If the stress distributions are assumed to be as shown in Figure 2.49a, then the average stress in the fibers may be obtained as follows.

Considering a differential section of the fiber as shown in Figure 2.49a, we obtain

$$F_1 = \sigma_f \frac{\pi d^2}{4}$$

$$F_2 = \left(\sigma_f + \frac{d\sigma_f}{dx}dx\right)\frac{\pi d^2}{4}$$

$$F_3 = \tau \pi d \, dx$$

For equilibrium,

$$F_1 = F_2 + F_3$$

so,

$$\sigma_f \frac{\pi d^2}{4} = \left(\sigma_f + \frac{d\sigma_f}{dx} dx\right) \frac{\pi d^2}{4} dx + \tau \pi d \, dx$$
$$\frac{d}{4} d\sigma_f = -\tau \, dx \tag{2.138}$$

Integrating gives

$$\frac{d}{4} \int_0^{\sigma_1} d\sigma_f = -\int_{l/2}^{x} \tau \, dx$$

$$\sigma_f = \frac{4\tau(l/2 - x)}{d} \tag{2.139}$$

Three cases may now be considered.

2.3.3.3 Fiber Length Less than l_c

In this case the peak stress occurs at $x=0$ (Figure 2.50a). So from Equation 2.139,

$$\sigma_f = \frac{2\tau l}{d}$$

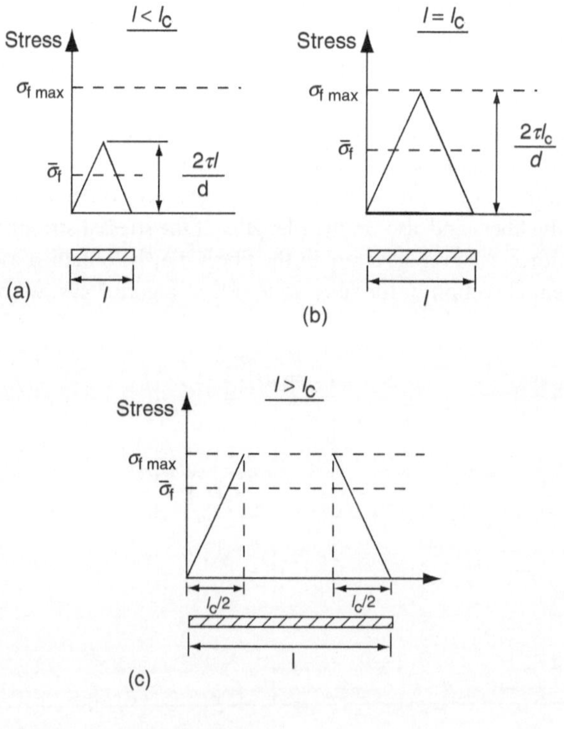

FIGURE 2.50 Stress variation for short and long fibers.

The average fiber stress is obtained by dividing the area of the stress-fiber length diagram by the fiber length; that is,

$$\bar{\sigma}_{\mathrm{f}} = \frac{(l/2)2\tau l/d}{l} = \frac{\tau l}{d}$$

The stress, σ_{c}, in the composite is now obtained from Equation 2.134

$$\sigma_{\mathrm{c}} = \sigma_{\mathrm{m}}(1-\Phi_{\mathrm{f}}) + \frac{\tau l k_1}{d}\phi_{\mathrm{f}} \qquad (2.140)$$

2.3.3.4 Fiber Length Equal to l_c

In this case the peak value of stress occurs at $x=0$ and is equal to the maximum fiber stress (Figure 2.50b). So

$$\sigma_{\mathrm{f}} = \sigma_{\mathrm{fmax}} = \frac{2\tau l_{\mathrm{c}}}{d} \qquad (2.141)$$

$$\text{Average fiber stress} = \bar{\sigma}_f = \frac{1}{2}\frac{l_{\mathrm{c}}(2\tau l_{\mathrm{c}}/d)}{l_{\mathrm{c}}}$$

i.e.,

$$\bar{\sigma}_{\mathrm{f}} = \frac{\tau l_{\mathrm{c}}}{d}$$

So from Equation 2.134,

$$\sigma_{\mathrm{c}} = \sigma_{\mathrm{m}}(1-\phi_{\mathrm{f}}) + k_1\left(\frac{\tau l_{\mathrm{c}}}{d}\right)\phi_{\mathrm{f}} \qquad (2.142)$$

2.3.3.5 Fiber Length Greater than l_c

1. For $l/2 > x > (l-l_{\mathrm{c}})/2$ (Figure 2.50c),

$$\sigma_{\mathrm{f}} = \frac{4\tau}{d}\left(\frac{1}{2}l - x\right)$$

2. For $(l-l_{\mathrm{c}})/2 > x > 0$ (Figure 2.50c),

$$\sigma_{\mathrm{f}} = \text{constant} = \sigma_{\mathrm{fmax}} = \frac{2\tau l_{\mathrm{c}}}{d}$$

The average fiber stress, from the area under the stress-fiber length graph, is

$$\bar{\sigma}_{\mathrm{f}} = \frac{(l_{\mathrm{c}}/2)\sigma_{\mathrm{fmax}} + (l-l_{\mathrm{c}})\sigma_{\mathrm{fmax}}}{l} = \left(1-\frac{l_{\mathrm{c}}}{2l}\right)\sigma_{\mathrm{fmax}} \qquad (2.143)$$

So from Equation 2.134,

$$\sigma_c = \sigma_m(1-\phi_f) + k_1\phi_f\left(1-\frac{l_c}{2l}\right)\phi_{fmax}. \quad (2.144)$$

It is evident from Equation 2.143 that to get the average fiber stress as close as possible to the maximum fiber stress, the fibers must be considerably longer than the critical length. At the critical length the average fiber stress is only half of the maximum fiber stress, i.e., the value achieved in continuous fibers (Figure 2.50c).

Equations such as Equation 2.144 give satisfactory agreement with the measured values of strength and modulus for polyester composites reinforced with chopped strands of glass fibers. These strength and modulus values are only about 20%–25% of those achieved by reinforcement with continuous fibers.

Example 2.5

Calculate the maximum and average fiber stresses for glass fibers of diameter 15 μm and length 2 mm embedded in a polymer matrix. The matrix exerts a shear stress of 40 kgf/cm^2 (3.9 MPa) at the interface, and the critical aspect ratio of the fiber is 50.

Answer.

$$l_c = 50 \times 15 \times 10^{-3} = 0.75 \text{ mm}$$

Since $l > l_c$, then

$$\sigma_{max} = \frac{2\tau l_c}{d} = 2 \times 40 \times 50 = 4 \times 10^3 \text{ kgf/cm}^2 (= 392 \text{ MPa})$$

Also,

$$\bar{\sigma}_f = \left(1-\frac{l_c}{2l}\right)\sigma_{f\,max} = \left(1-\frac{0.75}{2\times 2}\right)(4\times 10^3) = 3.25\times 10^3 \text{ kgf/cm}^2 (= 318 \text{ MPa})$$

2.3.4 Deformation Behavior of Fiber-Reinforced Plastic

As we have seen, the presence of fibers in the matrix has the effect of stiffening and strengthening it. The tensile deformation behavior of fiber-reinforced composites depends largely on the direction of the applied stress in relation to the orientation of the fibers, as illustrated in Figure 2.48. The maximum strength and modulus are achieved with unidirectional fiber reinforcement when the stress is aligned with the fibers (0°), but there is no enhancement of matrix properties when the stress is applied perpendicular to the fibers. With random orientation of fibers the properties of the composite are approximately the same in all directions, but the strength and modulus are somewhat less than for the continuous-fiber reinforcement.

In many applications the stiffness of a material is just as important as its strength. In tension the stiffness per unit length is given the product *EA*, where *E* is the modulus and *A* is the cross-sectional area. When the material is subjected to flexure, the stiffness per unit length is a function of the product *EI*, where *I* is the second moment of area of cross section (see Example 2.2). Therefore the stiffness in both tension and flexure increases as the modulus of the material increases, and the advantages of fiber reinforcement thus become immediately apparent, considering the very high modulus values for fibers.

2.3.5 Fracture of Fiber-Reinforced Plastics

Although the presence of the reinforcing fibers enhances the strength and modulus properties of the base material, they also cause a complex distribution of stress in the materials. For example, even under simple tensile loading, a triaxial stress system is set up since the presence of the fiber restricts the lateral contraction of the matrix. This system increases the possibility of brittle failure in the material. The type of fracture which occurs depends on the loading conditions and fiber matrix bonding.

2.3.5.1 Tension

With continuous-fiber reinforcement it is necessary to break the fibers before overall fracture can occur. The two different of fracture which can occur in tension are shown in Figure 2.51. It is interesting to note that when an individual fiber in a continuous-fiber composite breaks, it does not cease to contribute to the strength of the material, because the broken fiber then behaves like a long short fiber and will still be supporting part of the external load at sections remote from the broken end. In short-fiber composites, however, fiber breakage is not an essential prerequisite to complete composite fracture, especially when the interfacial bond is weak, because the fibers may then be simply pulled out of the matrix as the crack propagates through the latter.

2.3.5.2 Compression

In compression the strength of glass-fiber reinforced plastics is usually less than in tension. Under compressive loading, shear stresses are set up in the matrix parallel to the fibers. The fiber aligned in the loading direction thus promote shear deformation. Short-fiber reinforcement may therefore have advantages over continuous fibers in compressive loading because in the former not all the fibers can be aligned, so the fibers which are inclined to the loading plane will resist shear deformation. If the matrix-fiber bond is weak, debonding may occur, causing longitudinal cracks in the composite and buckling failure of the continuous fibers.

2.3.5.3 Flexure or Shear

In flexure or shear, as in the previous case of compression, plastics reinforced with short fibers are probably better than those with continuous fibers, because in the former with random orientation of fibers at least some of the fibers will be correctly aligned to resist the shear deformation. However, with continuous-fiber reinforcement if the shear stresses are on planes perpendicular to the continuous fibers, then the fibers will offer resistance to shear deformation. Since high volume fraction (ϕ_f) can be achieved with continuous fibers, this resistance can be substantial.

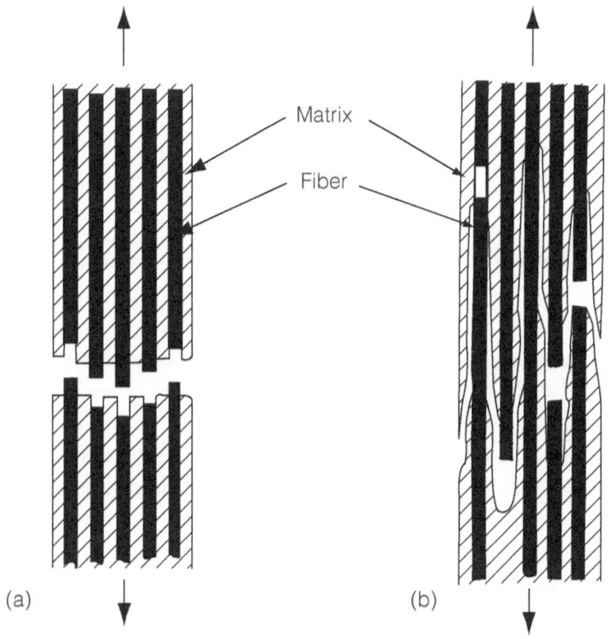

FIGURE 2.51 Typical fracture modes in fiber-reinforced plastics. (a) Fracture due to strong interfacial bond. (b) Jagged fracture due to weak interfacial bond. (After Crawford, R. J. 1981. *Plastics Engineering*, Pergamon, London.)

2.3.6 Fatigue Behavior of Reinforced Plastics

Like unreinforced plastics, reinforced plastics are also susceptible to fatigue. There is, however, no general rule concerning whether glass reinforcement enhances the fatigue endurance of the base material. In some cases the unreinforced plastic exhibits greater fatigue endurance than the reinforced material; in other cases, the converse is true.

In short-fiber glass-reinforced plastics, cracks may develop relatively easily at the interface close to the ends of the fibers. The cracks may then propagate through the matrix and destroy its integrity long before fracture of the composite takes place. In many glass-reinforced plastics subjected to cyclic tensile stresses, debonding many occur after a few cycles even at modes stress levels. Debonding may be followed by resin cracks at higher stresses, leading eventually to the separation of the component due to intensive localized damage. In other modes of loading, e.g., flexure or torsion, the fatigue endurance of glass-reinforced plastics is even worse than in tension. In most cases the fatigue endurance is reduced by the presence of moisture.

Plastics reinforced with carbon and boron, which have higher tensile moduli than glass, are stiffer than glass-reinforced plastics and are found to be less vulnerable to fatigue.

2.3.7 Impact Behavior of Reinforced Plastics

Although it might be expected that a combination of brittle reinforcing fibers and a brittle matrix (e. g., epoxy or polyester resins) would have low impact strength, this is not the case, and the impact strengths of the fibers or the matrix. For example, polyester composites with chopped-strand mat have impact strengths from 45 to 70 ft-lbf/in^2 (94–147 kJ/m^2), whereas a typical impact strength for polyster resin is only about 1 ft-lbf/in^2 (2.1 kJ/m^2).

The significant improvement in impact strength by reinforcement is explained by the energy required to cause debonding and to overcome friction in pulling the fibers out of the matrix. It follows from this that impact strengths would be higher if the bond between the fiber and the matrix is relatively weak, because if the interfacial bond is very strong, impact failure will occur by propagation of cracks across the matrix and fibers requiring very little energy.

It is also found that in short-fiber-reinforced plastics the impact strength is maximum when the fiber length has the critical value. The requirements for maximum impact strength (i.e., short fiber and relatively weak interfacial bond) are thus seen to be contrary to those for maximum tensile strength (long fibers and strong bond). The structure of a reinforced plastic material should therefore be tailored in accordance with the service conditions to be encountered by the material.

2.4 Electrical Properties

The usefulness of an insulator or dielectric ultimately depends on its ability to act as a separator for points across which a potential difference exists. This ability depends on the dielectric strength of the material, which is defined as the maximum voltage gradient that the material withstands before failure or loss of the material's insulating properties occurs.

Besides permittivity (dielectric constant), dielectric losses, and dielectric strength, another property used to define the dielectric behavior of a material is the insulation resistance, i.e., the resistance offered by the material to the passage of electric current. This property may be important in almost all applications of insulators.

This resistivity (i.e., reciprocal of conductivity) of a plastic material with a perfect structure would tend to be infinite at low electric fields. However, the various types of defects which occur in plastics may acts as sources of electrons or ions which are free to contribute to the conductivity or that can be thermally activated to do so. These defects may be impurities, discontinuities in the structure, and interfaces between crystallites and between crystalline and amorphous phases. Common plastics therefore have finite, though very high, resistivities from 10^8 to 10^{20} ohm-cm. These resistivity values qualify them as electrical insulators.

Polymeric materials have also been produced which have relatively large conductivities and behave in some cases like semiconductors and even photoconductors [28]. For example, polyphenylacetylene, polyaminoquinones, and polyacenequinone radical polymers have been reported with resistivities from 10^3 to 10^8 ohm-cm. It has been suggested that the conductivity in these organic semiconductors is due to the existence of large number of unpaired electrons, which are free within a given molecule and contribute to the conduction current by hopping (tunneling) from one molecule to an adjacent one (see Electroactive Polymers in Chapter 2 of *Industrial Polymers, Specialty Polymers, and Their Applications*).

2.4.1 Dielectric Strength

Dielectric strength is calculated as the maximum voltage gradient that an insulator can withstand before puncture or failure occurs. It is expressed as volts (V) per unit of thickness, usually per mil (1 mil = 1/1,000 in.).

Puncture of an insulator under an applied voltage gradient results from small electric leakage currents which pass through the insulator due to the presence of various types of defects in the material. (Note that only a perfect insulator would be completely free from such leakage currents.) The leakage currents warm the material locally, causing the passage of a greater current and greater localized warming of the material, eventually leading to the failure of the material. The failure may be a simple puncture in the area where material has volatilized and escaped, or it may be a conducting carbonized path (tracking) that short circuits the electrodes.

It is obvious from the cause of dielectric failure that the measured values of dielectric strength will depend on the magnitude of the applied electric field and on the time of exposure to the field. Since the probability of a flaw and a local leakage current leading ultimately to failure increases with the thickness of the sample, dielectric strength will also be expected to depend on the sample thickness.

The measurement of dielectric strength (Figure 2.52a) is usually carried out either by the short-time method or by the step-by-step method. In the former method the voltage is increased continuously at a uniform rate (500 V/sec) until failure occurs. Typically, a 1/8-in. thick specimen requiring a voltage of about 50,000 V for dielectric failure will thus involve a testing period of 100 sec or so.

In step-by-step testing, definite voltages are applied to the sample for a definite time (1 min), starting with a value that is half of that obtained by short-time testing, with equal increments of 2,000 V until failure occurs. Since step-by-step testing provides longer exposure to the electric field, dielectric strength values obtained by this method are lower than those obtained by the short-time test. Conditions to step-by-step testing correspond more nearly with those met in service. Even so, service failure almost invariably occurs at voltages below the measured dielectric strength. It is thus necessary to employ a proper safety factor to provide for the discrepancy between test and service conditions.

Increase in thickness increases the voltage required to give the same voltage gradient, but the probability of a flaw and a local leakage current leading ultimately to failure also increases.

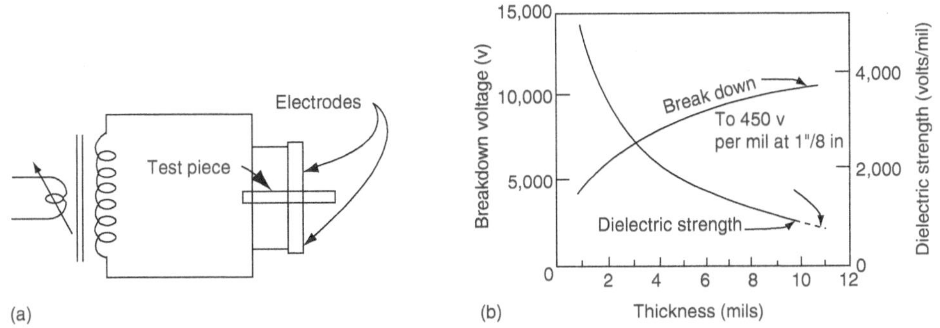

FIGURE 2.52 (a) Dielectric strength test. (b) Dependence of dielectric strength on thickness of sample.

The breakdown voltage increases proportionally less than thickness increases, and as a result the dielectric strength of a material decreases with the thickness of specimen (Figure 2.52b). For this reason, testing of insulation plastic should be done with approximately the thickness in which it is to be used in service.

It is seen from Figure 2.52b that the dielectric strength increases rapidly with decreasing thickness of the sample. A rule of thumb is that the dielectric strength varies inversely with the 0.4 power of the thickness. For example, if the dielectric strength of poly(vinyl chloride) plastic is 375 V/mil in a thickness of 0.075 in., it would be $375(75/15)^{0.4}$ or about 700 V/mil in foils only 15 mils thick. The fact that thin foils may have proportionally higher dielectric strength is utilized in the insulation between layers of transformer turns.

The dielectric strength of an insulation material usually decreases with increase in temperature and is approximately inversely proportional to the absolute temperature. But the converse is not observed, and below room temperature dielectric strength is substantially independent of temperature change.

Mechanical loading has a pronounced effect on dielectric strength. Since a mechanical stress may introduce internal flaws which serve as leakage paths, mechanically loaded insulators may show substantially reduced values of dielectric strength. Reductions up to 90% have been observed.

Dielectric strength of an insulating material is influenced by the fabrication detail. For example, flow lines in a compression molding or weld lines in an injection molding may serve as paths of least resistance for leakage currents, thus reducing the dielectric strength. Even nearly invisible minute flaws in a plastic insulator may reduce the dielectric strength to one-third its normal value.

2.4.2 Insulation Resistance

The resistance offered by an insulating material to the electric current is the composite effect of volume and surface resistances, which always act in parallel. Volume resistance is the resistance to leakage of the electric current through the body of the material. It depends largely on the nature of the material. But surface resistance, which is the resistance to leakage along the surface of a material, is largely a function of surface finish and cleanliness. Surface resistance is reduced by oil or moisture on the surface and by surface roughness. On the other hand, a very smooth or polished surface gives greater surface resistance.

A three-electrode system, as shown in Figure 2.53, is used for measurement of insulation resistance. In this way the surface and volume leakage currents are separated. The applied voltage must be well below the dielectric strength of the material. Thus, in practice, a voltage gradient less than 30 V/mil is applied. From the applied voltage and the leakage current, the leakage resistance is computed. Since the measured value depends, among other things, on the time during which the voltage is applied, it is essential to follow a standardized technique, including preconditioning of the specimen to obtain consistent results.

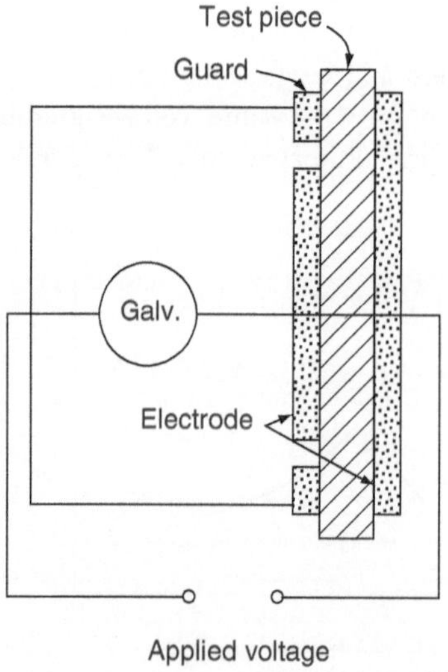

FIGURE 2.53 Insulation resistance test.

The insulation resistance of a dielectric is represented by its volume resistivity and surface resistivity. The volume resistivity (also known as specific volume resistance) is defined as the resistance between two electrodes covering opposite faces of a centimeter cube. The rang of volume resistivities of different materials including plastics is shown in Figure 2.54. Values for plastics range from approximately 10^{10} ohm-cm for a typical cellulose acetate to abut 10^{19} ohm-cm for a high-performance polystyrene.

The surface resistivity (also known as specific surface resistance) is defined as the resistance measured between the opposite edges of the

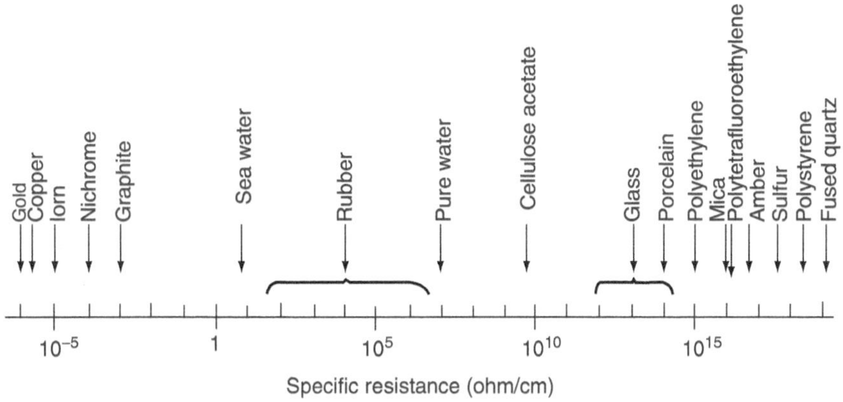

FIGURE 2.54 The resistivity spectrum.

surface of a material having an area of 1 cm^2 It ranges from 10^{10} ohm for cellulose acetate to 10^{14} ohm for polystyrene.

The insulation resistance of most plastic insulating materials is affected by temperature and the relative humidity of the atmosphere. The insulation resistance falls off appreciably with an increase in temperature or humidity. Even polystyrene, which has very high insulation resistance at room temperature, becomes generally unsatisfactory above 80°C (176°F). Under these conditions polymers like polytetrafluoroethylene and polychlorotrifluoroethylene are more suitable. Plastics that have high water resistance are relatively less affected by high humidities.

2.4.3 Arc Resistance

The arc resistance of a plastic is its ability to withstand the action of an electric are tending to form a conducting path across the surface. In applications where the material is subject to arcing, such as switches, contact bushes, and circuit breakers, resistance to arc is an important requirement. Arcing tends to produce a conducting carbonized path on the surface.

The arc resistance of an insulator may be defined as the time in seconds that an arc may play across the surface without burning a conducting path. A schematic of an arc-resistance test is shown in Figure 2.55.

Plastics that carbonize easily (such as phenolics) have relatively poor arc resistance. On the other hand, there are plastics (such as methacrylates) that do not carbonize, although they would decompose and give off combustible gases. There would thus be no failure in the usual sense. Special arc-resistant

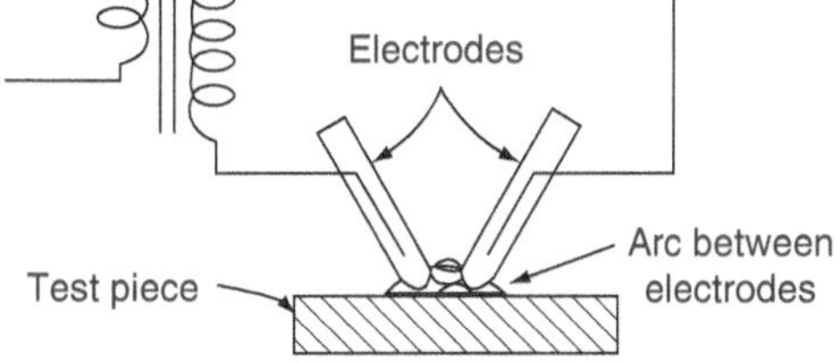

FIGURE 2.55 Arc-resistance test.

formulations involving noncarbonizing mineral fillers are useful for certain applications. But when service conditions are severe in this respect, ceramics ought to be used, because they generally have much better arc resistance than organic plastics.

Related to arc resistance is ozone resistance. This gas is found in the atmosphere around high-voltage equipment. Ignition cable insulation, for example, should be ozone resistant. Natural rubber is easily attacked and deteriorated by ozone. Fortunately, most synthetic resins have good ozone resistance and are satisfactory from this point of view.

2.4.4 Dielectric Constant

The effect of a dielectric material in increasing the charge storing capacity of a capacitor can be understood by considering the parallel-plate type sketched in Figure 2.56. If a voltage V is applied across two metal plates, each of area A m^2, separated by a distance, d m, and held parallel to each other in vacuum, the electric field established between the plates (Figure 2.56a) is

$$E = \frac{-V}{d} \tag{2.145}$$

The charge density, Q_0/A, where Q_0 is the total charge produced on the surface area A of each plate, is directly proportional to the electric field.

$$\frac{Q_0}{A} = -\varepsilon_0 E = \varepsilon_0 \frac{V}{d} \tag{2.146}$$

or

$$Q_0 = \frac{A\varepsilon_0}{d} V = C_0 V \tag{2.147}$$

The proportionality constant, ε_0, is called the dielectric constant (or permittivity) of a vacuum. It has units of

$$\varepsilon_0 = \frac{Q_0/A}{V/d} = \frac{\text{coul/m}^2}{\text{V/m}} = \frac{\text{coul}/V}{\text{m}} = \frac{\text{farad}}{\text{m}}$$

and a value of 8.854×10^{-12} farad/m.

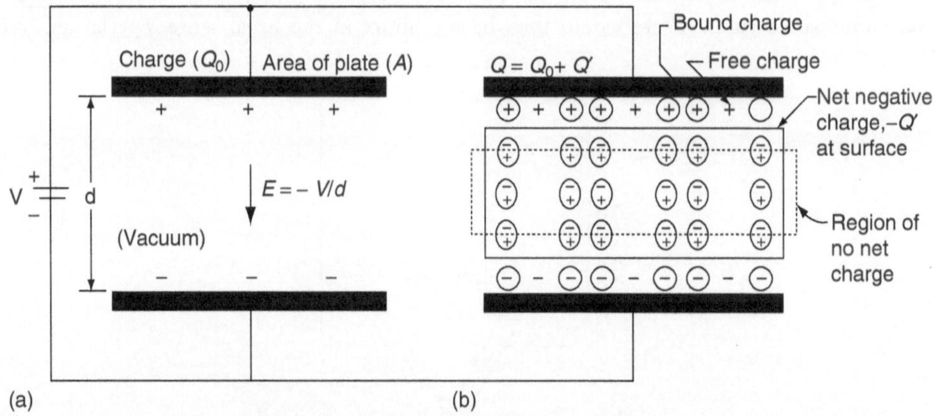

FIGURE 2.56 Schematic illustration of the effect of dielectric material in increasing the charge storing capacity of a capacitor.

The quality C_0 in Equation 2.147 is the capacitance of a capacitor (condenser) with a vacuum between its plates. It can be defined as the ratio of the charge on either of the plates to the potential differences between the plates.

Now if a sheet of a dielectric material is inserted between the plates of a capacitor (Figure 2.56b), an increased charge appears on the plates for the some voltage, due to polarization of the dielectric. The applied field E causes polarization of the entire volume of the dielectric and thus gives rise to induced charges, or bound charges, Q', at its surface, represented by the ends of the dipole chains. These induced charges may be pictured as neutralizing equal charges of opposite signs on the metal plates. If one assumes, for instance, that the induce charge $-Q'$ neutralizes an equal positive charge in the upper plate of the capacitor (Figure 2.56b), the total charge stored in the presence of the dielectric is $Q = Q_0 + Q'$.

The ratio of the total charge Q to the free charge Q_0 (which is not neutralized by polarization) is called the relative dielectric constant or *relative permittivity*, ε_r, and is characteristic of the dielectric material.

$$\varepsilon_r = \frac{\text{Total charge}}{\text{Free charge}} = \frac{Q}{Q_0} \tag{2.148}$$

Obviously, ε_r is always greater than unity and has no dimensions. For most materials ε_r exceeds 2 (Table 2.6).

Dividing both the numerator and denominator of Equation 2.148 by the applied voltage V and applying the definition of C from Equation 2.147, we obtain

$$\varepsilon_r = \frac{\varepsilon}{\varepsilon_0} = \frac{C}{C_0} \tag{2.149}$$

The relative dielectric constant or relative permittivity is thus defined as the ratio of the capacitance of a condense with the given material as the dielectric to that of the same condenser without the dielectric. (The dielectric constant for air is 1.0006. It is usually taken as unity—that is, the same as a vacuum—and the relative dielectric constant is referred to simply as the dielectric constant).

TABLE 2.6 Dielectric Properties of Electrical Insulators

Material	ε_r at 60 Hz	tan δ		Dielectirc Strength[a] (V/mil)
		60 Hz	10^6 Hz	
Ceramics				
Porcelain	6	0.010	—	—
Alumina	9.6	—	<0.0005	200–300
Zircon	9.2	0.035	0.001	60–290
Soda-lime	7	0.1	0.01	—
Fused silica	4	0.001	0.0001	—
Mica	7	—	0.0002	3,000–6,000 (1–3 mil specimen)
Polymers				
Polyethylene	2.3	<0.0005	—	450–1,000
Polystyrene	2.5	0.0002	0.0003	300–1,000
Polyvinyl chloride	7	0.1	—	300–1,000
Nylon-6, 6	4	0.02	0.03	300–400
Teflon	2.1	<0.0001	—	400

[a] Specimen thickness 1/8 in.

The dielectric susceptibility, χ, is defined as

$$\chi = \varepsilon_r - 1 = \frac{\varepsilon - \varepsilon_0}{\varepsilon_0} \tag{2.150}$$

It thus represents the part of the total dielectric constant which is a consequence of the material. From Equation 2.148,

$$\chi = \frac{\text{Bound charge}}{\text{Free charge}}$$

The magnitude of the induced or bound charge Q' per unit area is the polarization, P, which has the same units as charge density.

Therefore,

$$P = \frac{Q'}{A} = \frac{Q - Q_0}{A} \tag{2.151}$$

Substituting from Equation 2.146, Equation 2.148, and Equation 2.150, we obtain

$$P = \chi\varepsilon_0 E = \varepsilon_0(\varepsilon_r - 1)E \tag{2.152}$$

This equation, as we shall see later, provides a link between the permittivity, which is a macroscopic, measurable property of a dielectric, and the atomic or molecular mechanisms in the dielectric which give rise to this property.

2.4.4.1 Polarization and Dipole Moment

In terms of the wave-mechanical picture, an atom may be looked upon as consisting of a positively charged nucleus surrounded by a negatively charged cloud, which is made up of contribution from electrons in various orbitals. Since the centers of positive and negative charges are coincident (see Figure 2.57a), the net dipole moment of the atom is zero.

If an electric filed is applied, however, the electron cloud will be attracted by the positive plate and the nucleus by the negative plate, with the result that there will occur a small displacement of the center of gravity of the negative charge relative to that of the positive charge (Figure 2.57a). The phenomenon is described by the statement that the filed has induced an electric dipole in the atom; and the atom is said to have suffered electronic polarization—electronic because it arises from the displacement of the electron cloud relative to the nucleus.

The electric dipole moment of two equal but opposite charges, $+q$ and $-q$, at a distance r apart is defined as qr. For the atomic model of Figure 2.57a, it can be shown by balancing the opposite forces of the electric field and the coulombic attraction between the nucleus and the center of the electron cloud that the dipole moment, μ, induced in an atom by the field E is

$$\mu = (4\pi\varepsilon_0 R^3)E = \alpha_e E \tag{2.153}$$

where α_e, is a constant, called the electronic polarizability of the atom. It is proportional to the volume of the electron cloud, R^3. Thus the polarizability increases as the atoms become larger.

On the macroscopic scale we have earlier defined the polarization, P, to represent the bound charges induced per unit area on the surface of the material. Therefore, if we take unit areas on opposite faces of a cube separated by a distance d, the dipole moment due to unit area will be

$$\mu = Pd$$

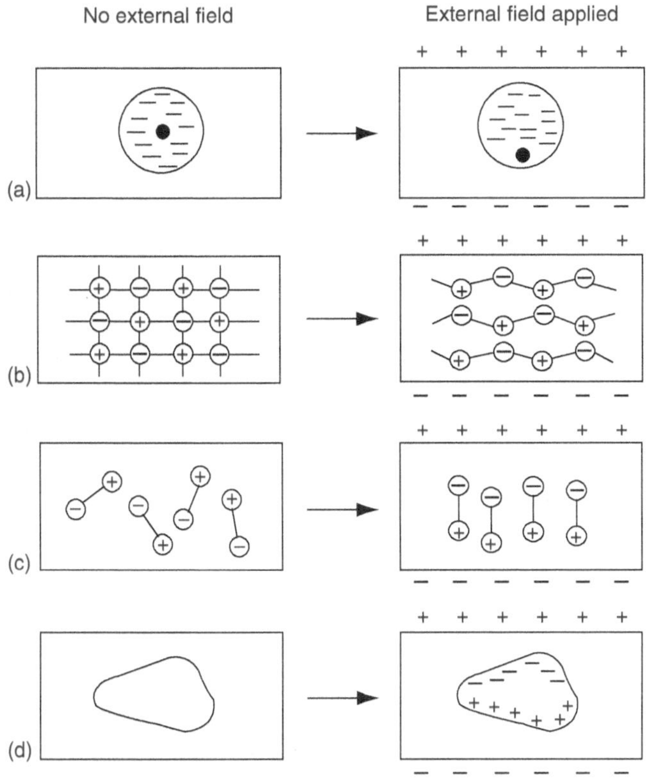

FIGURE 2.57 Schematic illustrations of polarization mechanisms. (a) Electronic displacement. (b) Ionic displacement. (c) Dipole orientation. (d) Space charge.

For $d=1$ (i.e., for unit volume) $\mu=P$. The polarization P is thus identical with the dipole moment per unit volume (check the units: coul/m^2=coul m/m^3).

A dipole moment may rise through a variety of mechanisms, any or all of which may thus contribute to the value of P. The total polarization may be represented as a sum of individual polarizations, each arising from one particular mechanism (Figure 2.57) or, more appropriately, as an integrated sum of all the individual dipole moments per unit volume.

$$P = P_e + P_i + P_0 + P_s = \frac{\sum \mu_e + \sum \mu_i + \sum \mu_0 + \sum \mu_s}{V} \tag{2.154}$$

Electronic polarization, P_e, as discussed previously, arises from electron displacement within atoms or ions in an electric field; it occurs in all dielectrics. Similarly, displacements of ions and atoms within molecules and crystal structures (Figure 2.57b) under an applied electric field give rise to ionic (or atomic) polarization, P_i. Orientation polarization, P_0, arises when asymmetric (polar) molecules having permanent dipole moments are present, since they become preferentially oriented by an electric field (Figure 2.57c). Interfacial (or space charge) polarization, P_s, is the result of the presence of higher conductivity phases in the insulating matrix of a dielectric, causing localized accumulation of charge under the influence of an electric field (Figure 2.57d).

Any or all of these mechanisms may be operative in any material to contribute to its polarization. A question to be discussed now is, which of the mechanisms are important in any given dielectric? The answer lies in studying the frequency dependence of the dielectric constant.

2.4.4.2 Dielectric Constant versus Frequency

Let us consider first a single dipole in an electric field. Given time, the dipole will line up with its axis parallel to the field (Figure 2.57c). If now the field is reversed, the dipole will turn 180° to again lie parallel to the field, but it will take a finite time; so if the frequency of the field reversal increases, a point will be reached when the dipole cannot keep up with the field, and the alteration of the dipole direction lags behind that of the field. For an assembly of dipoles in a dielectric, this condition results in an apparent reduction in the dielectric constant of the material. As the frequency of the field continues to increase, at some stage the dipoles will barely have started to move before the field reverses. Beyond this frequency, called the relaxation frequency, the dipoles make virtually no contribution to the polarization of the dielectric.

We may now consider the various mechanisms and predict, in a general way, the relaxation frequency for each one. Electrons with their extremely small mass have little inertia and can follow alterations of the electric field up to very high frequencies. Relaxation of electronic polarization is thus not observed until about 10^6 Hz (ultraviolet region). Atoms or ions vibrate with thermal energy, and the frequencies of these vibrations correspond to the infrared frequencies of the electromagnetic spectrum. The relaxation frequencies for ionic polarization are thus in the infrared range.

Molecules or groups of atoms (ions) behaving as permanent dipoles may have considerable inertia, so relaxation frequencies for orientation polarization may be expected to occur at relatively smaller frequencies, as in the radio-frequency range. Since the alternation of interfacial polarization requires a whole body of charge to be moved through a resistive material, the process may be slow. The relaxation frequency for this mechanism is thus low, occurring at about 10^3 Hz.

Figure 2.58 shows a curve of the variation of the dielectric constant (relative permittivity) with frequency for a hypothetical solid dielectric having all four mechanisms of polarization. Note that except at high frequencies the electronic mechanism makes a relatively low contribution to permittivity. However, in the optical range of frequencies, only this mechanism and the ionic mechanism operate; they therefore strongly influence the optical properties of materials.

2.4.4.3 Dielectric Constant versus Temperature

Liquids have higher dielectric constants than solids because dipole orientation is easier in the former. The effect is shown schematically in Figure 2.59a. After the abrupt change due to melting, the dielectric

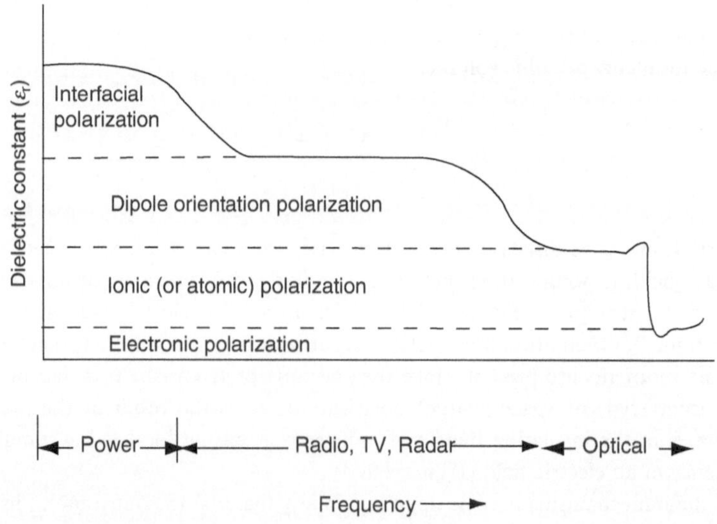

FIGURE 2.58 Dielectric constant versus frequency.

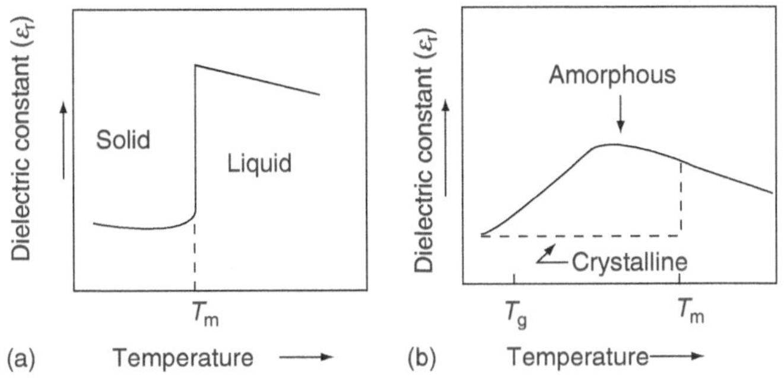

FIGURE 2.59 Variation of dielectric constant with temperature (schematic). (a) Crystalline material. (b) Amorphous polymer. A crystalline polymer containing polar group would behave as shown by dashed lines.

constant decreases as the temperature is increased, which is due to the higher atomic or molecular mobility and thermal collisions tending to destroy the orientation of dipoles.

Figure 2.59b shows the schematic variations of dielectric constants with temperature for amorphous solids, such as glasses and many polymers. Above the glass transition temperature (T_g), atoms and molecules have some freedom of movement, which allows orientation of permanent dipoles with the field, thereby increasing the dielectric constant. Since the polar groups which contribute to orientation polarization are not identically situated in an amorphous matrix, the dielectric constant changes over a temperature range rather than abruptly at a single temperature as in a crystalline material (cf. Figure 2.59a). The decrease in the dielectric constant after melting is again due to greater molecular mobility and thermal collisions.

2.4.4.4 Dielectric Losses

The behavior of a dielectric under an applied field has much in common with that of a material subjected to mechanical loading. The displacements of atoms and molecules within a material, when a mechanical force is applied, do not occur instantaneously but lag behind the force, resulting in elastic aftereffect and energy dissipation by mechanical hysteresis under an alternating force. Similarly, in dielectrics the lag of polarization behind the applied field produces energy dissipation by electrical hysteresis in an alternating field (Figure 2.60b). Such energy losses are related to the internal dipole friction. The rotation of dipoles with the field is opposed by the internal friction of the material, and the energy required to maintain this rotation contributes to the power loss in dielectrics.

Besides electrical hysteresis, leakage currents also contribute to dielectric losses. Leakage currents occur mainly by ionic conduction through the dielectric material and are usually negligible except at high temperatures. There are various ways of measuring energy losses by a dielectric.

A fundamental property of a capacitor is that if an alternating voltage is applied across it in a vacuum, the current that flows to and from it due to its successive charging and discharging is 90° out of phase with the voltage (Figure 2.60c), and no energy is lost. However, in real capacitors containing a dielectric, the lag of polarization causes a phase shift of the current (Figure 2.60d). The phase shift angle, δ, is called the loss angle, and its tangent ($\tan \delta$) is referred to as the loss tangent or dissipation factor. (An ideal dielectric would have a phase angle of 90°, and hence the loss angle would be zero.)

The sine of the loss angle ($\sin \delta$) or the cosine of the phase angle ($\cos \theta$) is termed the power factor. In electrical applications the power loss (PL) is defined as the rate of energy loss per unit volume and is derived to be

$$PL = \frac{\omega E^2 \varepsilon_0}{2} \varepsilon_r \tan \delta \qquad (2.155)$$

where E is the electric field and ω is the angular velocity; $\varepsilon_r \tan \delta$ is called the *loss factor.*

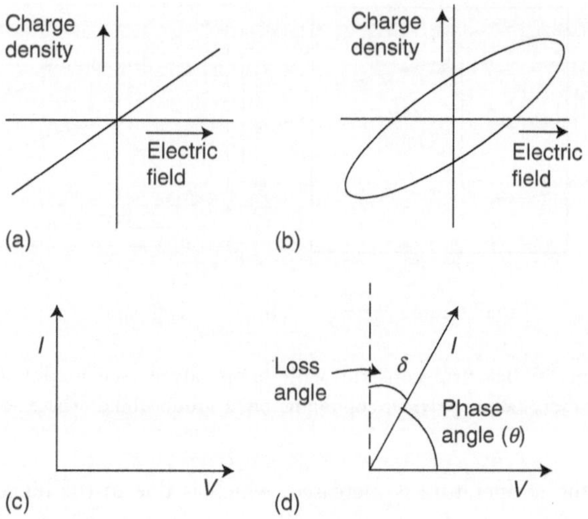

FIGURE 2.60 Charge density versus electric field. (a) Loss-free cycle. (b) High-loss cycle. (c) Phase shift in a perfect capacitor. (d) Phase shift in a real capacitor.

For most materials δ is very small; consequently, the three measures of energy dissipation—δ, tan δ, and sin δ ($=\cos\theta$)—are all approximately the same. Since these values vary somewhat with frequency, the frequency must be specified (see Table 2.6).

It is evident from this discussion that the power loss and heat dissipation in a dielectric will be aided by a high dielectric constant, high dissipation factor, and high frequency. Therefore, for satisfactory performance electrical insulating materials should have a low dielectric constant and a low dissipation factor but a high dielectric strength (Table 2.6) and a high insulation resistance.

Polyethylene and polystyrene with their exceptionally low dissipation factors (<0.0005) and low dielectric constant (2.3–2.5) are the most suitable for high-frequency applications, as in television and radar. For dielectrics used in capacitors, however, a high dielectric constant is desirable.

2.4.4.5 Dielectric Losses of Polar Polymers

When a polymer having polar groups (e.g., polymethyl methacrylate, polyvinyl chloride) is placed in an electric field, the polar groups behaving as dipoles tend to orient themselves in response to the field (Figure 2.61a). In an alternating field the friction of the dipoles rotating with the field contributes to the dielectric loss. This loss is small at low frequencies where the polar groups are able to respond easily to the field, and also at high frequencies where the polar groups are unable to change their alignment with the field. The loss is maximum in the transitional region where the polymer is passing from high response at low frequency to low response at high frequency (Figure 2.61b).

Since the friction of dipoles in an alternating field produces heat, polar polymers can be heated by the application of radio-frequency field in fabrication processes (see high-frequency welding, Chapter 2 of *Plastics Fabrication and Recycling*).

2.5 Optical Properties

Optical characteristics of plastics include color, clarity, general appearance, and more directly measurable properties, such as index of refraction [29]. For optical applications, however, other properties, including dimensional stability, scratch resistance, temperature limitation, weatherability, and water absorption, must be considered.

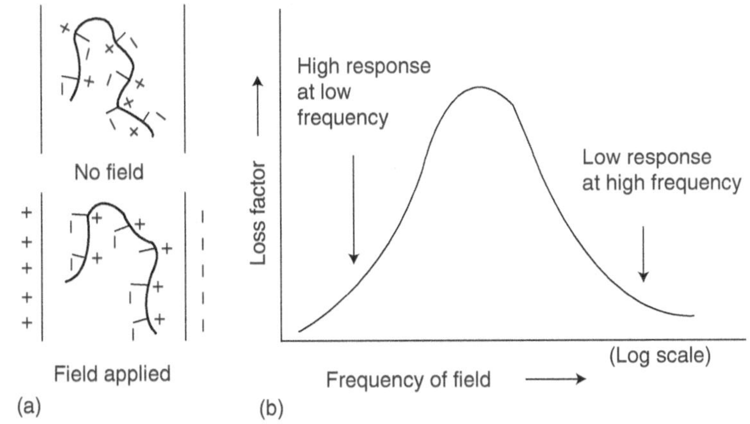

FIGURE 2.61 (a) Orientation of a polar polymer molecule in an electric field. (b) Dielectric response of a polar polymer in an alternating electric field.

2.5.1 Optical Clarity

Most resins by nature are clear and transparent. They can be colored by dyes and will become opaque as pigments or fillers are added. Polystyrene and poly (methyl methacrylate) are well known for their optical clarity, which even exceeds that of most glass. Optical clarity is a measure of the light transmitting ability of the material. It depends, among other things, on the length of the light path, which can be quantitatively expressed by the Lambert-Beer law, or log $(I/I_0) = -AL$, where I/I_0 is the fraction of light transmitted, L is the path length, and A is the absorptivity of the material at that wavelength. The absorptivity describes the effect of path length and has the dimension of reciprocal length.

Transparent colored materials are obtained by adding a dye to a water white resin. A color results when a dye removes part of the visible light traveling through the piece. The red color, for example, is produced by a dye which absorbs the blue, green, and yellow components of the light and transmits the red unchanged (Figure 2.62). However, for any dye to be effective it must be soluble in the plastic, and it is best incorporated in the plastic before molding. Fluorescent dyes absorb radiant energy at one wavelength, perhaps in the ultraviolet, and emit it as less energetic but more visible radiation.

Cast phenolics, allyls, cellulosics, and many other clear plastics show a natural tendency to absorb in the blue and to be yellowish. Ultraviolet photodecomposition is largely responsible for the development of

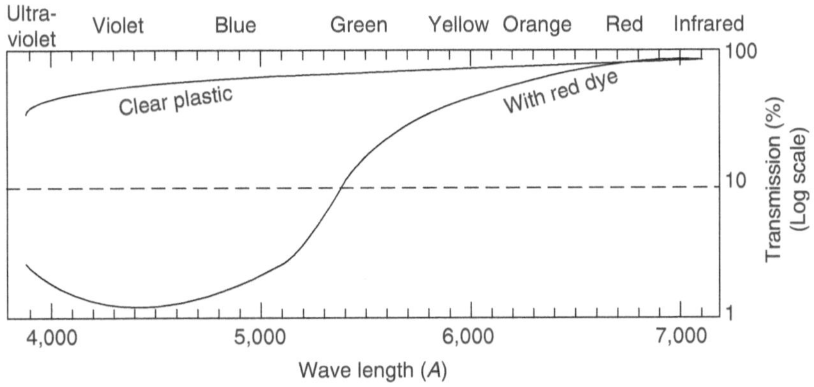

FIGURE 2.62 Light transmission diagram for plastic.

yellowish color in plastics exposed to sunlight. Incorporation of an invisible ultraviolet-absorbing material such as phenyl salicylate or dihydroxybenzophenone in the plastic greatly reduces photodecomposition.

Addition of a blue or green tint into the plastic can also mask the yellow color. In addition to yellowing, plastics show darkening due to outdoor exposure as the transmission curve shifts downward, and there is pronounced absorption at wavelengths shorter than about 5,000 A (Figure 2.63).

The optical clarity of a plastics specimen is measured by the lack of optical haze. Haze is arbitrarily defined as the fraction of transmitted light which is deviated $2\frac{1}{2}°$ or more from an incident beam by forward scattering. When the haze value is greater than 30%, the material is considered to be translucent. The forward scattering of the light beam, which is responsible for haze, occurs at internal interfaces such as caused by a dust particle, bubble, particles of a filler or pigment, or by density changes.

Due to scattering at interfaces, a crystalline plastic with myriads of crystallite regions bounded by interfaces is translucent. Crystalline polyethylene is thus translucent at room temperature, but, on warming, the crystallites disappear and the material becomes transparent. It can thus be inferred that plastics which are transparent at room temperature, such as polystyrene or poly(methyl methacrylate), are of the noncrystalline type and without fillers.

The effect of interfaces due to fillers depends to a large extent on the difference in the indices of refraction of plastic and filler. Thus transparent glass-filled polyester panels are obtained if the indices of refraction of glass and resin are identical and the glass is surface treated to enable the resin to wet the glass completely.

2.5.2 Index of Refraction

The index of refraction for any transparent material is the ratio of the sine of the angle of an incident ray to the sine of the angle of refraction (Figure 2.64). It also corresponds to the ratio of the speed of light in a vacuum (or air, closely) to its speed in the material. The refractive index values of several common plastics are compared with those of other materials in Table 2.7.

The refraction property makes possible the focusing action of a lens. Plastic lenses have the advantage that they are light in weight (about half as heavy as glass) and practically nonshattering. But they have the disadvantage of low scratch resistance and comparatively poor dimensional stability for temperature changes.

When tolerances are less critical, plastic lenses can be mass produced by virtue of the moldability of plastics; these lenses are quite satisfactory for inexpensive camera view finders, for example, but in applications where exacting tolerances are required, such as cameras, periscopes, or similar high-resolution devices, molded plastic lenses have not been suitable. If plastic lenses are to be used in these

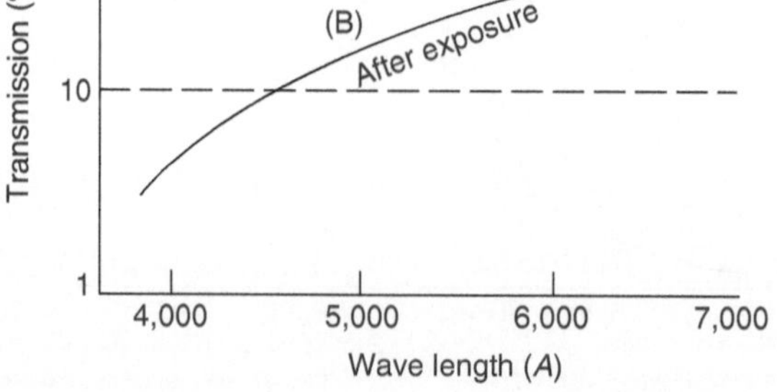

FIGURE 2.63 Yellowing of plastics on exposure.

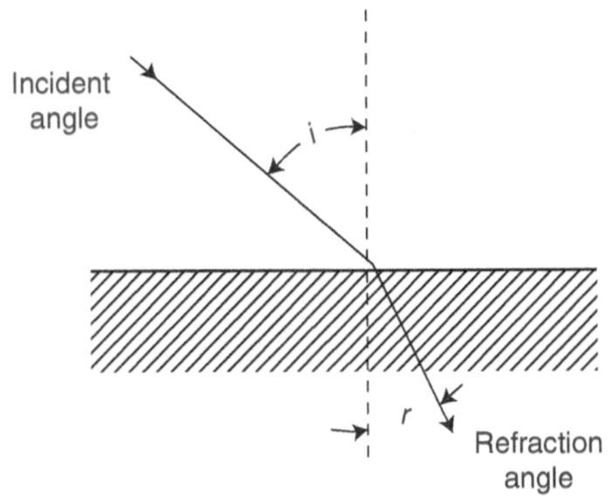

FIGURE 2.64 Refraction of light. Index of refraction: $n = (\sin i)/(\sin r)$.

applications, they should be ground and polished in much the same manner as in glass, though, of course, it is easier to do so (Figure 2.65).

The refractive index, n, of an isotropic material is given by the Lorentz–Lorenz relation

$$\frac{n^2 - 1}{n^2 + 2} = \frac{4}{3}\pi P$$

where P is the optical polarizability per unit volume of material. For a pure substance it is more convenient to write this equation in the form

$$\frac{(n^2 - 1)M}{(n^2 + 2)\rho} = \frac{4}{3}\pi N_A \alpha = R \qquad (2.156)$$

where M is the molecular weight of the substance, ρ is the density, N_A is Avogadro's number, α is *the optical polarizability* of a single molecule of the substance, and R is the *molar refraction.*

The quantity α represents the amount of polarization of the molecule per unit electric field caused by the alternating electric field associated with the light ray passing through the material. The polarization may be regarded as due to the shift of the center of charge of the more loosely bound electrons relative to the nucleus (see Figure 2.57).

The refractive index varies with wavelength of light and is measured by the optical dispersion—that is, the difference in the refractive indexes for different wavelengths. It is responsible for the spectrum-separating ability of a prism in a spectroscope. Most plastics have relatively low optical dispersion. This property makes them more suitable for eyeglasses and large lenses for projection television.

2.5.3 Piped Lighting Effect

The difference in indexes of refraction for air and for a transparent solid (plastic or glass) is responsible for the ability of the latter, used as a rod or plate, to bend or pipe light around a curve. Bending can be explained as follows. It is evident from Figure 2.64 that a light beam in air cannot enter a material with $i > 90°$. Since $\sin 90° = 1$, the maximum angle of refraction is that for which the sine is $1/n$, where n is the index of refraction. For poly(methyl methacrylate) $n = 1.49$, and the maximum angle of refraction becomes $\sin^{-1}(1/1.49)$ or 42° approximately. Obviously a light beam refracted at a greater angle cannot have come from the air, but it must have been reflected internally. A plastic plate with small curvature therefore gives internal reflections and can thus bend a light beam, as shown in Figure 2.65.

TABLE 2.7 Index of Refraction

Material	Index of Refraction (n)
Air	1.00
Water	1.33
Cellulose acetate	1.48
Poly(methyl methacrylate)	1.49
Common glass	1.52
Poly(vinyl chloride) copolymer	1.53
Polystyrene	1.59
Flint glass	1.65
Diamond	2.42

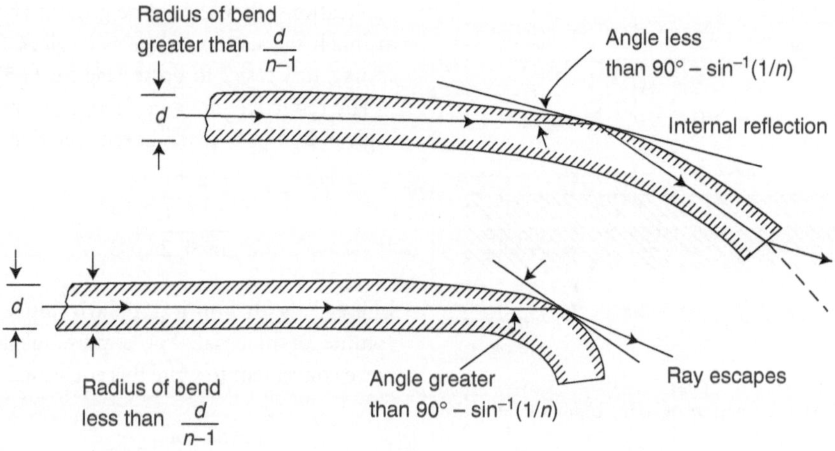

FIGURE 2.65 Principle of piped light.

Consideration of geometric optics shows the minimum radius of curvature for this piping of light to correspond to $1/(n-1)$ times the thickness.

Polymer has numerous advantages over glass [30]. First and foremost, it is more pliable. It is also nonshattering, lightweight, and easy to fabricate. Polymer optical fiber (POF) has a shorter bend radius (i.e., more flexibility) and is more resilient to damage and abuse than glass due to its intrinsic material characteristics. The bending radius depends largely on the diameter of the fiber, large fiber giving a larger bend radius. Thus, for an outside diameter of one millimeter (1,000 μm), the bend radius (damage threshold) is approximately one centimeter, while for single-mode POF with outside diameter of 125 μm, the bend radius is as low as 0.125 cm. The level of intensity loss from bending depends on the individual characteristics of the fiber, but intensity loss values have been observed at 10%–15% for bends close to the minimum (damage threshold) radius.

Attenuation is a measurement, in decibels (dB) per unit length, of how much reduction in light (due to absorption) is experienced per unit length of the fiber. Absorption varies largely with the wavelength of light. The attenuation of PMMA, generally used as POF core material, is approximately 100 dB/km in the visible region and increases rapidly at higher wavelength (Figure 2.66). This high attenuation of POF compared to the silica-based fiber limits the data link length. However, the field witnessed real progress in 1980s when low loss polymers were developed and drawn into fibers (see Chapter 2 of *Industrial Polymers, Specialty Polymers, and Their Applications*).

The light bending ability of POF is made use of in a wide variety of applications ranging from lighting scheme for pools and fountains to telecommunications and consumer electronics. In these applications (see more details in Chapter 2 of *Industrial Polymers, Specialty Polymers, and Their Applications*), the POF serves as a medium for transmitting light for illumination or imaging, or for transporting and/or controlling information that is encoded on a beam of light [31].

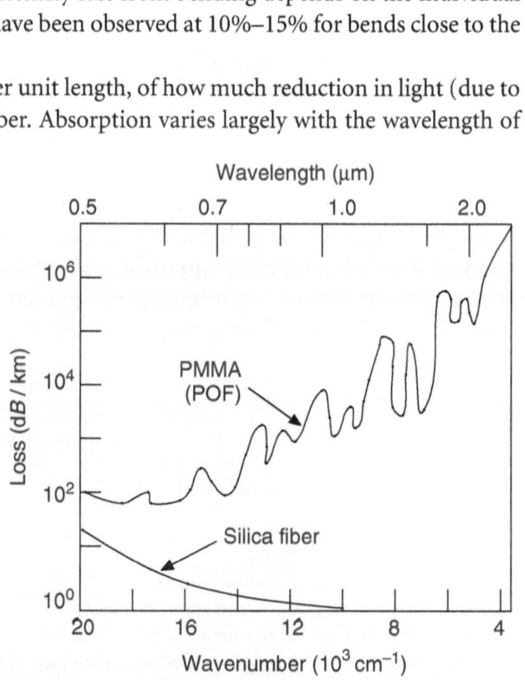

FIGURE 2.66 Variation of attenuation of PMMA and silica-based fibers with wavelength.

2.5.4 Stress-Optical Characteristics

An important phenomenon observed in amorphous plastics (also observed in optical glass) is the development of optical anisotropy due to stress. The stress-optical characteristic of transparent plastics is the basis of the important technique of photoelasticity by which stress and strain in complicated shapes, for which no analytical solution is readily available, can be determined experimentally and simply.

The amount of strain in various parts of a transparent plastic model of a machine part, subjected to loads simulating those in actual operation, is determined by measuring the anisotropy with polarized light; therefore from this measurement the distribution of stress and strain in the actual metal part can be deduced. Very complicated shapes, including complete structures such as bridges or aircraft wings, have been successfully analyzed in this manner, employing plastic models.

The same stress-optical characteristic also permits examination of locked-in stresses in a molded plastic part. The part is examined under polarized light, and the amount of stress is indicated by the number of fringes or rings that become visible. Illumination with white light gives colorful patterns involving all the colors of the spectrum. Monochromatic light is, however, used for stress analysis because it permits more precise measurements.

In general, a single ray of light entering an anisotropic transparent material, such as crystals of sufficiently low symmetry and strained or drawn polymers, is propagated as two separate rays which travel with different velocities and are polarized differently. Both the velocities and the state of polarization vary with the direction of propagation. This phenomenon is known as birefringence or double refraction, and the material is said to be birefringent.

The stress-optical characteristic of plastics arises from this phenomenon of birefringence, induced by strains due to applied stress. The applied stress produces different densities along different axes. The stress-optical coefficient of most plastics is about 1,000 psi (70 kgf/cm^2 or 6.9 Mpa) per inch (2.54 cm) thickness per fringe. This means that a 1-in. thick part when illuminated with monochromatic light and viewed through a polarizing filter will show a dark fringe or ring for a stress of 1,000 psi. This sensitivity is many times higher than that shown by glass and makes transparent plastics useful in photoelastic stress analysis.

In one variation of the process the deforming stress is applied to a warm, soft plastic model which is then quenched to room temperature. The resulting locked-in stress may then be analyzed at leisure or, perhaps more conveniently, by cutting sections from the model and examining them separately. For this application, however, the plastic should have stress-optical stability. In this regard, unplasticized transparent plastics such as poly(methyl methacrylate), polystyrene, and cast poly(allyl phthalate) are superior to plasticized materials such as cellulose acetate.

2.6 Thermal Properties

The useful thermal properties of plastics include specific heat, thermal expansion, thermal conductivity, and thermal softening [32].

2.6.1 Specific Heat

For most plastics the specific heat value (calories per gram per °C) lies between 0.3 and 0.4. On a weight basis this value is much higher than that of most metals. Both iron and copper, for example, have specific heats of about 0.1 at ordinary temperatures. However, along a volume basis, the specific heats of plastics are lower than those of common metals, because of the substantially lower density of plastics.

Knowledge of specific heat of a material helps to determine the energy requirement for increasing its temperature. In compression molding and injection molding the theoretical heat requirement can be calculated as the sum of this direct heat and any latent heat of melting minus any energy released by chemical reaction. But this heat requirement is not large compared to the heat loss by radiation and conduction from the press. The proportion of losses is substantially lower for injection molding than for

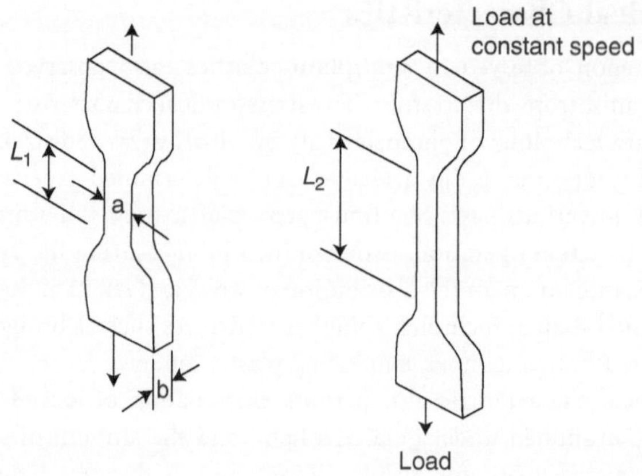

FIGURE 2.67 Tests for tensile strength (stress at fracture of the specimen) and elongation (extension of materials under load) of plastics. The first provides a measure of the breaking strength of the material but is radically affected by the rate of loading and the ambient temperature. Tensile strength in lbf/in.^{2} or kgf/cm^2=load (lbf or kgf)/a×b (in.^{2} or cm^2). Standard test methods: BS 2782 method 301, ASTM D638, ISO R527. Elongation %=$(L_2-L_1)\times 100/L_1$. Standard test methods: BS 2782 method 301, ASTM D638.

compression molding. It is nevertheless customary in both cases to establish the heating requirements as well as mold cooling requirements by a process of trial and error.

In dielectric preheating, however, the specific heat of a molding powder is of more direct concern; this knowledge along with the amount of material and time enables one to calculate the required amount of radio-frequency power. For example, to raise the temperature of a 1-kg preform of specific heat 0.35 through 80°C in 1 min requires 1×0.35×80 or 28 kcal/min or about 2 kW.

2.6.2 Thermal Expansion

Linear (thermoplastic) polymers have very high thermal expansion coefficients since they are weakly bonded materials and need less input of thermal energy to expand the structure. This applies to all polymers of the vinyl type which have expansion coefficients of about 90×10^{-6}/°C. Network (thermosetting) polymers having a three-dimensional framework of strong covalent bonds exhibit less thermal expansion and have expansion coefficients in the range of 30–70$\times10^{-6}$/°C. These may be

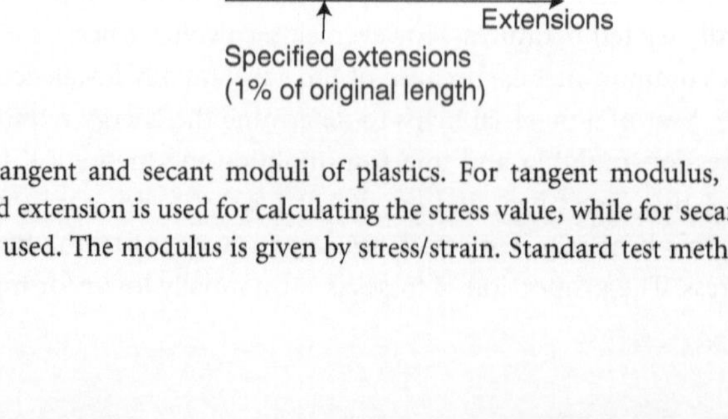

FIGURE 2.68 Test for tangent and secant moduli of plastics. For tangent modulus, load on tangent to load-extension curve at specified extension is used for calculating the stress value, while for secant modulus load on secant to load-extension curve is used. The modulus is given by stress/strain. Standard test methods: BS 2782 method 302, ASTM D638, ISO R527.

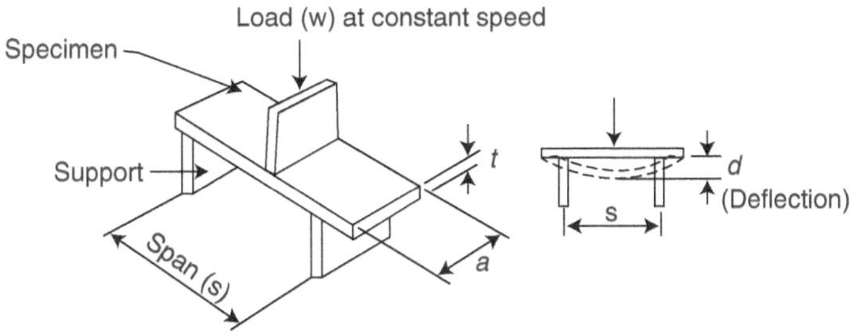

FIGURE 2.69 Tests for flexural properties of plastics. Flexural strength $= 3ws/2at$ (lbf/in.^{2} or kgf/cm^2). Standard test method: ASTM D790. Modulus in flexure $= s^3w/4at^3d$ (lbf/in.^{2} or kgf/cm^2). Standard test methods: BS 2782 method 302D, ASTM D790, ISO R178.

compared with values of 11×10^{-6}/°C for mild steel, 17×10^{-6}/°C for ordinary brass, and less than 10×10^{-6}/°C for ceramics. In spite of high thermal expansion, plastics do not easily undergo thermal cracking, because they also have very low elastic moduli and large strains do not induce high stresses.

One area where the high expansion of polymers plays a significant role is the molded dimensions of plastic parts. The linear mold shrinkage due to thermal contraction from molding to room temperature is usually about 1/2%–1%. Polyethylene and certain other materials exhibit even a higher shrinkage. As a result, plastic parts with close tolerances are difficult to make. It is also possible during the molding operation that different parts are not all at a uniform temperature. This may lead to differential shrinkage during cooling and produce warping, locked-up internal stresses, and weakening of the part.

Warping may be prevented by a suitable mounting. Internal stresses in plastics that are more subject to creep and cold flow tend to relieve themselves slowly; warming the part accelerates the process. Internal stresses in polystyrene and other brittle plastics can be removed by an annealing process similar to the one used in glass manufacture. It involves a controlled heating-cooling cycle, and for plastics it is obtained by immersing the parts in a liquid held at the proper temperature, followed by very slow cooling. Polystyrene for example can be heated in water at 80°C and then slowly cooled to 65°C. It is then cooled in undisturbed air.

The relatively high thermal expansion of plastics poses a problem in the use of molded metal inserts which are sometimes required for electrical contacts, screw thread mountings, or increased strength. To minimize stresses due to inserts, manufacturers usually use plastics with low coefficient of expansion. Phenolics and ureas are commonly used because their coefficient of expansion are among the lowest of the common plastics.

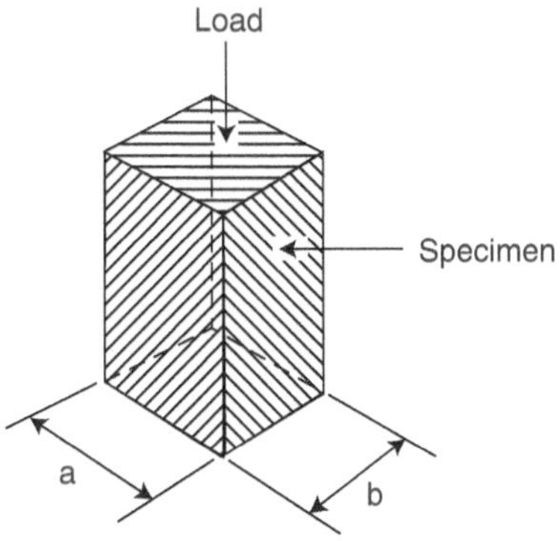

FIGURE 2.70 Test for compressive strength of plastics. Compressive strength (lbf/in.^{2} or kgf/cm^2) = load (lbf or kgf)/$a \cdot b$ (in.^{2} or cm^2). Standard test methods: ASTM D795, BS 2782 method 303, ISO R604.

For many plastics the use of molded metal inserts is not satisfactory because of the excessive stress produced. As an example, polystyrene with brass inserts on cooling from the molding temperature of 160 to 20°C produces a strain of $(70-17) \times 10^{-6} \times (160-20)$ or 0.0074. For an elastic modulus of 0.46×10^6 psi (3.2 Gpa), the internal stresses become 3,400 psi (23 Mpa). The presence of this much internal stress renders the part useless, since the tensile strength of polystyrene is only 3,600 psi (25 Mpa).

However, by use of an appropriate filler the thermal expansion of plastics and, hence, the internal stresses can be reduced. Addition of 11%

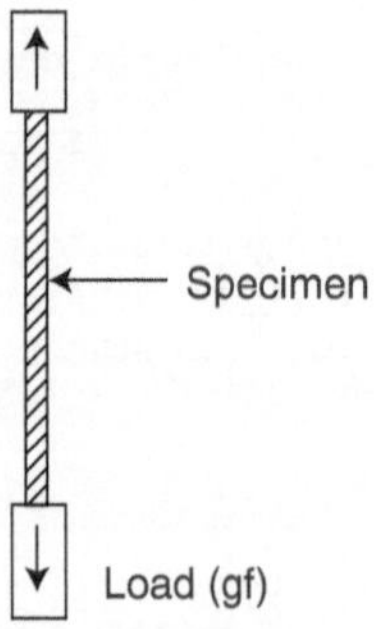

FIGURE 2.71 Test for tenacity of filaments, cords, twines, etc. Tenacity (gf/denier) = breaking load (gf)/denier of specimen. (Denier of filaments, cords, twines, etc., is equal to the weight in grams of 9,000 meters of the sample.)

by weight of aluminum oxides in polystyrene gives a mixture whose thermal expansion is identical to that of brass. Brass inserts in this alumina-polystyrene composite show no evidence of internal stresses and are quite satisfactory.

When synthetic resins are used as cements, the thermal expansion needs to be matched; this matching is done by judiciously choosing the filler. In the production of panels of aluminum alloy overlaid on phenolic laminate, a filler consisting of a glass fiber-starch mixture is used to reduce the expansion characteristic of the resin and to match its expansion exactly with that of aluminum. The aluminum panel then adheres completely and cannot be removed without tearing the metal or rupturing the phenolic core.

There are cases where the high thermal expansion of plastics is used to advantage. One example is the shrink fitting of handles of cellulose nitrate on screwdrivers. Another example is in mold design where the shrinkage on cooling is sufficient to permit a small undercut.

2.6.3 Thermal Conductivity

Thermal conductivities of plastics are relatively low and approximate 0.0004 (cal-cm)/(°C-cm^2-sec). The corresponding values are 0.95 for copper, 0.12 for cast iron, 0.002 for asbestos, 0.0008 for wood, and 0.0001 for cork (SI values in J/s-m-k are obtained by multiplying by 418.7). Because of their low thermal conductivities, plastics are used for handles to cooking utensils and for automobile steering wheels. The low thermal conductivity is also responsible for the pleasant feel of plastic parts. Quite hot or quite cold objects can be handled with less difficulty if they are made of plastic, since the thermal insulation afforded by the plastic prevents a continuous rush of heat energy to (or from) the hand.

Both thermal conductivity and temperature resistance of plastics have to be considered for their use at high temperatures. As an example consider a teapot handle; it must not deform even at 100°C. Therefore,

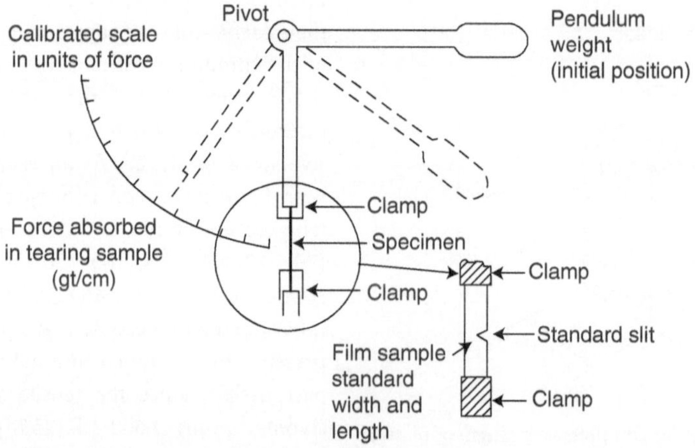

FIGURE 2.72 Test for tear propagation resistance of plastic film and thin sheeting. Tear resistance (gf/cm/mil) = force required to tear sample (gf/cm)/thickness of film (mil). Standard test methods: ASTM D1922, BS 2782 method 308B.

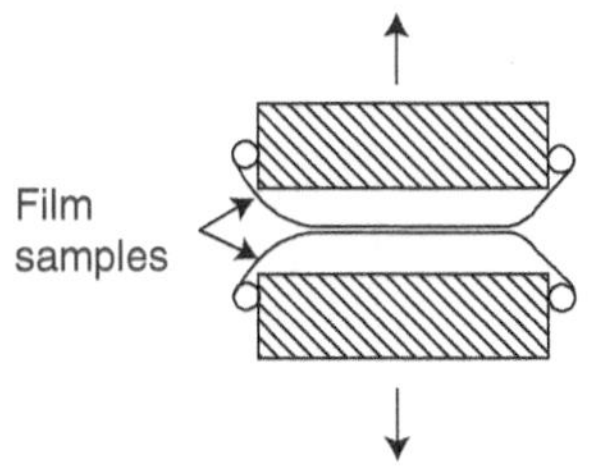

FIGURE 2.73 Tests for blocking of plastic film. Blocking force (lbf/in.^{2} or kgf/cm^2) = load (lbf or kgf)/initial area of films in contact (in.^{2} or cm^2). Standard test method: ASTM D1893.

common thermoplastics such as cellulose acetate are ruled out, but both a wood-flour-filled and an asbestos-filled phenolic might be considered.

The thermal conductivity of a mixture is nearly proportional to the volume percentage of each component. Wood-flour-filled phenolic has a higher thermal conductivity than the pure resin, but the conductivity of this composite is still low enough to justify its use as the handle to a teapot. This composite can also withstand temperatures up to 100°C sufficiently well to give the handle a reasonable service life. For parts subjected to higher temperatures, asbestos-filled phenolic is a better choice. It can be used as the insulating connection to an electric iron, for example.

Design of a handle determines to a great extent its service life. Quite often handles are found fastened directly to the hot object without regard for any temperature limitation; at the junction the plastic becomes brittle because of high temperature, and failure occurs. If the handle can be separated from the heated part and some cooling arrangement is included as part of the design, improved performance is to be expected.

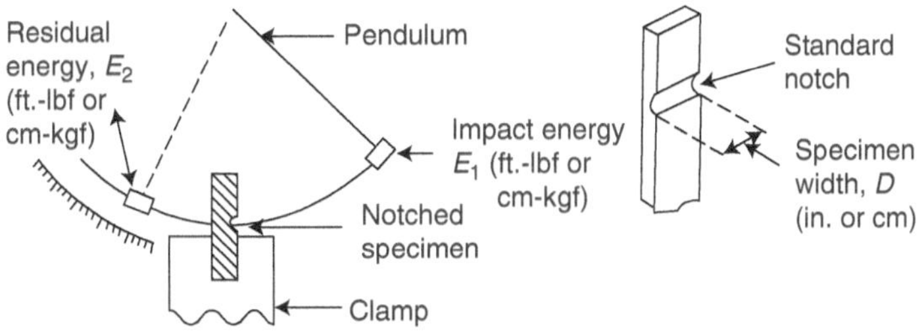

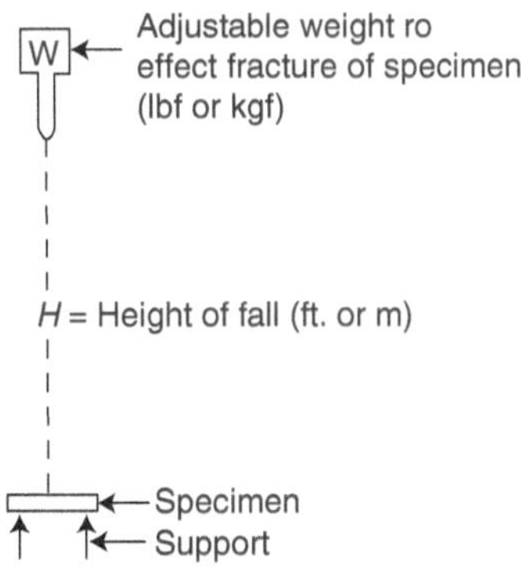

FIGURE 2.74 Tests for impact resistance of plastics. Izod impact strength = $(E_1 - E_2)/D$ ft-lbf/in. of notch or cm-kgf/cm of notch. Standard test methods: ASTM D256; BS 2782 method 306A, ISO R180. Falling weight impact strength = $W \cdot H$ ft-lbf or m-kgf. (F_{50} is the energy required to fracture 50% of the specimens.) Standard test method: BS 2782 method 306 B.

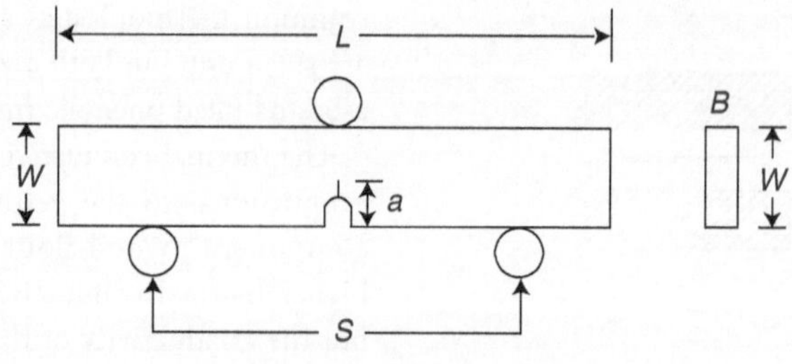

FIGURE 2.75 Test for fracture toughness of plastics. Specimen in three-point bending configuration, as shown, is loaded in a testing machine that has provision for autographic recording of load applied to the specimen. Specimen dimensions: L, length (mm); W, width (mm); B, thickness (mm); S, span length (mm); a, initial crack length (mm). Plane strain fracture toughness, $K_{Ic}=PSf\,(a/W)/BW^{3/2}$, where P=load (N) determined from load displacement record; $f\,(a/W)$=geometric factor. Example: L=64 mm, W=10 mm B=3 mm, S=40 mm, a=3.2 mm, P= 50.88 N, $f\,(a/W)$=1.61 (from ASTM standard), K_{Ic}=34.3 N $mm^{-3/2}$=1.09 MPa $m^{1/2}$. Standard test method: ASTM E399-74.

2.6.4 Transition Temperatures and Temperature Limitations

Both first- and second-order transitions are observed in polymers. Melting and allotropic transformations are accompanied by latent-heat effects and are known as first-order transitions. During second-order transitions, changes in properties occur without any latent-heat effects. Below the second-order-transition temperature (glass transition temperature) a rubberlike material acts like a true solid (see Chapter 1). Above this temperature the fixed molecular structure is broken down partially by a combination of thermal expansion and thermal agitation. The glass transition temperature of polystyrene is 100°C; below 100°C polystyrene is hard and brittle, and above 100°C it is rubberlike and becomes easily deformed.

The design engineer often requires to know the maximum temperature for which a polymer can be used in a given application. This depends largely on two independent factors: (1) the thermal stability of the polymer, particularly in air; and (2) the softening behavior of the polymer.

Thermal stability testing requires the study of the change in properties on aging at various service temperatures. The relationship between structure and stability was considered briefly in Chapter 1 and will be considered again in Chapter 2 of *Industrial Polymers, Specialty Polymers, and Their Applications.* The use of additives for improving such stability is discussed in Chapter 1.

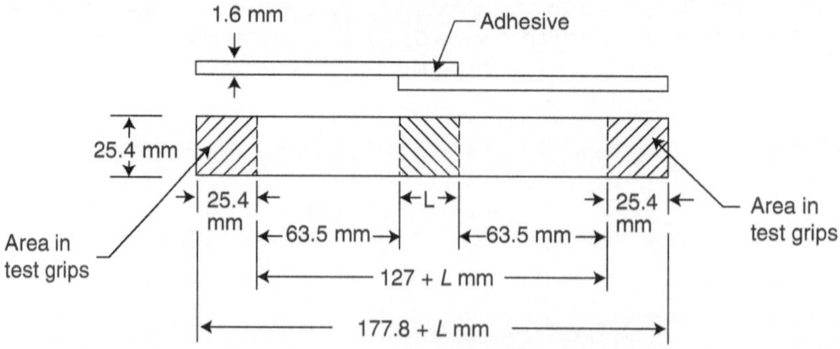

FIGURE 2.76 Form and dimensions of test specimen for single-lap-joint shear test to determine shear strengths of adhesives for bonding metals (standard test method: ASTM D1002-94). Recommended length of overlap (L) for most metals of 1.62 mm in thickness is 12.7 mm. Specimen is placed in grips of a testing machine so that outer 25.4 mm (1 in.) of each end is in contact with the jaws. Load is applied at the rate of 80–100 kg/cm^2 (1,200–1,400 psi) of shear area per min and continued till failure occurs.

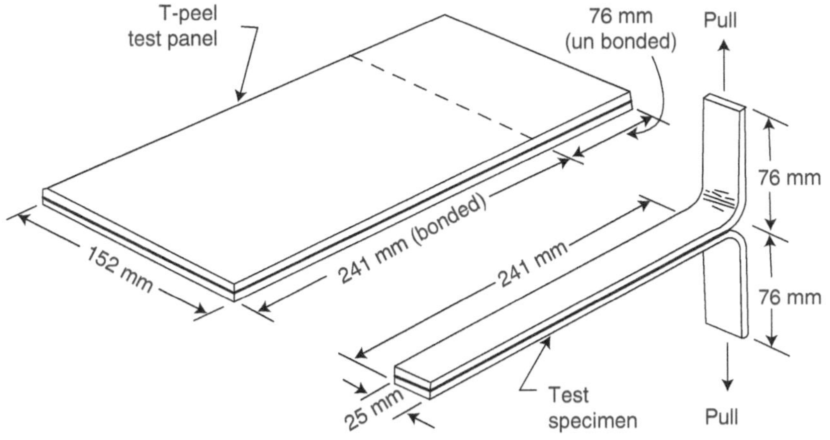

FIGURE 2.77 Test panel and T-type test (T-peel test) specimen for peel resistance of adhesives (standard test method: ASTM D1876-95). The bent, unbonded ends of test specimen are clamped in test grips of tensile testing machine and load applied at a constant head speed of 254 mm (10 in.). Average peeling load (in pounds per inch of specimen width) is determined for the first 127 mm (5 in.) of peeling after the initial peak.

The rigidity of a polymer is determined by the ease with which polymer molecules are deformed under load. Young's modulus is a fundamental measure of the rigidity or stiffness of a material. It is thus clear from Figure 1.13 that the softening point is associated with the T_g for amorphous polymers and with the T_m for highly crystalline polymers. In addition, there are many polymers that soften progressively between T_g and T_m and the value determined for the softening point can depend significantly on the test method used.

Two particular test methods have become very widely used. These are the Vicat softening point test and the test widely known as the heat distortion temperature test (also called the deflection temperature under load test). In the Vicat softening point test a sample of polymer is heated at a specified rate temperature increase and the temperature is noted at which a needle of specified dimensions indents into the polymer a specified distance under a specified load.

The heat-distortion temperature is defined as the temperature at which the midpoint of a beam 1/2 by 1/2 by 5 in. supported 1/2-in. from each end, shows a net deflection of 0.01 in. when loaded centrally with 2.5 kg and heated at the specified rate of 2°C/min. Testing is also done at one-quarter of this load. For most materials the two heat-distortion temperatures are within 10°.

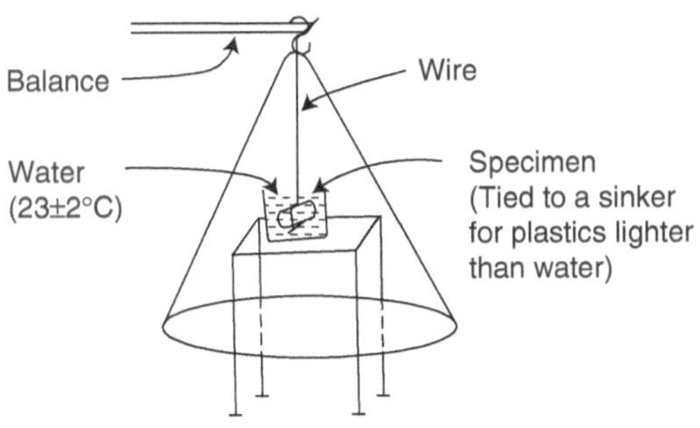

FIGURE 2.78 Test for specific gravity and density of plastics. Sp. gr. = $a/(a-w-b)$ where a = *wt.* of specimen without wire, b = *wt.* of specimen completely immersed and of the wire partially immersed in water, and w = *wt.* of partially immersed wire. Standard test method: ASTM D792.

A large difference in these two heat distortion temperature indicates a material sensitive to temperature change, and for such materials at any elevated temperature stress should be minimum up to the heat-distortion temperature. A plastic is expected to maintain its shape under load, and hence this temperature represents an upper limiting point at which a plastic may be used.

Recommendations of upper temperature limits for plastics are usually based on general experience, although some consideration is given to high-load and low-load heat-distortion temperatures. The size and shape of the part as well as its molding

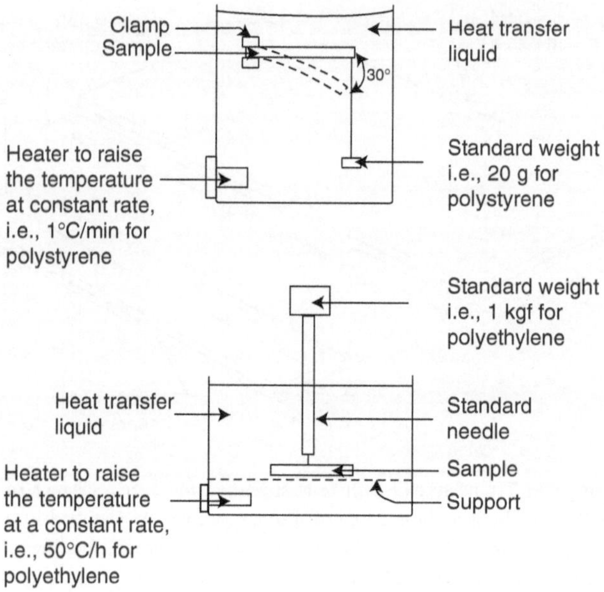

FIGURE 2.79 Tests for softening temperatures of plastics. Cantilever softening point is the temperature at which the sample bends through 30°. Standard test method: BS 2782 method 102 C. Vicat softening point is the temperature at which the needle penetrates 1 mm into the sample. Standard test methods: ASTM D1525, BS 2782 method 102D, ISO R306.

conditions govern to a certain extent the maximum permissible temperature; service conditions such as temperature variations and humidity are also important.

2.6.5 Standard Properties of Plastics

Figure 2.67 through Figure 2.93 illustrate schematically (in simplified form) the bases of some of the standard properties of plastics [33]. Where standard test methods have been developed, these have been included. The principles of the different tests are shown, and the properties measured are indicated.

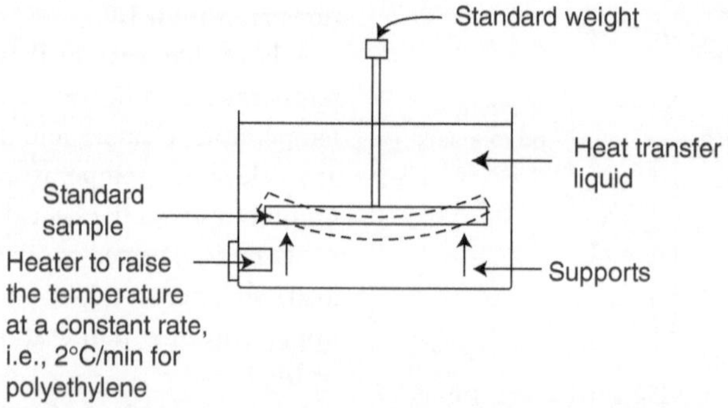

FIGURE 2.80 Test for deflection temperature of plastics under flexural load. Heat distortion temperature is the temperature at which a sample deflects by 0.1 in. (2.5 mm). Two measurements are made and quoted: (a) with a stress of 66 lbf/in.^{2} (4.6 kgf/cm^2) and (b) with a stress of 264 lbf/in.^{2} (18.5 kgf/cm^2). Standard test methods: ASTM D648, BS 2782 method 102, ISO R75.

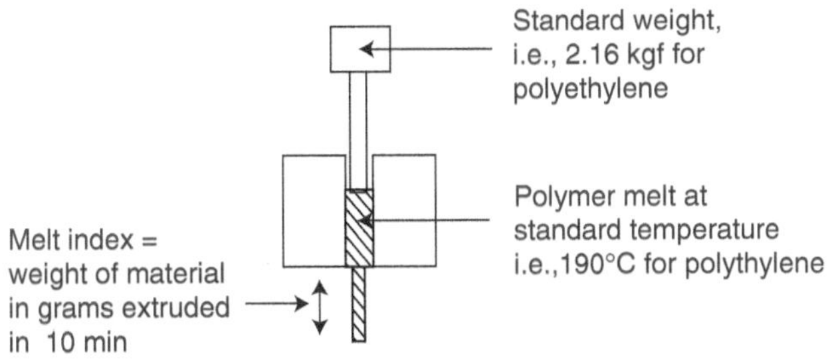

FIGURE 2.81 Melt index of plastics. The test measures the rate of flow of polymer melt. It provides an indication of the ease of processing. Standard test methods: ASTM D 1238, BS 2782 method 105C, ISO R292.

Note that the property values of plastics are highly dependent on the specimen preparation, equipment, and testing techniques used. For this reason it is essential to refer to the appropriate official standard test method when executing the work.

2.7 Identification of Common Plastics

The first step in the identification of polymers is a critical visual examination. While the appearance of the sample may indicate whether it is essentially a raw polymer or a compounded and processed item, learning about its form, feel, odor, color, transparency or opacity, softness, stiffness, brittleness, bounce, and surface texture may be important in the process of the identification of the polymer. For example, polystyrene, the general purpose polymer, is transparent and brittle, and produces a characteristic metallic tinkle when objects molded from it are dropped or struck.

Identification of plastics [34] is carried out by a systematic procedure: preliminary test, detection of elements, determination of characteristic values, and, finally, specific tests. For an exact identification, however, the test sample should first be purified so that it contains no additives (plasticizers, fillers, pigments, etc.) that may affect the results of an analysis. Purification is achieved by solvent extraction; either the material is dissolved out and polymer is obtained by reprecipitation or evaporation of the solvent, or the pure polymer remains as the insoluble residue. The solvent varies, and a general method cannot be given. However, for many materials particularly for those in which additives do not interfere, the unpurified material can be investigated and qualitative preliminary tests used.

In general, the following series of investigations should be carried out for preliminary tests of the material: (1) behavior on heating in an ignition tube and in the flame; (2) qualitative detection of heteroelements such as N, S, and halogens; (3) determination of properties such as density, refractive index, and melting point or range; (4) determination of solubility as an aid to polymer identification; and (5) ash or sulfated ash determination as a test for inorganic additives (fillers, pigments, stabilizers).

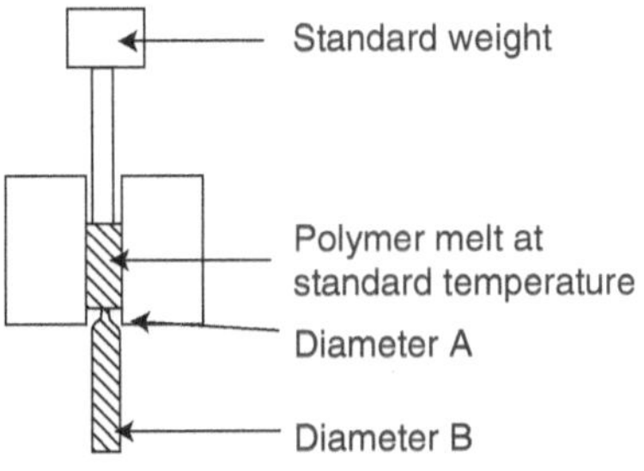

FIGURE 2.82 Measurement of swelling of die extrudate. Die swell ratio = B/A.

2.7.1 Behaviors on Heating and Ignition

Many polymers can be roughly identified by their behavior when carefully heated and ignited. Nitrocellulose and plastics containing this (e.g., celluloid) burn with explosive violence and other materials such as poly(vinyl chloride) or fluoro-hydrocarbons decompose with the evolution of poisonous or irritating vapors. Only

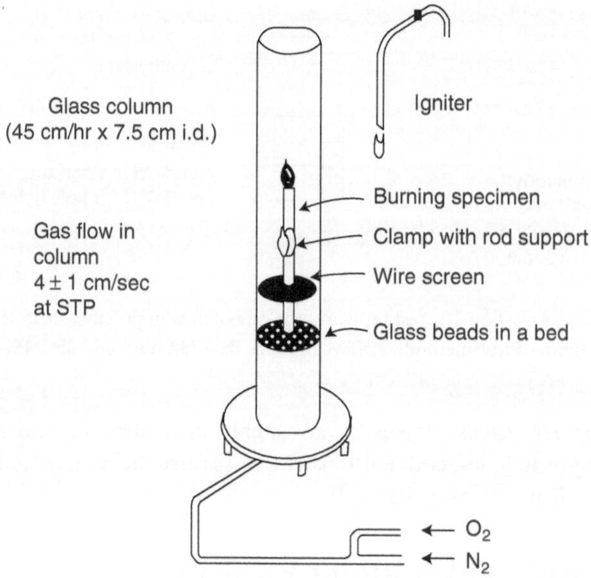

FIGURE 2.83 Determination of oxygen index. Oxygen index, $n\% = 100 \times O_2/(O_2 + N_2)$, where O_2, volumetric flow rate of oxygen, cm^3/sec, at the minimum concentration necessary to support flaming combustion of a material initially at room temperature and N_2, corresponding flow rate of nitrogen, cm^3/sec. Standard test method: ASTM D2863.

small quantities of material should therefore be taken for heating tests. The heating should be done gently, because if the heating is too rapid or intense, the characteristic changes may not be observed.

In a typical procedure, a small piece or amount (0.1 g) of the test sample is placed on a cleaned glass or stainless steel spatula, previously heated to remove any traces of combustible or volatile materials, and then warmed gently over a small colorless Bunsen flame until the sample begins to fume. The decomposing sample is removed from the flame and the nature of the fume or gas is examined with respect to color, odor, inflammability, and chemical identity including acidity, alkalinity, etc.

The sample is next moved to the hottest zone of the small Bunsen flame and note is taken of the following: (1) if the material burns and if so, how readily; (2) the nature and color of the flame as the material burns; (3) whether the material is self-extinguishing or continues to burn after removal from the flame, and (4) the nature of the residue.

Observations on heating and ignition of some common polymers are listed in Table 2.8.

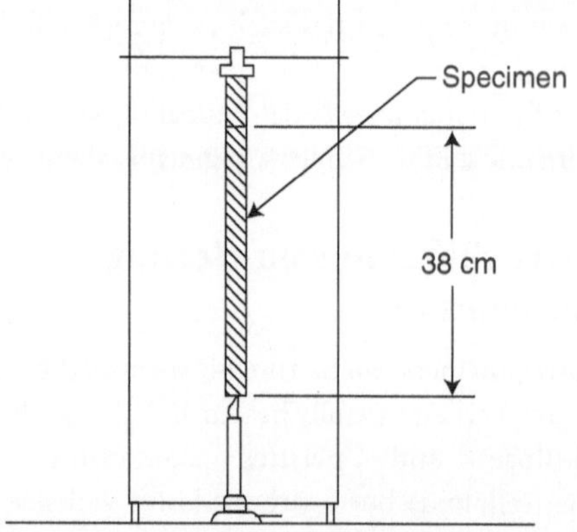

FIGURE 2.84 Test for rate of burning. Burning rate = 38 (cm)/t (min). Standard test method: ASTM D568.

2.7.2 Tests for Characteristic Elements

The results of tests for characteristic elements such as nitrogen, sulfur and halogens may serve to roughly indicate the nature of the unknown material, including the nature of the base polymer and additives, if present. The following tests may be performed for qualitative detection of N, S, and halogens.

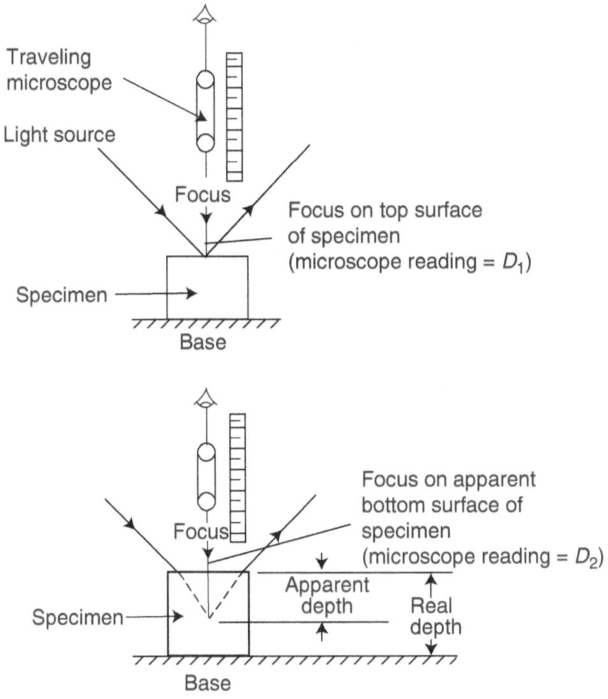

FIGURE 2.85 Test for index of refraction of transparent plastics. Refractive index = real depth (measured with vernier)/apparent depth ($D_2 - D_1$). Standard test method: ASTM D542, ISO R489.

1. *Nitrogen.* Above 50 mg of the test material is heated carefully in an ignition tube with twice the amount of freshly cut sodium, until the sodium melts. A further small amount of material is added and the tube is heated to red heat. It is then dropped carefully into water (20 ml) in a mortar. The solid is powdered and the solution filtered.

   A little freshly prepared ferrous sulfate is added to the filtrate and the latter boiled for 1 min. It is acidified with dilute hydrochloric acid, and 1–2 drops of ferric chloride solution added. A deep blue coloration or precipitate (Prussian blue) indicates nitrogen (if very little nitrogen is present, a green–blue color is formed initially). This test sometimes fails with substances containing nitro-groups and with nitrogen-containing heterocyclic compounds.
2. *Sulfur.* The fusion is carried out as for nitrogen. The filtrate (2 ml) is acidified with acetic acid and a few drops of an aqueous solution of lead acetate is added. A black precipitate of lead sulfide indicates the presence of sulfur.
3. *Halogens.* The fusion is carried out as for nitrogen. The filtrate (2 ml) is acidified with dilute nitric acid, boiled in the fume cupboard to remove H_2S and HCN, and then a few drops of silver nitrate solution is added. A while precipitate, turning grey–blue, indicates chlorine, a yellowish precipitate indicates bromine and a yellow precipitate indicates iodine.

In Beilstein's test, which provides a simple means of detecting halogens, a minute quantity of the test material is placed on a copper wire (initially heated in a nonluminous bunsen flame until the flame appears colorless and then cooled) and heated in the outer part of the flame. Carbon burns first with a luminous flame. A subsequent green to blue–green color, produced by volatilized copper halide, indicates halogen (chlorine, bromine, iodine).

The different polymers may be classified into several groups according to the element present as shown in Table 2.9. The focus of identification may be further narrowed down on the basis of other preliminary observations, e.g., fusibility or otherwise, melting point or range, heat distortion temperature, flame tests

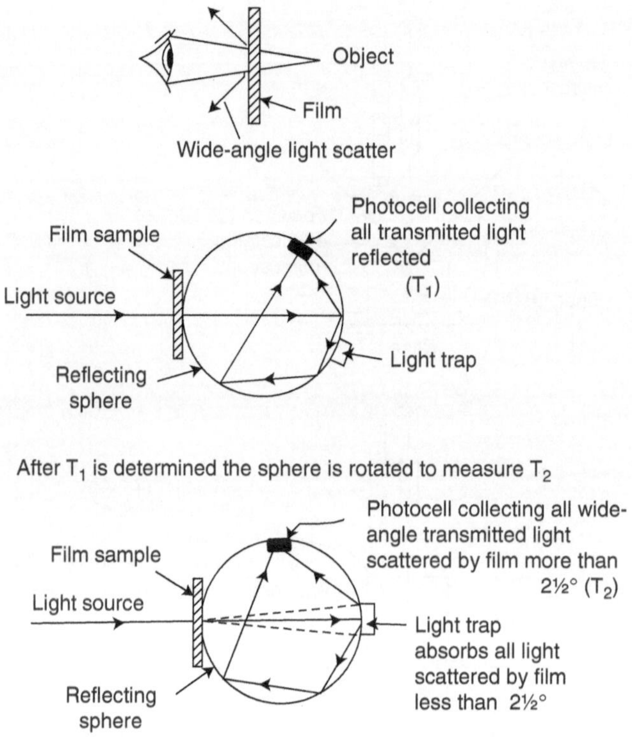

FIGURE 2.86 Test for haze of transparent plastics. Haze, % = 100 × T_2/T_1. A low haze value is important for good short distance vision. Standard test method: ASTM D1003.

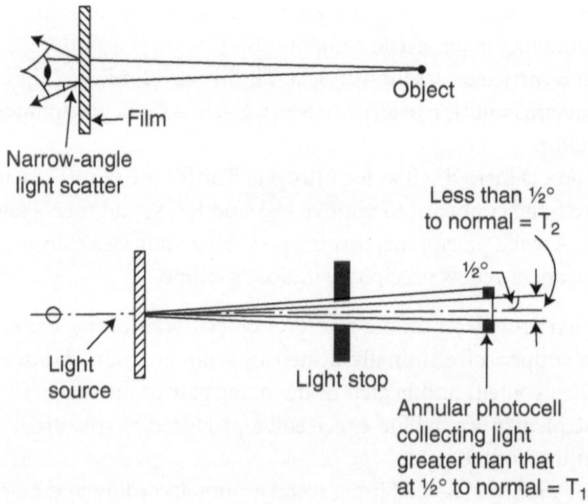

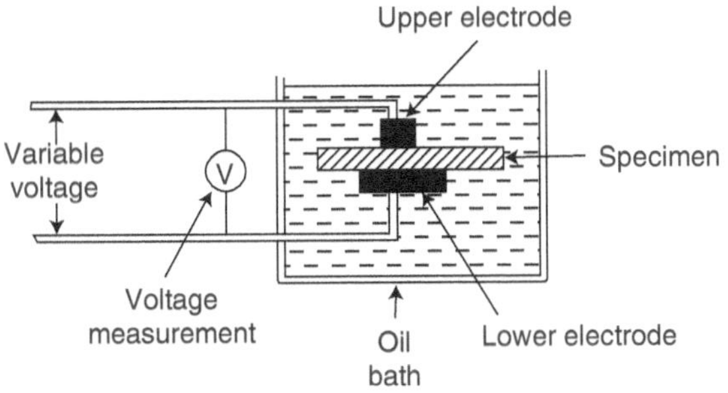

FIGURE 2.88 Test for dielectric strength of solid insulating materials. Dielectric strength (V/mil) = maximum voltage before failure (V)/thickness of specimen (mil). Standard test methods: ASTM D149, BS 2782 method 201.

and tests for thermal degradation, and solubility or extractability in water or different organic solvents. The solubility behaviors of common polymers are compared in Table 2.10.

2.7.3 Specific Tests

When the observations and results of preliminary tests have been considered and most of the possible structures for the polymer base eliminated, an exact identification can be made by carrying out specific tests [35]. Some specific tests for ready identification of specific polymers are described below.

1. *Tests for polystyrene and styrenic copolymers.* The test depends on the aromatic rings of the styrene units in the polymer chain. The polymer sample (0.1 g) is heated under reflux in concentrated nitric acid and the clear mixture is poured into water (25 ml) to yield a pale yellow precipitate which is then extracted with ether (2×25 ml). The ether extracts are combined and washed thoroughly with water (2×5 ml), then extracted with dilute sodium hydroxide solution (2×5 ml) and finally with water (5 ml).

   The alkaline extracts and the aqueous extract are combined and the nitro compounds present in the mixture are reduced by adding granulated zinc (1 g) and concentrated hydrochloric acid and warming gently. The solution is cooled, filtered and then diazotized with a dilute solution

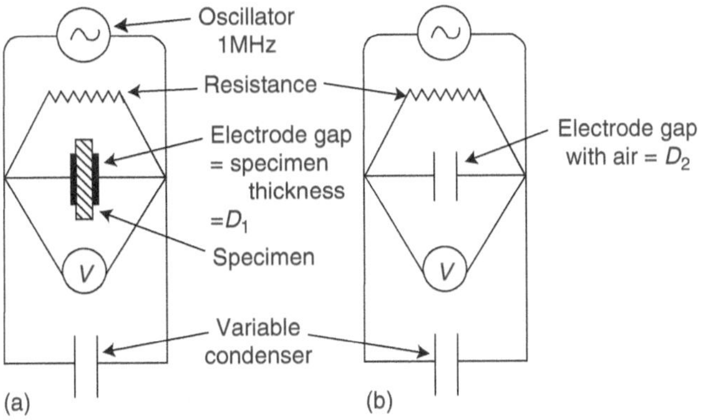

FIGURE 2.89 Test for permittivity (dielectric constant) of insulating materials. (a) Position of maximum voltage obtained with sample by adjusting variable capacitor. Electrode gap = specimen thickness = D_1; (b) position of maximum voltage obtained with air by adjusting electrode gap to D_2 [variable capacitor remains as set in (a)]. Dielectric constant = D_1 (in. or mm)/D_2 (in. or mm). Standard test methods: ASTM D150, BS 2782 method 207A.

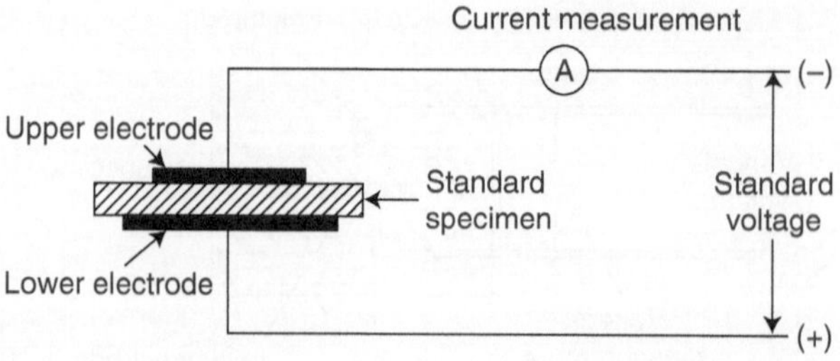

FIGURE 2.90 Test for DC resistance of insulating materials. Electrical resistance of specimen (ohm) = applied voltage (V)/current measured (A). Volume resistivity (ohm-cm) = resistance of specimen (ohm) × arc of upper electrode (cm^2)/specimen thickness (cm). Standard test methods: ASTM D257, BS 2782 method 202.

(5 ml) of sodium nitrite under ice-cooled condition. The solution is finally poured into excess of alkaline β-naphthol solution producing a deep red color.

2. *Test for poly(vinyl chloride)*. Specific tests for chlorine containing polymers are performed only when the presence of chlorine is confirmed by preliminary test. The simplest method of chlorine determination is the Beilstein test, described previously. Plasticizers are removed from the test material by ether extraction and the Beilstein test for chlorine is repeated to make certain that chlorine is still present. The material is then dissolved in tetrahydrofuran, filtered and the polymeric product reprecipitated by adding methanol.

   On addition of a few drops of a ~10% methanolic solution of sodium hydroxide to 2–5 ml of ~5% solution of the plasticizer free material, the mixture changes with time from colorless to light yellow–brown, dark brown, and finally to black.

   In another test, poly(vinyl chloride) and vinyl chloride polymers readily form brown coloration and eventually dark brown precipitates when their pyridine solutions are boiled and treated with a few drops of methanolic sodium hydroxide (5%).

3. *Test for poly(vinylidene chloride)*. Poly(vinylidene chloride), when immersed in morpholine, develops a brown color both in the liquid and the partially swollen polymer. The change takes place faster if the mixture is warmed in a water bath.

4. *Test for poly(vinyl alcohol) and poly(vinyl acetate)*. When a small volume of iodine solution (0.2 g iodine and 1.0 g KI dissolved in 20 ml of a 1:1 alcohol–water mixture and diluted to 100 ml using 2N hydrochloric acid solution) is added to an equal volume of 0.25% neutral or acidic solution of poly(vinyl alcohol), a blue color develops almost immediately or on addition of a pinch of borax into the solution. Poly (vinyl acetate), however, turns deep brown on contact with the dilute iodine solution.

5. *Test for polyacrylonitrile and acrylonitrile copolymers.* When a strip of cupric acetate paper that is freshly moistened with a dilute solution of benzidine in dilute acetic acid is held in the pyrolytic vapors of polyacrylonitrole or its copolymers, a bright blue color develops readily. In another test, if the condensed pyrolyzate of polyacrylonitrile or its copolymers is made alkaline, boiled with a trace of ferrous sulfate, and then acidified, Prussian blue precipitate is readily produced.

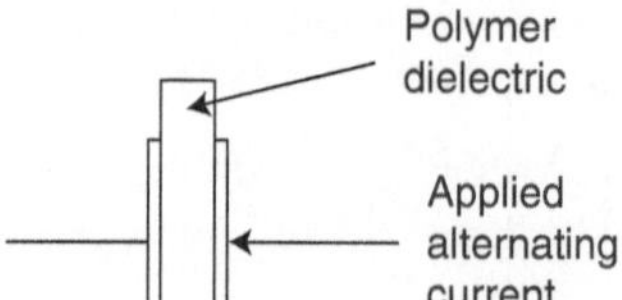

FIGURE 2.91 Test for AC loss characteristics of solid insulating materials. Power factor = $W/V \times I$, where W = power loss in watts and $V \times I$ = effective sinusoidal voltage × current in volt-amperes. Standard test methods: ASTM D150, BS 2782 method 207A.

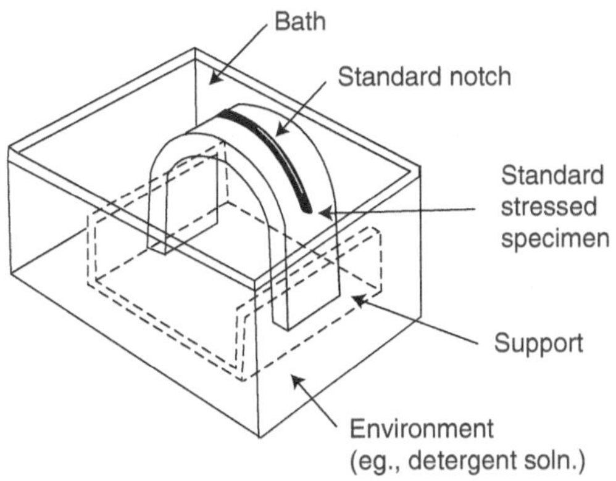

FIGURE 2.92 Test for environmental stress cracking of ethylene plastics. Stress cracking resistance (F_{50}) = time taken for 50% of the specimens to fail (h). Standard test method: ASTM D1693.

6. *Test for phenolic resins.* The test material (dry) is heated in an ignition tube over a small flame. The mouth of the tube is covered with a filter paper, prepared by soaking it in an ethereal solution of 2,6-dibromoquinone-4-chloro-imide and then drying it in air. After the material has been pyrolyzed for about a minute, the paper is removed and moistened with 1–2 drops of dilute ammonia solution. A blue color indicates phenols (care must be taken with plastics that contain substances yielding phenols on pyrolysis, e.g., phenyl and cresyl phosphate, cross-linked epoxide resins, etc.).
7. *Test for formaldehyde condensate resins.* Formaldehyde enters into the composition of several resins and polymers. Common examples are phenol-, urea-, and melamine-formaldehyde resins, polyoxymethylene and poly (vinyl formal). Formaldehyde is evolved when these are thermally treated or boiled with water in the presence of an acid (H_2SO_4).

   The aqueous extracts or acid distillates are treated with chromotropic acid (1,8-dihydroxynaphthalene-3, 6-disulfonic acid). A few drops of 5% aqueous chromotropic acid solution are added to the aqueous test solution and then an excess of concentrated sulfuric acid is added, and the mixture preferably warmed to nearly 100°C for a few minutes. In the presence of formaldehyde, the solution turns violet/dark violet.

   Phenol, urea, and melamine are also obtained as intermediates on acid hydrolysis of the corresponding formaldehyde condensate resins and appropriate tests for phenol, urea, and melamine may be employed for identification purposes.

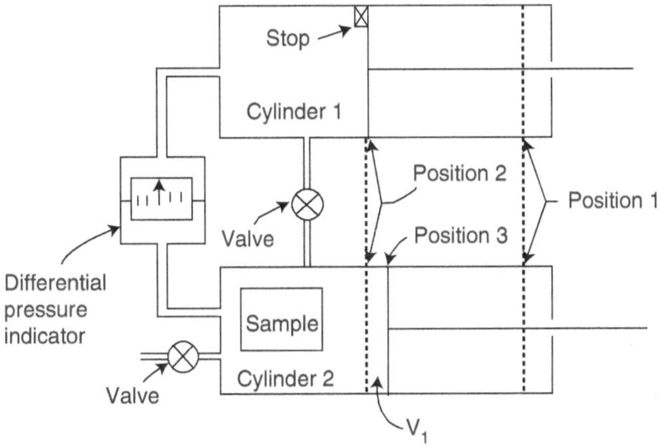

FIGURE 2.93 Determination of open cell content of rigid cellular plastics. Open cell content (approx.) = 100 ($V-V_1$)/V, where V_t, displacement volume of specimen, cm^3 and V, geometric volume of specimen, cm^3. Standard test method: ASTM D2856.

TABLE 2.8 Heating Tests of Some Common Polymers

Polymer	Color of Flame	Odor of Vapor	Other Notable Points
The material burns but self-extinguishes on removal from flame			
Poly(vinyl chloride)	Yellow–orange, green bordered	Resembles hydrochloric acid and plasticizer (usually ester like)	Strongly acidic fumes (HCl), black residue
Poly(vinylidene chloride)	As above	Resembles hydrochloric acid	As above
Polychloroprene	Yellow, smoky	As above	As above
Phenol-formaldehyde resin	Yellow, smoky	Phenol, formaldehyde	Very difficult to ignite, vapor reaction neutral
Melamine-formaldehyde resin	Pale yellow, light	Ammonia, amines (typically fish like), formaldehyde	Very difficult to ignite, vapor reaction alkaline
Urea-formaldehyde resin	As above	As above	As above
Nylons	Yellow–orange, blue edge	Resembling burnt hair	Melts sharply to clear, flowing liquid; melt can be drawn into a fiber; vapor reaction alkaline
Polycarbonate	Luminous, sooty	Phenolic	Melts and chars
Chlorinated rubber	Yellow, green bordered	Acrid	Strongly acidic fumes, liberation of HCl; swollen, black residue
The material burns and continues burning on removal from flame			
Polybutadiene (BR)	Yellow, blue base, smoky	Disagreeable, sweet	Chars readily; vapor reaction neutral
Polyisoprene (NR, gutta percha, synthetic)	Yellow, sooty	Pungent, disagreeable, like burnt rubber	As above
Styrene-butadiene rubber (SBR)	Yellow, sooty	Pungent, fruity smell of styrene	As above
Nitrile rubber (NBR)	Yellow, sooty	Like burnt rubber/ burnt hair	As above
Butyl rubber (IIR)	Practically smoke free candle like	Slightly like burnt paper	Melt does not char readily
Polysulfide rubber (polymer itself emits unpleasant, mercaptan like odor)	Smoke-free, bluish	Pungent; smell of H_2S	Yellow, acidic (SO2) fumes
Cellulose (cotton, cellophane, viscose rayon, etc.)	Yellow	Burnt paper	Chars, burns without melting
Cellulose acetate	Yellow–green, sparks	Acetic acid, burnt paper	Melts, drips, burns rapidly, chars, acidic fumes
Cellulose acetate butyrate	Dark yellow (edges slightly blue), somewhat sooty, sparks	Acetic acid/butyric acid, burnt paper	Melts and forms drops which continue burning
Cellulose nitrate (plasticized with camphor)	Yellow	Camphor	Burns very fast, often with explosion
Methyl cellulose	Yellow, luminous	Burnt paper	Melts and chars
Ethyl cellulose	Pale yellow with blue–green base	Slightly sweet, burnt paper	Melts and chars
Polyacrylonitrile	Yellow	Resembling burnt hair	Dark residue; vapor reaction alkaline
Poly(vinyl acetate)	Yellow, luminous, sooty	Acetic acid	Sticky residue, acidic vapor
Poly(vinyl alcohol)	Luminous, limited smoky	Unpleasant, charry smell	Burns in flame, self extinguishing slowly on removal; black residue

(continued)

TABLE 2.8 (Continued)

Polymer	Color of Flame	Odor of Vapor	Other Notable Points
Poly(vinyl butyral)	Bluish (yellow edge)	Like rancid butter	Melts, forms drops
Poly(vinyl acetal)	Purple edge	Acetic acid	Does not drip like poly (vinyl butyral)
Poly(vinyl formal)	Yellow–white	Slightly sweet	Does not drip like poly (vinyl butyral)
Polyethylene	Luminous (blue center)	Like paraffin wax (extinguished candles)	Melts, forms drops; droplets continue burning
Poly(α-olefins) (PP, EPR, etc.)	As above	As above	As above
Polystyrene	Luminous, very sooty	Sweet (styrene)	Softens, easily ignited
Poly(methyl methacrylate)	Luminous, yellow (blue, edge), slightly sooty, crackling	Sweet, fruity	Softens, chars slightly

Source: From Krause, A. and Lange, A. 1969. *Introduction to Chemical Analysis of Plastics.* Iliffe Books Ltd, London; Ghosh, P. 1990. *Polymer Science and Technology of Plastics and Rubber.* Tata McGraw-Hill, New Delhi, India.

a. *Phenol.* On addition to the extract (10 ml), approximately 0.5N potassium hydroxide (8–10 ml) and 2 ml of diazotized *p*-nitroaniline (5% sodium nitrite solution added to an ice-cold solution of 2–5 mg of *p*-nitroaniline dissolved in 500 ml of approximately 3% hydrochloric acid, until the solution becomes just colorless), a red or violet color develops indicating phenol. No differentiation between phenol and its homologs is possible by this test.
b. *Urea and melamine.* The aqueous extract is divided into two parts. (solutions 1 and 2). Solution 1 is made alkaline with dilute sodium hydroxide, and 1 ml of sodium hypochlorite solution is added. Solution 2 is treated with freshly prepared furfural reagent (5 drops of pure, freshly distilled furfural, 2 ml of acetone, 1 ml of concentrated hydrochloric acid, and 2 ml of water).

TABLE 2.9 Classification of Most Common Polymers According to Elements Present

Group	Element Found	Rubbers[a]	Plastics/Fibers
Group 1	N	Nitrile (NBR), polyurethane (ester/ether urethanes)	Cellulose nitrate, silk, polyamides, polyimides, polyurethanes, polyacrylonitrile, SAN and ABS resins, urea-formaldehyde resins, melamine-formaldehyde resins, etc.
Group 2	S	S-vulcanized diene rubbers (NR, IR, SBR, BR, CR, IIR, EPDM, NBR), polysulfide rubbers and chlorosulfonated polyethylenes.	Wool, polysulfones, ebonite, etc.
Group 3	Cl	Polychloroprene, chlorosulfonated polyethylenes, etc.	Poly(vinyl chloride), poly(vinylidene chloride) and related copolymers, polychlorotrifluoroethylene, chlorinated or hydrochlorinated rubber, etc.
Group 4	Absence of N, S, Cl, etc.	Peroxide cross-linked or unvulcanized hydrocarbon rubbers (NR, IR, SBR, BR, IIR, EPDM, EPR, etc.)	Petroleum resins, coumaroneindene resins, cellulose and cellulosics other than cellulose nitrate, polyesters, polyethers or acetal resins, polyolefins, polystyrene, poly(methyl methacrylate), poly(vinyl acetate/alcohol), etc.

[a] See Appendix A2 for abbreviations.

TABLE 2.10 Solubility Behavior of Some Common Plastics

Resin	Soluble In	Insoluble In
Polyolefins		
Polyethylene	Dichloroethylene, xylene, tetralin, decalin (boiling)	Alcohols, esters, halogenated hydrocarbons
Polypropylene	Chloroform, trichloroethylene, xylene, tetralin, decalin (boiling)	Alcohols, esters
Polyisobutylene	Ethers, petroleum ether	Alcohols, esters
Poly(vinyl chloride)	Dimethyl formamide, tetrahydrofuran, cyclohexanone	Alcohols, hydrocarbons, butyl acetate
Poly(vinylidene chloride)	Butyl acetate, dioxane, ketones, tetrahydrofuran	Alcohols, hydrocarbons
Polytetrafluoroethylene	Insoluble	All solvents
Polychlorotrifluoroethylene	*o*-Chlorobenzotrifluoride (above 120°C)	All solvents
Polystyrene	Benzene, methylene chloride, ethyl acetate	Alcohols, water
ABS	Chlorinated hydrocarbons, eg., *p*-dichlorobenzene	Alcohols, water
Polybutadiene	Benzene	Aliphatic hydrocarbons, alcohols, esters, ketones
Polyisoprene	Benzene	Alcohols, esters, ketones
Acrylics		
Polyacrylonitrile	Dimethylformamide and nitrophenol	Alcohols, esters, ketones, hydrocarbons
Polyacrylamide	Water	Alcohols, esters, hydrocarbons
Esters of polyacrylic acid	Aromatic hydrocarbons, esters, chlorinated hydrocarbons, ketones, tetrahydrofuran	Aliphatic hydrocarbons
Esters of polymethacrylic acid	Aromatic hydrocarbons, chlorinated hydrocarbons, esters, ketones, dioxane	Aliphatic hydrocarbons alcohols, ethers
Poly(vinyl acetate)	Aromatic hydrocarbons, chlorinated hydrocarbons, ketones, methanol	Aliphatic hydrocarbons
Poly(vinyl alcohol)	Formamide, water	Aliphatic and aromatic hydrocarbons, alcohols, ethers, esters, ketones
Poly(vinyl acetals)	Esters, ketones, tetrahydrofuran, (butyrals in 9:1 chloroform-methanol mixture)	Aliphatic hydrocarbons, methanol
Poly(vinyl ethers)		
Methyl ether	Water, alcohol, benzene, chlorinated hydrocarbons, ethers, esters	Petroleum ether
Ethyl ether	Petroleum ether, benzene, chlorinated hydrocarbons, alcohols, ethers, esters, ketones	Water
Polyesters	Benzyl alcohol, nitrated hydrocarbons, phenols	Alcohols, esters, hydrocarbons
Polycarbonate	Chlorinated hydrocarbons, cyclohexanone, dimethyl formamide, cresol	(Only swelling in usual solvents)
Nylon	Formic acid, phenols, trifluoroethanol	Alcohols, esters, hydrocarbons
Molded phenolic resins	Benzylamine (at 200°C)	All common solvents
Molded amino resins (urea, melamine)	Benzylamine (at 160°C)	All common solvents

(continued)

TABLE 2.10 (Continued)

Resin	Soluble In	Insoluble In
Polyurethanes		
Noncross-linked	Methylene chloride, hot phenol, dimethylformamide	Petroleum ether, benzene, alcohols, ethers
Cross-linked	Dimethylformamide	Common solvents
Polyoxymethylene	Insoluble	All solvents
Poly(ethylene oxide)	Alcohols, chlorinated hydrocarbons water	Petroleum ether
Epoxy resins		
Uncured	Alcohols, ketones, esters, dioxane, benzene, methylene chloride	Aliphatic hydrocarbons, water
Cured	Practically insoluble	
Cellulose, regenerated	Schweizer's reagent	Organic solvents
Cellulose ethers		
Methyl	Water, dil. Sodium hydroxide	Organic solvents
Ethyl	Methanol, methylene chloride	Water, aliphatic and aromatic hydrocarbons
Cellulose esters (acetate, propionate)	Ketons, esters	Aliphatic hydrocarbons
Cellulose nitrate	Esters (ethyl acetate, butyl acetate, etc.), ketones (acetone, methyl ethyl ketone, etc.), mixtures (eg. 80% methyl isobutyl ketone +20% isopropyl alcohol, 80% butyl acetate+ 20% isopropyl alcohol)	Aliphatic hydrocarbons (hexane, heptane, etc.) methyl alcohol, water
Natural rubber	Aromatic hydrocarbons, chlorinated hydrocarbons	Petroleum ether, alcohols ketones, esters
Chlorinated rubber	Esters, ketones, linseed oil (80–100°C), carbon tetrachloride, tetrahydrofuran	Aliphatic hydrocarbons
Styrene-butadiene rubber	Ethyl acetate, benzene, methylene chloride	Alcohols, water

Source: From Krause, A. and Lange, A. 1969. *Introduction to Chemical Analysis of Plastics.* Iliffe Books Ltd., London.

For urea, solution 1 remains colorless and solution 2 becomes orange to red after 3–5 h. For melamine, solution 1 slowly becomes orange and solution 2 remains colorless.

Detection of urea with urease also provides a useful differentiation between urea and melamine resins. For this test, the powdered sample (0.25 g) is placed in a 100 ml Erlenmeyer flask and boiled with 5% sulphuric acid until the smell of formaldehyde has disappeared. The mixture is neutralized with sodium hydroxide (phenolphthalein as indicator). Then 1 drop of 1 N sulphuric acid and 10 ml of 10% urease solution is added, a strip of red litmus paper is placed in the vapors and the flask is stoppered. The appearance of a blue color in the litmus paper after a short time indicates urea and thus the presence of urea resin.

8. *Test for cellulose esters.* Cellulose esters respond to the Molisch test for carbohydrates. The sample is dissolved in acetone and treated with 2–3 drops of 2% ethanolic solution of α-naphthol; a volume of 2–2.5 ml of concentrated H_2SO_4 is so added as to form a lower layer. A red to red–brown ring at the interface of the liquids indicates cellulose (glucose). A green ring at the interface indicates nitrocellulose and differentiates it from other cellulose esters.

   A more sensitive test for nitrocellulose is provided by an intense blue color reaction when a drop of a solution of diphenylamine in concentrated H_2SO_4 (5% w/v) is added to the sample in the absence of other oxidizing agents.

9. *Test for polyamides.* When a strip of filter paper soaked in a fresh saturated solution of *o*-nitrobenzaldehyde in dilute sodium hydroxide is held over the pyrolytic vapors of polyamides containing adipic acid, a mauve-black color is readily developed. Pyrolytic vapors of polyamides from diacids other than adipic acid produce a grey color when tested similarly.
10. *Tests for natural rubber and synthetic rubbers.* Rubbers may be identified by testing the pyrolytic vapors from test samples. A strip of filter paper soaked in an ethereal solution containing *p*-dimethylaminobenzaldehyde (3%) and hydroquinone (0.05%) and then moistened with a 30% solution of trichloroacetic acid in isopropanol produces different color reactions in the presence of pyrolytic vapors of different rubbers.

Vapors from natural rubber (NR) produce a deep blue or blue–violet color, and those from styrene-butadiene rubber (SBR) pyrolysis turn the paper green or blue with a distinct green tinge. Polyisobutylenes and butyl rubber resemble NR, and silicone rubbers resemble SBR, in this color reaction. Pyrolytic vapors from nitrile rubbers, on the other hand, give a brown or brown–yellow color and those from polychloroprenes (neoprene) turn the test paper grey with a yellow tinge.

In another test, pyrolytic vapors from polyisobutylenes and butyl rubbers produce a bright yellow color on filter paper freshly soaked in a solution obtained by dissolving yellow mercuric oxide (5 g) in concentrated sulfuric acid (15 ml) and water (85 ml) on boiling. Other rubbers may either produce little change or give an uncharacteristic brown color.

Chromic acid oxidation provides a simple test for polyisoprenes. About 0.1 g of sample is gently heated in a test tube in the presence of chromic acid solution (5 ml) prepared by dissolving chromium trioxide (2 g) in water (5 ml) and adding concentrated sulfuric acid (1.5 ml). 1,4-Polyisoprenes (the synthetic equivalent of natural rubber) yield acetic acid, which can be readily identified by its odor.

If the above test is positive, the sample may be further subjected to a modified Weber color test for polyisoprenes. For this test, an acetone-extracted polymer (0.05 g) is dissolved or suspended in carbon tetrachloride, treated with a little solution of bromine in carbon tetrachloride and heated in a water bath to remove excess bromine. The residue is then warmed with a little phenol. In the presence of polyisoprenes, the solid or the solution turns violet or purple and gives a purple solution with chloroform. The test is also positive for natural rubber and butyl rubber.

References

1. Ogorkiewicz, R. M. 1970. *Engineering Properties of Thermoplastics,* John Wiley, New York.
2. Williams, J. G. 1973. *Stress Analysis of Polymers,* Longmans, London.
3. Turner, S. 1973. *Mechanical Testing of Plastics,* CRC Press, Cleveland, OH.
4. Crawford, R. J. 1981. *Plastics Engineering,* Pergamon, London.
5. Ferry, J. D. 1970. *Viscoelastic Properties of Polymers, 2nd Ed.,* John Wiley, New York.
6. Tobolsky, A. V. 1960. *Properties and Structure of Polymers,* John Wiley, New York.
7. Bartenev, G. M. and Zuyev, Y. S. 1968. *Strength and Failure of Viscoelastic Materials,* Pergamon, London.
8. McLoughlin, J. R. and Tobolsky, A. V. 1952. *J. Colloid Sci.,* 7, 555.
9. Williams, M. L., Landel, R. F., and Ferry, D. 1955. *J. Am. Chem. Soc.,* 77, 3701.
10. Nielsen, L. E. 1974. *Mechanical Properties of Polymers and Composites, Vol. 5,* Marcel Dekker, New York.
11. Ward, I. M. 1971. *Mechanical Properties of Solid Polymers,* Interscience, New York.
12. Operations Manual of Rheometric Dynamic Spectrometers, RDS-7700, Rheometrics, Inc., New Jersey.
13. Lenk, R. S. 1978. *Polymer Rheology,* Applied Science Publishers, London.
14. Graessley, W. W. 1967. *J. Chem. Phys.,* 47, 1942.
15. Bueche, F. and Harding, W. S. 1958. *J. Polym. Sci.,* 32, 177.
16. Herzog, R. and Weissenberg, K. 1928. *Colloid Z.,* 46, 227.
17. Nason, H. K. 1945. *J. Appl. Phys.,* 16, 338.

18. Tordella, J. P. 1969. *Tordella, Rheology—Theory and Applications, Vol. 5,* F. R. Eirich, ed., p. 355. Academic Press, San Diego, Chap. 2.
19. van Wazer, J. R., Lyons, J. W., Kim, K. Y., and Colwell, R. E. 1963. *Viscosity and Flow Measurements,* Interscience Publishers, New York.
20. Bagley, E. B. 1961. *Trans. Soc. Rheol.,* 5, 355.
21. Rabinowitsch, B. 1929. *Z. Physik. Chem.,* 145A, 1.
22. Vincent, P. I. 1970. *Impact Tests and Service Performance of Thermoplastics,* Plastics Institute, London.
23. Braden, M. and Gent, A. N. 1960. *J. Appl. Polym. Sci.,* 3, 90.
24. Gaube, E. 1959. *Kunststoffe,* 49, 446.
25. Holloway, L. 1977. *Glass Reinforced Plastics in Construction,* Surrey University Press, Surrey.
26. Richardson, M. O. W. 1977. *Polymer Engineering Composites,* Applied Science Publishers, London.
27. Johnston, A. F. 1979. *Engineering Design Properties of GRP,* British Plastics Federation, England.
28. Brophy, J. J. and Buttrey, J. W. 1962. *Organic Semiconductors,* Macmillan, New York.
29. Saunders, D. W. 1965. Optical properties of high polymers, In *Physics of Plastics,* P. D. Ritchie, ed., Van Nostrand Rheinhold, New York.
30. Hornak, L. 1991. *Polymers for Lightwave and Integrated Optics,* Marcel Dekker, New York.
31. Daum, W., Krauser, J., Zamzow, P. E., and Ziemann, O. 2002. *Polymer Optical Fibers for Data Communication,* Springer, Berlin, Heidelberg, New York.
32. Gordon, M. 1965. Thermal properties of high polymers, In *Physics of Plastics,* P. D. Ritchie, ed., Van Nostrand Rheinhold, New York.
33. Glanville, A. B. 1971. *The Plastics Engineer's Data Book,* Industrial Press, New York.
34. Krause, A. and Lange, A. 1969. *Introduction to Chemical Analysis of Plastics,* Iliffe Books Ltd, London.
35. Ghosh, P. 1990. *Polymer Science and Technology of Plastics and Rubber,* Tata McGraw-Hill, New Delhi, India.

A1
Conversion of Units

SI Units and Conversion Factors

Physical Quantity	Name of SI Unit	Symbol for SI Unit	Definition of SI Unit
Length	Meter	m	Basic unit
Mass	Kilogram	kg	Basic unit
Time	Second	s	Basic unit
Force	Newton	N	$kg\ m\ s^{-2}$ $(=J\ m^{-1})$
Pressure	Pascal	Pa	$kg\ m^{-1}\ s^{-2}$ $(=N\ m^{-2})$
Energy	Joule	J	$kg\ m^2\ s^{-2}$
Power	Watt	W	$kg\ m^2\ s^{-3}$ $(=J\ s^{-1})$

Physical Quantity	Customary Unit	SI Unit	To Convert from Customary Unit to SI Units Multiply by
Length	in.	m	2.54×10^{-2}
Mass	lb	kg	4.5359237×10^{-1}
Force	dyne	N	1×10^{-5}
	kgf	N	9.80665
	lbf	N	4.44822
Pressure	dyne/cm^2	Pa or N/m^2	1×10^{-1}
	atm	Pa or N/m^2	1.01325×10^5
	mm Hg	Pa or N/m^2	1.33322×10^2
	lbf/in.^{2} or psi	Pa or N/m^2	6.89476×10^3
Energy	erg	J	1×10^{-7}
	Btu	J	1.055056×10^3
	ft.-lbf	J	1.35582
Area	in.^{2}	m^2	6.4516×10^{-4}
	ft.^{2}	M2	9.290304×10^{-2}
Density	lb/ft.^{3}	kg /m^3	1.6018463×10
Viscosity	poise	kg /(m.s) or N s/m^2	1×10^{-1}

A2

Index of Trade Names, Manufacturers, and Suppliers of Antioxidants

The following index of trade names, manufacturers, and suppliers is based on a choice of representative antioxidants for thermoplastics. No claim is made for completeness. Detailed lists can be found in the source cited.

The letters A to M symbolize chemical classes of compounds with the following meanings: A = alkylphenols; B = alkylidene-bisphenols (molecular weight 300–600); C = alkylidene-bisphenols (molecular weight >600); D = thiobisphenols (molecular weight 300–600); E = hydroxybenzyl compounds (molecular weight 300–600); F = hydroxy-benzyl compounds (molecular weight >600); G = Acylaminophenols (4-hydroxyanilides); H = octadecyl-3-(3,5-di-*tert*-butyl-4-hydroxyphenyl)-propionate; I = poly-hydroxyphenylpropionates (molecular weight >600); J = amines; K = thioethers; L = phosphites and phosphonites; M = zinc dibutyldithiocarbamate.

	Chemical Class													
Trade Name	A	B	C	D	E	F	G	H	I	J	K	L	M	Manufacturer/Supplier
Age Rite										X			X	R.T Vanderbilt Norwalk, CT, U.S.A.
Vanox		X				X			X	X	X	X		
Anox	X							X	X	X				Ente Nazionale Idrocarburi, Rome, Italy
CAO	X	X		X										Ashland Chemical Ltd., Columbus, OH, U.S.A.
Carstab									X	X				Morten Thiokol Inc., Danvers, MA, U.S.A.
Cyanox		X				X					X			American Cyanamid Co., Wayne, NJ, U.S.A.
Ethanox		X		X		X		X						Ethyl Corporation, Baton Rouge, LA, U.S.A.
Good-rite						X			X	X				B.F. Goodrich Chemical Group, Cleveland, OH, U.S.A.
Hostanox			X								X	X		Hoechst AG, Frankfurt, Germany; American Hoechst Corp., Somerville, NJ, U.S.A.

(continued)

Trade Name	Chemical Class A	B	C	D	E	F	G	H	I	J	K	L	M	Manufacturer/Supplier
HPM			X											Societe Francaise d'Organo-Synthese, Neville-sur-seine, France
Garbefix		X		X							X	X		
Ionol	X													Shell Nederland Chemie BV, Gravenhage, Netherlands
Ionox						X								
Irgafos												X		Ciba–Geigy AG, Basel, Switzerland; Ciba–Geigy Corp., Hawthorne, NY, U.S.A.
Irganox		X		X	X	X	X	X	X	X	X			
Isonox	X	X												Schenectady Chemicals Inc., Schenectady, NY, U.S.A.
Keminox	X	X	X	X		X								Chemipro Kasei Ltd., Chuo-Ka, Kobe, Japan
Lowinox	X	X		X				X	X	X	X	X		Chemische Werke Lowi GmbH, Waldkraiburg, Germany
Mark										X		X		Argus Chemical S.A.-N.V., Brussels, Belgium; Witco Chem. Co., New York, NY, U.S.A.
Mixxim		X										X		Fairmount Chem. Co., Newark, NJ, U.S.A.
Naugard		X	X					X	X	X	X	X	X	Uniroyal Ltd., Bromsgrove., Worcs., U.K.; Uniroyal
Naugawhite		X												
Polygard												X		Chem., Naugatuck, CT, U.S.A.
Nonox	X	X								X				ICI Ltd., London, U.K.
Topanol	X	X								X				ICI Americas, Wilmington, DE, U.S.A.
Negonox			X											
Oxi-chek		X						X						Ferro Corp. Chem. Div., Walton Hills, OH, U.S.A.
Perkanox	X													Akzo Chemie, Nederland NV., Amsterdam, Netherlands
Permanax	X	X												Vulanax International, Saint Cloud, France
Samilizer		X		X				X	X		X	X		Sumitomo Chemical Co. Ltd., Osaka, Japan
Antigene										X				
Santonox				X										Monsanto Europe SA., Brussels, Belgium; Monsanto Co., Akron., OH, U.S.A.
Santowhite	X	X		X						X				
Santoflex										X				
Santicizer				X								X		
Seenox		X		X		X					X			Shipro Kasei Ltd., Kita-ku, Osaka, Japan
Tominox								X	X					Yoshitomi Pharmaceutical Ind. Ltd., Osaka, Japan
Yoshinox		X		X										
Vulkanox	X	X								X	X			Mobay Chem. Corp., Pittsburgh, PA, U.S.A.
Weston												X		Borg-Warner Chemicals., Parkersburg, WV, U.S.A.

(continued)

	Chemical Class													
Trade Name	A	B	C	D	E	F	G	H	I	J	K	L	M	Manufacturer/Supplier
Ultranox	X	X		X				X		X		X		
Wing Stay	X			X						X				Goodyear SA, Rueil-Malmaison, France; Goodyear, Atkron, OH, U.S.A.
Wytox	X	X								X		X		Otlin Corp., Stamford, CT, U.S.A.

Source: Henman, T. J. 1982. *World Index of Polyolefin Stabilizers*, Kogan Page, London.

A3

Index of Trade Names, Manufacturers, and Suppliers of Metal Deactivators

The following index of trade names, manufacturers, and suppliers is based on a choice of representative metal deactivators for thermoplastics. No claim is made for completeness. Detailed lists can be found in the source cited.

Trade Name	Chemical Class	Manufacturer/Supplier
Cyanox	Not disclosed	Cyanamid GmbH, Wolfratschausen, Germany American Cyanamid Co., Wayne, NJ, U.S.A.
Eastman Inhibitor OABH	Dibenzalhydrazone	Eastman Chemical Products, Kingsport, TN, U.S.A.
Hostanox	Phosphorous acid ester	Hoechst AG, Augsburg, Germany
Irganox MD	Diacylhydrazine	Ciba–Geigy AG, Division KA, Basel, Switzerland Ciba–Geigy Crop., Plastics and Additives Div., Hawthorne, NY, U.S.A.
Mark	Not disclosed	Argus Chemical SA, Drogenbos, Belgium Argus Chemical Corp., Brooklyn, NY, U.S.A.
Naugard	Oxalamide derivative	Uniroyal Chemical, Naugatuck, CT, U.S.A.

Source: Henman, T. J. 1982. *World Index of Polyolefin Subilizers*, Kogan Page Ltd., London.

A4

Index of Trade Names, Manufacturers, and Suppliers of Light Stabilizers

The following index of trade names, manufacturers, and suppliers is based on a choice of representative light stabilizers for thermoplastics. No claim is made for completeness. Detailed lists can be found in the source cited.

The letters A to J symbolize chemical classes of compounds with the following meanings: A = benzophenones; B = benzotriazoles; C = nickel compounds; D = salicylates; E = cyanocinnamates; F = benzylidene malonates; G = benzoates; H = oxanilides; I = sterically hindered amiens and J = polymeric sterically hindered amines.

Trade Name	Compound Type	Manufacturer
Aduvex	A	Ward Blenkinsop and Co. Ltd., Wembley (Middx.),
Anti-UV	A, D	Societe Francaise D'Organo Synthese. Neuilly-sur-Seine, France
Carstab	A	Cincinnati Milacron Chemicals Inc., New Brunswick, NY, U.S.A.
Chimassorb	A, C, J	Chimosa S.p.A., Pontecchio Marconi, Italy
Cyasorb UV	A, B, C, E, F, I, J	American Cyanamid Co., Wayne, NJ, U.S.A.
Eastman Inhibitor	A	Eastman Chemical Product Inc. Kingsport, TN, U.S.A.
UV-Chek AM	A, C, G	Ferro Corp., Cleveland, OH, U.S.A.
Goodrite UV	I	B. F. Goodrich, Chemical Group, Cleveland, OH, U.S.A.
Hostavin	A, I, K	Hoechst AG, Frankfurt/M. Germany
Interstab	G	Interstab Ltd., Liverpool, U.K.
Irgastab	C	Ciba–Geigy AG, Basel Switzerland
Mark	A, I	Argus Chemical Corp., Brooklyn, NY, U.S.A.
Mixxim	I, J	Fairmount Chemical Co., Newark, NJ, U.S.A.
Rylex	A, C	E. I. Du Pont de Nemours and Co., Wilmington, DE, U.S.A.
Salol	A, D	Dow Chemical Co., Midland, MI, U.S.A.
Sanol LS	I	Sankyo Co. Ltd., Tokyo, Japan
Seesorb	A, B, C, D, E, G	Sun Chemical Corp., Tokyo, Japan
Sumisorb	A, D	Sumitomo Chemical Co., Ltd., Osaka, Japan
Tinuvin	B, H, I, J	Ciba–Geigy AG. Basel, Switzerland
UV-Absorber Bayer	A, E	Bayer AG, Leverkusen, Germany
Uvinul	A, E	BASF AG, Ludwighafen, Germany
Viosorb	A, B, D	Kyodo Chemical Co. Ltd., Tokyo, Japan

Source: Henman, T. J. 1982. *World Index of Polyolefin Stabilizers*, Kogan Page Ltd., London.

A5
Formulations of Flame-Retarded Selected Thermoplastics

In the following formulations the fire performance properties of flame-retarded thermoplastics are compared with those of the base resin, the properties of the latter being given in brackets. The test methods used are: ASTM-D 633-77, the value for average time of burning (ATB) given in seconds and that for average extent of burning (AEB) in millimeters; ASTM 2863-77, limiting oxygen index (LOI) in % O2.

Low-Density Polyethylene (LDPE)

90.5%	LDPE
6.0%	Chlorinated paraffin (70% chlorine)
3.5%	Antimony trioxide

Properties

ASTM	ATB	<5	(160)
	AEB	18	(100)
	LOI	23.75	(17.5)

High-Density Polyethylene (HDPE)

89.9%	HDPE
5.0%	Octabromodiphenyl
1.6%	Chlorinated paraffin (70% chlorine)
3.5%	Antimony trioxide

Properties

ASTM	ATB	<5	(210)
	AEB	20	(100)
	LOI	25.75	(16.75)

Polypropylene

94.5%	Polypropylene
3.5%	Bis-dibromopropylether of tetrabromobisphenol A
2.0%	Antimony trioxide

Properties

ASTM	ATB	5	(195)
	AEB	14	(100)
	LOI	27.5	(18.25)

Polystyrene

a. *Crystal Polystyrene*

97%	Polystyrene
3%	Hexabromocyclododecane

Properties

ASTM-D635	ATB	<5	(117)
	AEB	28	(100)

b. *Impact Polystyrene*

80%	Impact polystyrene
15%	Octabromoldiphenyl
5%	Antimony trioxide

Properties

ASTM	ATB	<5	(150)
	AEB	17	(100)
	LOI	25.0	(16.5)

c. *Polystyrene Foam*

82%	Polystyrene
13%	Decabromodiphenyl ether
5%	Antimony trioxide

Properties

ASTM	ATB	<5	(90)
	AEB	19	(100)

Plasticized PVC

53.5%	Suspension PVC
27.7%	Dioctylphthalate
4.9%	Epoxidized oleic acid ester
4.9%	Pentabromodiphenyl ether
8.3%	Tricresyl phosphate
0.5%	Lubricant
0.4%	Stabilizer (butyltin carboxylate)

[The formulation passes the FMVSS 302 (Federal Motor Vehicle Safety Standard) test used for the assessment of the burning behavior of materials used for vehicle interiors. A specimen of 35.5×10.0× max. 13 cm is mounted horizontally and ignited for 15 sec. The rate of flame spread should not exceed 10 cm/min.]

ABS Terpolymer

82%	ABS
13%	Octabromodiphenyl ether
5%	Antimony trioxide

Properties

ASTM	ATB	<5	(195)
	AEB	10	(100)
	LOI	24	(18)

Saturated Polyester

a. *Polyester*

88%	Poly (butylene terephthalate) (PBTP)
8%	Octabromodiphenyl
4%	Antimony trioxide

Properties

ASTM	ATB	<5	(240)
	AEB	16	(100)
	LOI	28.5	(22.0)

b. *Polyester with Glass Fiber Reinforcement*

87%	PBTP (30% glass fiber)
9%	Decarbromodiphenyl
4%	Antimony trioxide

Properties

LOI	33	(19)

Polycarbonates

Flame-ratarded polycarbonates preferably contain tetrabromobisphenol-A as flame-retardant added before polycondensation. Untreated polycarbonate can be processed to a flame-retardant product as follows:

93.1%	Polycarbonate
5.0%	Decabromodiphenyl ether
1.9%	Antimony trioxide

Properties

ASTM	ATB	<5	(17)
	AEB	12	(24)
	LOI	40	(25)

Polyamides

a. *Nylon-6*

84%	Nylon-6
10%	Decabromodiphenyl ether
6%	Antimony trioxide

Properties

ASTM	ATB	<5	(178)
	AEB	17	(60)
	LOI	22.75	(21.75)

b. *Fiberglass-Reinforced Nylon-6*

53%	Nylon-6
22%	Fiberglass
18%	Dechlorane 602
7%	Antimony trioxide

Properties

LOI	32	(21)

A6

Index of Trade Names and Suppliers of Flame Retardants

The following index of trade names and suppliers is based on a choice of representative flame retardants for thermoplastics. No claim is made for completeness. Detailed lists can be found in the source cited.

Trademark	Chemical Class	Producer
Amgard	Phosphoric acid ester	Albright and Wilson, London, U.K.
Chlorowax	Chlorinated paraffins	Diamond Shamrock Plastics Div., Cleveland OH, U.S.A.
Cereclor	Chlorinated paraffins	ICI, London, U.K.
Dechlorane	Chlorine containing cycloaliphatics	Occidental Chemical Corp., Niagara Falls, NY, U.S.A.
Disflamoll	Organic phosphates and halogenated organic phosphates	Bayer AG, Leverkusen, Germany
Exolit	Chlorinated paraffins, organic phosphates	Hoechst AG, Frankfurt/M, Germany
Firebrake	Inorganic flame retardants	US Borax & Chem. Corp., Los Angeles, CA, U.S.A.
Fire Fighters	Organic bromine compounds	Great Lakes Chem. Corp., West Lafayette, IN, U.S.A.
Fireguard	Organic bromine compounds	Teijin Chem. Co., Nishishinbasi, Tokyo, Japan
Firemaster	Organic bromine compounds	Great Lakes Chem. Corp., West Lafayette, IN, U.S.A.
Fire-Shield	Organic bromine compounds, antimony trioxide	PPG Industries Inc., Chicago, IL, U.S.A.
Firex	Organic halogenated compound with antimony trioxide	Dr. Th. Bohme KG. Geretstried, Germany
Flammastik	Flame retardant on the basis of antimony	Chem. Fabrik Grunau GmbH, Illertissen, Germany
Hydral	Hydrated alumina	Aluminum Company of America, Pittsburgh, PA, U.S.A.
Ixol	Organic bromine and chlorine compounds	Kali-Chemie AG, Hannover, Germany
Martinal	Hydrated alumina	Martinswerk GmbH, Bergheim, Germany
Nonnen	Organic bromine compounds	Marubeni Corp., Osaka, Japan
Nyacol	Flame retardant on the basis of antimony	Nyacol Products Inc., Ashland, MA, U.S.A.
Phosflex	Organic phosphates	Stauffer Chem. Co., Specialty Chem. Div., Westport, CT, U.S.A.
Proban	Phosphoric acid esters, quaternary phosphonium compounds	Albright and Wilson, London, U.K.
Pyro-Check	Organic bromine compounds	Ferro Corp., Chem. Div., Bedford, OH, U.S.A.
Pyrovatex	Organic phosphates	Ciba–Geigy AG, Basel, Switzerland

(continued)

Trademark	Chemical Class	Producer
Sandoflam	Flame retardants based on organic bromine compounds	Sandoz AG, Basel, Switzerland
Santicizer	Phosphoric acid esters, halogenated organic compounds	Monsanto Co., St. Louis, MO, U.S.A.
Saytex	Organic bromine compounds	Ethyl Corp., Chemical Group, Baton Rouge, LA, U.S.A.
Tardex	Organic bromine compounds	ISC-Chemicals, Avonmouth, Bristol, England
Timonox	Antimony oxides	Cookson Ltd., Stoke-on-Trent, England
Zerogen	Hydrated alumina	Solem Industries, Inc., Norcross, GA, U.S.A.
—	Smoke reducer based on molybdenum	Climax Molybdenum Co. Ltd., London, U.K.

Source: Troitzsch, J. 1983. *International Plastics Flammability Handbook,* Carl Hanser Verlag, Munchen.

A7

Harmonic Motion of a Maxwell Model

Trigonometric Notation

Starting with a sinusoidal input of strain in a Maxwell element (see Chapter 2), we derive the resulting sinusoidal stress. First we let the strain ϵ be a function of a maximum or peak strain ϵ_0 and time t with a frequency ω:

$$\epsilon = \varepsilon_0 \sin \omega t \tag{A7.1}$$

For the Maxwell element:

$$\frac{d\epsilon}{dt} = \frac{1}{E}\frac{d\sigma}{dt} + \frac{\sigma}{\lambda E} \tag{A7.2}$$

Differentiating Equation A7.1:

$$\frac{d\epsilon}{dt} = \omega\varepsilon_0 \cos \omega t \tag{A7.3}$$

Rearranging Equation A7.2:

$$\frac{d\sigma}{dt} + \frac{\sigma}{\lambda} = \omega\epsilon_0 E \cos \omega t \tag{A7.4}$$

This is a simple linear differential equation of the form

$$\frac{dy}{dx} + Py = Q$$

The general solution for such an equation, when P and Q are functions of x only, is

$$y \exp(\psi) = \int \exp(\psi) Q dx + C, \quad \psi = \int P dx \tag{A7.5}$$

For Equation A7.4, the analogy is

$$\psi = \frac{t}{\lambda} \tag{A7.6}$$

$$\sigma \exp\left(\frac{t}{\lambda}\right) = \omega\epsilon_0 E \int \exp\left(\frac{t}{\lambda}\right)\cos \omega t \, dt + C \tag{A7.7}$$

$$\sigma \exp\left(\frac{t}{\lambda}\right) = \frac{\omega\epsilon_0 E\lambda}{1+\omega^2\lambda^2}(\cos \omega t + \omega\lambda \sin \omega t)\exp\left(\frac{t}{\lambda}\right) + C \tag{A7.8}$$

or

$$\sigma = \frac{\omega\lambda}{1+\omega^2\lambda^2}\epsilon_0 E(\cos \omega t + \omega\lambda \sin \omega t) + C\exp\left(\frac{-t}{\lambda}\right) \tag{A7.9}$$

The second term on the right is a transient one which drops out in the desired steady-state solution for $t/\lambda >> 1$.

Let us now define an angel δ by

$$\tan \delta = \frac{1}{\omega\lambda} = \frac{\sin \delta}{\cos \delta} \text{ and } \sin \delta = \frac{1}{(1+\omega^2\lambda^2)^{1/2}} \tag{A7.10}$$

Then, making use of trigonometric identities:

$$\cos \omega t + \omega\lambda \sin \omega t = \frac{\cos \omega t(\sin \delta)}{\sin \delta} + \frac{\sin \omega t(\cos \delta)}{\sin \delta} \tag{A7.11}$$

$$= \frac{\sin(\omega t + \delta)}{\sin \delta} \tag{A7.12}$$

$$= (1+\omega^2\lambda^2)^{1/2}\sin(\omega t + \delta) \tag{A7.13}$$

Finally, combining Equation. A7.13 and Equation A7.9 with the transient term dropped, one arrives at

$$\sigma = \frac{\omega\lambda}{(1+\omega^2\lambda^2)^{1/2}}\epsilon_0 E \sin(\omega t + \delta) \tag{A7.14}$$

Complex Notation

Starting with a complex strain, the real part of which is the actual strain:

$$\epsilon^* = \epsilon_0 \exp(i\omega t) \tag{A7.15}$$

The motion of the Maxwell element, in terms of a complex stress and strain, is

$$\frac{d\epsilon^*}{dt} = \frac{1}{E}\frac{d\sigma^*}{dt} + \frac{\sigma^*}{\lambda E} \tag{A7.16}$$

Differentiating Equation A7.15:

$$\frac{d\epsilon^*}{dt} = i\omega\epsilon_0 \exp(i\omega t) \tag{A7.17}$$

Rearranging Equation A7.16 and Equation A7.17:

$$\frac{d\sigma^*}{dt} + \frac{\sigma^*}{\lambda} = E\frac{d\epsilon^*}{dt} = i\omega\epsilon_0 E \exp(i\omega t) = Q \tag{A7.18}$$

As in Equation A7.5, the general solution is

$$\sigma^* \exp\left(\frac{t}{\lambda}\right) = \int \exp\left(\frac{t}{\lambda}\right) Q dt + C \tag{A7.19}$$

$$\begin{aligned} \int \exp\left(\frac{t}{\lambda}\right) Q dt &= i\omega\varepsilon_0 E \int \exp\left(i\omega t + \frac{t}{\lambda}\right) dt \\ &= \frac{i\omega\varepsilon_0 E \exp(i\omega t + t/\lambda)}{i\omega + 1/\lambda} \end{aligned} \tag{A7.20}$$

Substitution and rearrangement yields

$$\sigma^* = \frac{i\omega\varepsilon_0 \lambda E \exp(i\omega t)}{i\omega\lambda + 1} + C \exp\left(-\frac{t}{\lambda}\right) \tag{A7.21}$$

Once again, the second term on the right-hand side is a transient term that drops out at $t/\lambda >> 1$. Multiplying both numerator and denominator by $1 - i\omega\lambda$ and substituting ϵ^* for its equivalent, $\epsilon_0 \exp(i\omega t)$ gives

$$\sigma^* = \frac{\omega^2\lambda^2\epsilon^* E + i\omega\lambda\epsilon^* E}{1 + \omega^2\lambda^2} \tag{A7.22}$$

Rearranging gives

$$\frac{\sigma^*}{\epsilon^*} = \frac{E\omega^2\lambda^2}{1 + \omega^2\lambda^2} + \frac{i(E\omega\lambda)}{1 + \omega^2\lambda^2} \tag{A7.23}$$

The definition of complex E^* is

$$E^* = E' + iE'' = \frac{\sigma^*}{\epsilon^*} \tag{A7.24}$$

Comparing Equation A7.23 and Equation A7.24 one concludes that

$$E' = \frac{E\omega^2\lambda^2}{1 + \omega^2\lambda^2} \text{ and } E'' \frac{E\omega\lambda}{1 + \omega^2\lambda^2} \tag{A7.25}$$

The dynamic modulus E', which is the real component of E^*, is associated with energy storage and release in the periodic deformation and is therefore called the storage modulus. The imaginary part of the modulus, E'', is associated with viscous energy dissipation and is called the loss modulus (see Chapter 2 for more details).

Index

D

E

F

G

H

I

K

L

M

N

O

P

Q

R

S

T

U

V

W

X

Y

Z